Construction Technology 2:
INDUSTRIAL AND COMMERCIAL BUILDING

THIRD EDITION

Mike Riley

Director of the School of the Built Environment,
Liverpool John Moores University, UK

and

Alison Cotgrave

Deputy Director of the School of the Built Environment,
Liverpool John Moores University, UK

palgrave
macmillan

First edition published 2002
Second edition published 2009
This edition first published 2014 by
PALGRAVE MACMILLAN

Palgrave Macmillan in the UK is an imprint of Macmillan Publishers Limited,
registered in England, company number 785998, of Houndmills, Basingstoke,
Hampshire RG21 6XS.

Palgrave Macmillan in the US is a division of St Martin's Press LLC,
175 Fifth Avenue, New York, NY 10010.

Palgrave Macmillan is the global academic imprint of the above companies
and has companies and representatives throughout the world.

Palgrave® and Macmillan® are registered trademarks in the United States,
the United Kingdom, Europe and other countries.

ISBN: 978–1–137–37169–0

This book is printed on paper suitable for recycling and made from fully
managed and sustained forest sources. Logging, pulping and manufacturing
processes are expected to conform to the environmental regulations of the
country of origin.

A catalogue record for this book is available from the British Library.

Typeset by Wileman Design, Farnham, Surrey

Printed in Great Britain by
Lavenham Press Ltd., Lavenham, Suffolk

Contents

List of comparative studies and case studies

Acknowledgements

The authors would like to express their thanks to the following people for their support and contribution to the book:

Dave Tinker: for his time and effort in creating the illustrations contained within the book

Robin Hughes: for his advice and commentary on the preliminary drafts

Carillion, David McLean Property and Development Group, HHL: for their kind permission to use examples of their buildings for illustrating the case studies and for granting unlimited access to their construction sites

Bill Atherton: for assistance with structural elements of the book

Peter Williams: for writing the section of the book relating to designing for Health and Safety

Alice Fursdon: of Sheffield Hallam University

Paul Kenny: for providing excellent photographs

Julie, Steve and Sam: for their support and understanding during the writing of this text

Dianne Marsh: for writing the section on BIM

Laurie Brady and Derek King: for writing Chapter 14.

Figures 14.1, 14.2, 14.3 and 14.5 have been adapted from CIBSE publications with their kind permission.

The publishers and author would like to thank the organisations and people listed below for permission to reproduce the following material:

Macmillan Education and Getty for the chapter openers for Chapters 1 and 3
Macmillan Education for the chapter opener for Chapter 2
Macmillan Education and Brand X for the chapter opener for Chapter 14
Fotolia and flik47 for the chapter opener for Chapter 4
Fotolia and Chris Hill for the chapter opener for Chapter 5
Fotolia and Sergejs Katkovskis for the chapter opener for Chapter 6
Fotolia and rabbit75_fot for the chapter opener for Chapter 7
Fotolia and patleem for the chapter opener for Chapter 8
Fotolia and miket for the chapter opener for Chapter 10
Fotolia and Stephen Finn for the chapter opener for Chapter 11
Fotolia and photo 5000 for the chapter opener for Chapter 12
Fotolia and Lev for the chapter opener for Chapter 13
Cadassist Ltd for permission to reprint Figure 2.19.

Preface

There are many texts available in the area of commercial and industrial building technology, each with their own merits. Each new book in this complex and wide-ranging field of study attempts to provide something extra, something different to set it apart from the rest. The result is an ever-expanding quantum of information for the student to cope with. While such texts are invaluable as reference sources, they are often difficult to use as learning vehicles. This text seeks to provide a truly different approach to the subject of construction technology primarily associated with framed buildings. Rather than being a reference source, although it may be used as such, the book provides a learning vehicle for students of construction and property-related subjects. This volume builds upon the subject matter that was introduced in the first book of the series, *Construction Technology 1: House Construction*, although it is also a valuable standalone learning resource. This third edition is further updated to take into account changes in legislation and sustainable technology.

A genuine learning text

The text is structured so that it provides a logical progression and development of knowledge from basic introductory material to more advanced concepts relating to the technology of commercial and industrial buildings. The content is aimed at students wishing to gain an understanding of the subject matter without the need to consult several different and costly volumes. Unlike reference texts that are used for selectively accessing specific items of information, this book is intended to be read as a continuous learning support vehicle. Whilst the student can access specific areas for reference, the major benefit can be gained by reading the book from the start, progressing through the various chapters to gain a holistic appreciation of the various aspects of commercial building construction.

Key learning features

The learning process is supported by several key features that make this text different from its competitors. These are embedded at strategic positions to enhance the learning process.

- Case studies include photographs and commentary on specific aspects of the technology of framed buildings. Thus students can visualise details and components in a real situation.
- Reflective summaries are included at the end of each section to encourage the reader to reflect on the subject matter and to assist in reinforcing the knowledge gained.

- Review tasks aid in allowing the reader to consider different aspects of the subject at key points in the text.
- Comparative studies allow the reader to quickly compare and contrast the features of different details or design solutions, and are set out in a simple-to-understand tabular format.
- The Info points incorporated into each chapter identify key texts or sources of information to support the reader's potential needs for extra information on particular topics.
- In addition, the margin notes are used to expand on certain details discussed in the main text without diverting the reader's attention from the core subject matter.

New to the third edition

This latest edition incorporates sustainability as a key theme in each aspect of technology as opposed to having a bespoke chapter at the end of the book. The authors have taken this approach because sustainability is now being seen as mainstream as a criterion for the choice of materials and systems used in commercial and multi-storey buildings. A useful addition is the chapter on building services installations which should enable the reader to gain a better understanding of modern services installations and how they integrate with the building form and enhance function.

Website

A website supporting this book and designed to enhance the learning process can be found at www.palgrave.com/engineering/riley2. This contains outline answers to the Review tasks, plus further photographs from the Case studies in the book, with accompanying commentary.

Summary

The overall aim of the text and website is to inspire students of construction and property-related disciplines to develop an understanding of the principles of construction associated with industrial and commercial buildings. The format is intended to be easily accessible and the support features will allow the reader to work through the text in a progressive manner without undue complexity. The text is written in a clear and concise fashion and should provide all that students require to support the development of their learning in the area of industrial and commercial building construction.

This book is intended to be an accessible tool to support the learning process; it is not an expansive reference book. With the aid of the book it is envisaged that students will be able to navigate their way through the typical syllabus of a construction technology programme dealing with the construction of high-rise and long-span buildings. The need to buy several expensive texts should be a thing of the past.

MIKE RILEY
ALISON COTGRAVE

Preparing to build

1 Functions and requirements of industrial and commercial buildings

AIMS

After studying this chapter you should be able to:

■ Appreciate the main physical functions of buildings
■ Describe the factors that must be considered in creating an acceptable living and working environment
■ Discuss links between these factors and the design of modern commercial and industrial high-rise and long-span buildings
■ Recognise the sources and nature of loadings applied to building elements and the ways in which they affect those elements
■ Appreciate the influence of the choice of materials and the selection of design features on building performance

This chapter contains the following sections:

1.1   Physical and environmental functions of industrial and commercial buildings
1.2   Forces exerted on and by buildings
1.3   Structural behaviour of elements

INFO POINT

■ Building Regulations Approved Document A: Structures (2004 including 2010 amendments)
■ BS 648: Schedule of weights of building materials (1964 – withdrawn)
■ BS 5250: Code of practice for control of condensation in buildings (2011)
■ BS 6399 Part 1: Loadings for buildings. Code of practice for dead and imposed loads (1984)
■ BS EN ISO 7730: Ergonomics of the thermal environment – Analytical determination and interpretation of thermal comfort using calculation of the PMV and PPD indices and local thermal comfort criteria (2005)
■ BS ISO 6243: Climatic data for building design (1997)
■ BRE Digest 426: Response of structures to dynamic loads (2004)
■ BRE Report 487: Designing quality buildings (2007)
■ *Durability of materials and structures in building and civil engineering*, WHITTLES (2006)
■ London District Surveyors Association (LDSA) *Sustainability Guidance*, *Reducing, recycling, and reusing demolition and construction waste* (2010)
■ http://www.natural-building.co.uk/

1.1 | Physical and environmental functions of industrial and commercial buildings

Introduction

- After studying this section you should be aware of the nature of buildings as environmental enclosures.
- You should appreciate the nature of the building user's need to moderate the environment and understand how buildings have evolved to allow this to be achieved.
- You should also have an awareness of the link between environmental needs and the form of the building fabric.
- You should be able to recognise the key features of building form that affect the internal environment.

Overview

The ways in which the internal environments of buildings are controlled have become very sophisticated as the needs of occupiers have evolved. The extent to which the internal environment can be controlled through the building fabric and building services is immense. However, it is easy to overlook the importance of the external fabric with regard to the internal environment. Most buildings are designed to be aesthetically pleasing and many of the details associated with building styles have evolved in order to satisfy functional requirements. As buildings have developed, the role of building services to control heat, light and ventilation has become more significant, and it is easy to forget that these services rely on an appropriate building envelope in order to achieve required levels of performance. The main functions of the building envelope are therefore to protect occupiers from the elements; to provide a suitable enclosure for building services that will enable a suitable internal environment to be provided; and to be aesthetically pleasing.

The building as an environmental envelope

Historically, people have sought to modify and control the environment in which they live. In prehistoric times caves and other naturally occurring forms of shelter were used as primitive dwellings, providing protection from the external environment. As civilisation has developed, so the nature of human shelter has become more refined and complex, developing from caves and natural forms of shelter to simple artificial enclosures, such as those used throughout history by nomadic peoples worldwide. The way in which the structures created by humans have developed has depended upon the nature of the climate in specific locations and the form of building materials available locally.

The ability to transport building materials over relatively large distances is a recent development. In Britain this was limited prior to the Industrial Revolution by the lack of effective transport networks. Hence **vernacular architecture** has arisen to cope with specific environmental demands, using the materials available locally. Examples of vernacular architecture are found throughout the world. In areas such as the Middle

Vernacular architecture is that which is native to a particular country or region, and develops due to the culture and climate of the area.

Eastern desert regions, where diurnal temperatures vary considerably, being very hot during the day and cool at night, buildings of massive construction are common. Such buildings are referred to as 'thermally heavy' structures. The intense heat of the day is partly reflected by the use of white surface finishes. That which is not reflected is absorbed by the building fabric, rather than being transmitted into the occupied space. As a result of the slow thermal reaction of the building, this stored heat is released at the times of day when the external temperatures may be very low, acting as a form of storage heater. The effects of direct solar gain are reduced by the use of a limited number of small window openings. Conversely, in areas where the climate is consistently warm and humid, such as the Far East, a very different approach to building design is required. In such situations, rare breezes may be the only cooling medium which can remove the oppressive heat and humidity of the internal environment. Since this cooling and dehumidifying effect takes place over a short period, the building must be able to react quickly to maximise any potential benefit. Hence a thermally light structure is essential to transmit external changes to the interior with minimal delay. The nature of buildings in such areas reflects these requirements, with lightweight building fabric and many large openings to allow cooling breezes to pass through the building. Figure 1.1 illustrates the differing properties of thermally light buildings, with fast response to external changes, and thermally heavy buildings, which insulate the interior from external changes as a result of slow reaction times.

The use of protective structures or enclosures is not the only method utilised in the moderation of the human environment. Since fire was first discovered and used by primitive people to provide light and heat, the use of energy to aid in environmental moderation has been fundamental. Although the use of built enclosures can moderate the internal environment and reduce the effects of extremes in the external climate, the active control and modification of the internal environment requires the input of energy. The use of buildings to house people, equipment and processes of differing types, exerting differing demands in terms of the internal environment, has resulted in the

Figure 1.1
Differing properties of thermally light and heavy buildings.

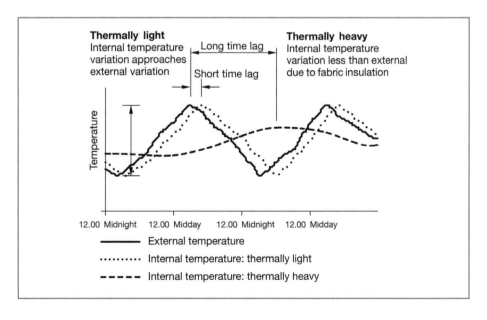

development of buildings and associated services capable of moderating the internal conditions within desired parameters with great accuracy.

The nature of people's perception of comfort within buildings has also developed; the simple exclusion of rain and protection from extreme cold or heat are no longer sufficient to meet human needs. The provision of an acceptable internal condition relies on a number of factors, which may be summarised as follows:

- Thermal insulation and temperature control
- Acoustic insulation
- Provision of light (natural or artificial)
- Control of humidity
- Exclusion of contaminants
- Heat gain due to equipment such as photocopiers and computers.

Figure 1.2 illustrates these requirements and the ways in which they are met in modern construction forms.

Figure 1.2
The building as an environmental modifier.

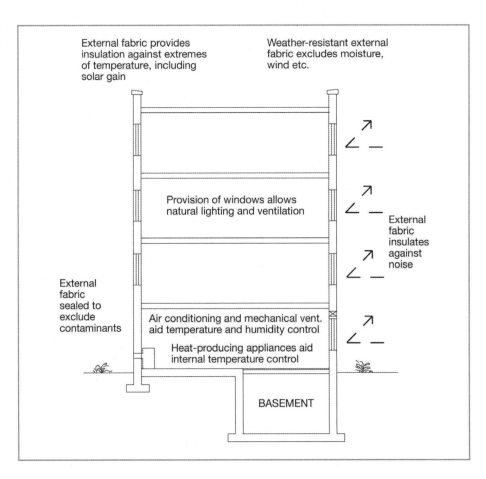

Performance requirements of the building fabric

The requirement to provide an acceptable internal environment is simply one of the performance requirements of modern buildings. The level of performance of buildings depends upon several factors, and the emphasis which is placed upon these individual performance requirements varies from situation to situation.

However, **minimum standards** are set out by statutes and guidelines, such as the Building Regulations, which must be achieved in any instance. The increasing role of the building as an asset has also affected the ways in which buildings have been designed to maximise the long-term value and minimise the maintenance costs of the structure and fabric.

The performance requirements of buildings may be summarised as follows.

Minimum standards that are acceptable for buildings are set out by statutes or guidelines such as the Building Regulations. These aid the designer in ensuring that the building is designed effectively.

Structural stability

In order to satisfactorily fulfil the functions required of it a building must be able to withstand the loadings imposed upon it without suffering deformation or collapse. This necessitates the effective resistance of loadings or their transfer through the structure to the ground.

Durability

The long-term performance of the structure and fabric demands that the component parts of the building are able to withstand the vagaries and hostilities of the environment in which they are placed, without deterioration. The ability of the parts of the building to maintain their integrity and functional ability for the required period of time is fundamental to the ability of the building to perform in the long term. This factor is particularly affected by the occurrence of fires in buildings.

Thermal insulation

The need to maintain internal conditions within fixed parameters and to conserve energy dictate that the external fabric of a given building provides an acceptable standard of resistance to the passage of heat. The level of thermal insulation which is desirable in an individual instance is, of course, dependent upon the use of the building, its location and so on.

Exclusion of moisture and protection from weather

The passage of moisture from the exterior, whether in the form of ground water rising through capillary action, precipitation or other possible sources, should be resisted by the building envelope. The ingress of moisture to the building interior can have several undesirable effects, such as the decay of timber elements, deterioration of surface finishes and decorations and risks to health of occupants, in addition to effects upon certain processes carried out in the building. Hence details must be incorporated to resist the passage of moisture, from all undesirable sources, to the interior of the building. The exclusion of wind and water is essential to the satisfactory performance of any building fabric.

Acoustic insulation

The passage of sound from the exterior to the interior, or between interior spaces, should be considered in building construction. The level of sound transmission which is acceptable in a building will vary considerably, depending upon the nature of the use of the building and its position.

Flexibility

In industrial and commercial buildings in particular, the ability of the building to cope with and respond to changing user needs has become very important. Hence the level of required future flexibility must be taken into account in the initial design of the building; this is reflected, for example, in the trend to create buildings with large open spaces, which may be subdivided by the use of partitions that may be readily removed and relocated.

Aesthetics

Aesthetics relate to the principles of art and taste. People have very different ideas about what they consider aesthetically pleasing.

The issue of building **aesthetics** is subjective. However, it should be noted that in some situations the importance of the building's aesthetics is minimal, while in others, of course, it is highly important.

For example, the appearance of a unit on an industrial estate is far less important than that of a city centre municipal building. The extent to which aesthetics are pursued will have an inevitable effect on the cost of the building.

This summary is not a definitive list of the performance requirements of all building components in all situations. However, it is indicative of the factors which affect the design and performance of buildings and their component parts (Figure 1.3).

Sustainability

Since the authors wrote the first and second editions of this book, the concept of what sustainability means in the context of construction work has slowly evolved. The basic principles are the same, but how the industry can achieve these principles has become better understood.

There is a wide acceptance that sustainability integrates, at least, three dimensions:

- Social dimension
- Economic dimension
- Environmental dimension.

In order to address sustainability, all three dimensions need to be considered, and most importantly they need to be considered in a local context.

There are a number of different models that are used to illustrate these three dimensions but the authors believe that the three pillars/triple bottom line approach is the easiest to understand and conceptualise in the context of construction.

Figure 1.3
Modern, sophisticated buildings require careful design of structure, fabric and services to ensure that they meet user requirements.

Figure 1.4 illustrates the concept of sustainability using the three pillars model. It also shows how this links to the triple bottom line view of sustainability. All three pillars and all three aspects of the triple bottom line need to be addressed if a sustainable development is to be achieved, but for the purpose of this book which is about construction technologies, the one that will be discussed predominantly is the ecological pillar and planet bottom line. Construction work affects the environment from an ecological perspective, i.e. it impacts on the environment.

Figure 1.4
The three pillars of sustainability and the triple bottom line.

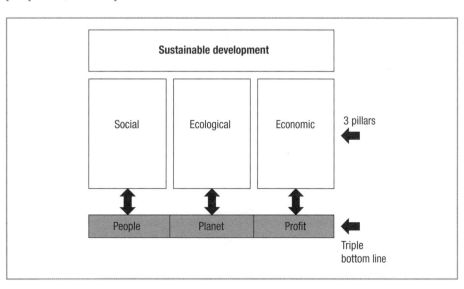

Impact of construction work and buildings on the environment

Construction work and buildings have a major impact on the environment in terms of land use, natural resource usage, carbon dioxide emissions and energy requirements. For example, buildings in use account for about 50 per cent of total energy used in the UK and the construction of buildings accounts for another 5–10 per cent. Figure 1.5 illustrates this graphically.

Figure 1.5
Percentages of total energy use in the UK.

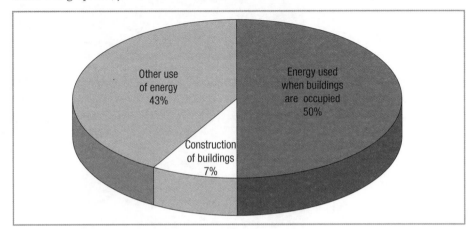

How is all this energy used? Figure 1.6 illustrates how much energy is used for particular activities in occupied buildings.

The highest percentage by far is the 57 per cent used for heating and consequently this is an area where a great deal of research has been undertaken to try to reduce this value, and a number of new technologies have been developed to address energy efficiency. Although good design of the different elements that make up a building can help to improve this figure significantly, even then there can be problems arising from heat loss through the structure of the building and air leakage from the building. It has been estimated that two-thirds of heat loss from a house built to current Building Regulations will be through the structure and one-third from air leakage. In super-insulated buildings the heat loss through the fabric will be much less but the air leakage loss will be about

Figure 1.6
Percentages of total energy used in buildings for particular functions.

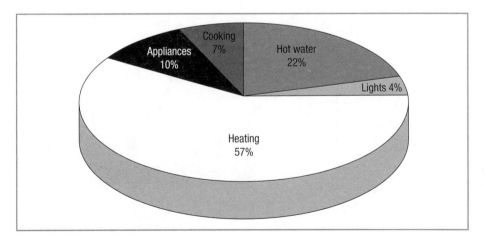

the same. Ironically, if heat loss from buildings through the structure and air leakage are improved significantly then a different problem may arise, which is overheating. If this occurs, the inclusion of air conditioning in buildings in the UK will increase which requires energy to run and also produces pollutants that can adversely affect the atmosphere. This is a good example of the conundrum of sustainable buildings – you can do something that will improve a particular aspect, but this may increase the demand for a counter system that could be more damaging environmentally in the long term.

Some 5–10 per cent of the total energy is used in the UK during the construction phase. This is a lot of energy and it surprises many people that such a significant amount of energy is used during this phase. This is because of the embodied energy required to manufacture building components. **The embodied energy of a material is the amount of energy required for its manufacture**. Materials such as steel require a huge amount of energy to manufacture and this energy use has to be calculated into the overall energy requirement of a building during construction and in use. However even this does not tell the whole story. In order to gain a greater understanding of the 'real energy requirement' you would need also to calculate the energy required to transport the material to its place of installation, the amount of energy required to install it, perhaps also the energy required for the operatives installing the material or system to travel to the site and even maybe the amount of energy used by those operatives powering the water boiler to make their tea!

The other major impact of construction work on the environment is waste. Reducing waste in construction is the intervention that could have the biggest impact on the reduction of mineral extraction and the overuse of materials. The construction industry generates approximately 70 million tonnes of waste annually. At £15 per tonne for disposal this equates to just over £1 billion annually. To put this into context, a reasonably sized and complex hospital would cost around £100 millon to build, so £1 billion would build 10 hospitals. Wembley Stadium cost £798 million to build – with the money spent on disposing waste another Wembley could be built with plenty of money to spare.

Waste is generated in different ways:

- *Design waste*: this occurs when designs change during the construction phase and new materials are required. The original materials if already delivered are then redundant.
- *Process waste*: this occurs when materials are ordered but the standard order size means that there will always be waste. For example, timber is ordered in multiple lengths of 300 mm. Therefore if a 2.5 m piece of timber is required, a standard length of 2.7 m will be ordered and 200 mm will be waste. More of this type of waste is generated in wet trades undertaken on-site such as brickwork and plastering, so using system building techniques more frequently could be a major waste saver.
- *Estimating waste*: this occurs when materials are overordered, delivered to site and not used. This accounts for approximately 20 per cent of all construction waste. It needs to be considered whether this waste can be used on other sites rather than being placed in skips.
- *Damage waste*: this occurs when materials are delivered and not stored properly. The materials become damaged and unusable.

Reducing waste is, consequently, viewed as one of the most important factors that needs to be addressed in making construction work more sustainable. In many areas of the UK there are increasing numbers of facilities to sort waste into different categories and then to sell on for recycling. This is a reflection on what is happening with domestic waste which many Local Authorities require to be sorted into recyclable and non-recyclable bins. The types of materials that are recyclable are growing. However, this needs to be managed carefully on-site as different skips/bins can easily become contaminated if the wrong materials are placed in them. Demolition industry contractors already have an excellent record for recycling waste (over 90 per cent of demolition 'waste' was reused or recycled in 2005–06 – NFDC figures). However, to date, the construction industry as a whole has paid less attention to the possibilities of recycling, reuse and reclaiming of materials, but this has to change.

In this third edition of *Construction Technology 2*, technologies that have been developed to reduce mineral extraction, increase energy and reduce waste are introduced in the relevant chapters. Some of the technologies, such as timber framing for multi-storey buildings to reduce the need for steel and concrete and the use of MMC techniques to reduce waste, were included in the previous editions, but the growing acceptance of these techniques as more 'sustainable' practices is emphasised further. In addition, construction technologies that could be badged as 'green' are integrated into chapters. The authors believe that by integrating these green technologies as opposed to having a 'green technologies' chapter, their choice as potentially realistic solutions will be better evidenced. If these technologies are dealt with separately then potentially they will only be used when a building is required to be show-cased as being green. This approach leads to tokenism rather than a holistic view of the building from a sustainable perspective.

REFLECTIVE SUMMARY

With reference to the building as an environmental shelter, remember:

- Heat needs to be preserved, while allowing light into a building.
- Thermally heavy structures tend to be heavy and intercept heat by absorption.
- Aesthetics of buildings are very subjective and will depend very much on the nature and location of a building.
- A large amount of heat is generated by equipment in industrial and commercial buildings, and this needs to be seriously considered in the design of these types of building.
- The increasing recognition of the impact of construction on the environment has led to a recognition of *sustainability* as a core function of building.

REVIEW TASK

- What are the advantages of *thermally light buildings* and *thermally heavy buildings* with regard to the performance requirements of multi-storey buildings?
- Visit the companion website at www.palgrave.com/engineering/riley2 to view sample outline answers to the review task.

1.2 Forces exerted on and by buildings

Introduction

- After studying this section you should be aware of the forces that act upon the structure of a building.

- You should have developed a knowledge of the origins of these forces and the ways in which they act upon the structural elements of a building.
- You should also have an intuitive knowledge of the magnitude of the different forces and be able to distinguish between the forces acting on a building and the forces exerted by it.

Overview

The forces applied to buildings derive from a variety of sources and act in many different ways. However, a number of basic principles of structural behaviour can be considered to encompass all of these applications and effects. The ways in which the structure and fabric of a building behave will depend upon their ability to cope with the inherent and applied loadings. If the building is able to withstand the loadings imposed upon it, it will remain static – in such a state it is considered to be stable. Any force acting upon a building must be considered as a loading, whether it be from the actions of wind on the building, the positioning of furniture, equipment or people, or simply the effect of the self-weight of the structure.

In order to withstand such forces two basic structural properties must be provided by the building:

- The component parts of the building must possess adequate strength to carry applied loads.
- Applied forces must be balanced, to resist the tendency for the building to move. That is, the structure must be in equilibrium.

The forces, or loadings, applied to buildings can be considered under two generic classifications, *dead loads* and *live loads*. Dead loads would normally include the self-weight of the structure, including floors, walls, roofs, finishes, services and so on. Live loads would include loadings applied to the building in use, such as the weight of people, furniture, machinery and wind loadings. Such loadings are normally considered as acting positively on the building; however, in the case of wind loadings, suction zones may be created, i.e. negative loadings; this effect is often illustrated by the action of roofs being lifted from buildings in high wind conditions. Hence buildings must be designed to cope with forces acting in a variety of ways.

The ability of the materials used in the construction of buildings to withstand these loadings is termed *strength*. In considering whether a building has sufficient strength, the nature of loadings must be considered.

Stress

When subjected to forces, all structural elements tend to deform, and this deformation is resisted by stresses, or internal forces within the element. If these stresses do not exceed the level which can be satisfactorily withstood by the material then the building will remain structurally sound.

Types of stress

The formula used to calculate stress is W/A, where W = load and A = cross-sectional area, and **stress** is measured in N/mm^2 or kN/m^2.

Compressive stress

Compressive stress (Figure 1.7) is the internal force set up within a structural element when an external applied force produces a tendency for the member to be compressed. An element is said to be in compression. Some materials, such as concrete, are very good at resisting compressive stress, whereas others, such as steel, are not so good. The amount of stress in a structural member will increase if the load is increased due to an increase in dead loading caused by the weight of the structural elements.

Figure 1.7
Compressive stress.

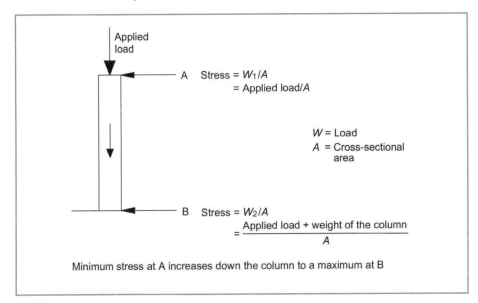

Tensile stress

Tensile stress (Figure 1.8) is the internal force induced within an element which resists an external loading that produces a tendency to stretch the component. When such a force is applied the member is said to be in tension.

Steel is very good at withstanding tensile stress, whereas concrete is not. As different materials are better at withstanding the different stresses, the use of two materials to

Figure 1.8
Tensile stress.

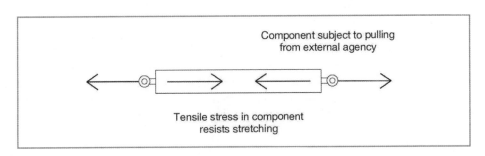

Figure 1.9
Stresses induced in reinforced concrete beams.

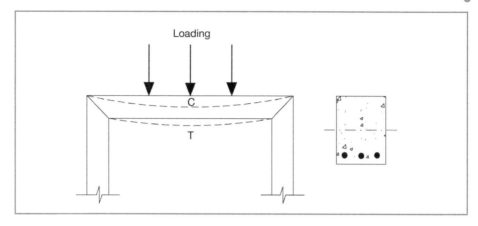

produce structural elements is common. Reinforced concrete beams (Figure 1.9) are an example of this.

The top of the beam is in compression, so the fibres are being crushed together; the bottom of the beam is in tension, and the fibres are being pulled apart. The cross-section of the beam shows that steel has been placed at the bottom to withstand the tension, whilst the top half of the beam is concrete to withstand the compression. This makes for a very economical beam section, taking advantage of the properties of both materials.

Shear stress

Shear stress (Figure 1.10) is the internal force created within a structural element which resists the tendency, induced by an externally applied loading, for one part of the element to slide past another.

Figure 1.10
Shear stress.

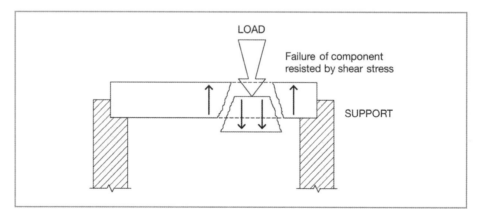

Torsional stress

Torsional stress (Figure 1.11) is the internal force created within a structural element that resists an externally applied loading which would cause the element to twist.

Strain

The effect of a tensile or compressive stress on an element is to induce an increase or decrease in the length of the element. The magnitude of such a change in length depends

Figure 1.11
Torsional stress.

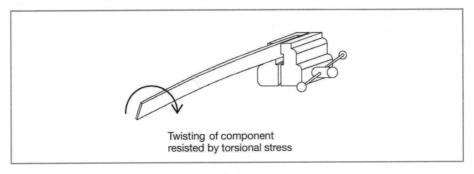

Twisting of component
resisted by torsional stress

upon the length of the unit, the loading applied and the stiffness of the material. The relationship between this change in length and the original length of the component gives a measure of *strain* (*e*):

$$e = \delta l / L$$

where δl = change in length and L = original length.
Strain has no unit.

This effect is also evident in materials subject to shear stress, although the deformation induced in such cases tends to distort the element into a parallelogram shape.

The relationship between stress and strain (subject to loading limits) is directly proportional and is a measure of the material's stiffness. The ratio of stress to strain is known as the *Young's modulus of elasticity*.

To calculate the value of a **moment** about a point, use the formula:

Force × distance the force acts at 90° from that point.

Moments

The application of a force can in certain instances induce a tendency for the element to rotate. The term given to such a tendency is **moment** (Figure 1.12). The magnitude of such a moment depends on the extent of the force applied and the perpendicular distance between the point of rotation and the point at which the loading is applied. As a result

Figure 1.12
Moments applied to a
building element.

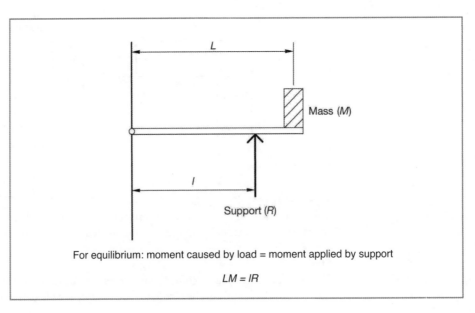

L

Mass (*M*)

l

Support (*R*)

For equilibrium: moment caused by load = moment applied by support

$$LM = lR$$

of the effects of leverage, relatively small loadings applied at considerable distances from the point of action can induce rotational forces. Moments upon structures must be in equilibrium in order for the structure to remain stable, i.e. clockwise moments (+) must be equalled by anticlockwise moments (–). The magnitude of a moment is the product of the force applied and the distance from the point of action at which it is applied (the lever arm) and is expressed in newton millimetres.

It is necessary to calculate the maximum bending moment that will be created in a structural element by the predicted loadings in order for the element and its supports to be designed.

REFLECTIVE SUMMARY

■ Loads derive from a variety of sources including:
  - Dead and live loads
  - Horizontal, vertical and oblique directions
  - Point loads, axial loads and uniformly distributed loads.
■ There is a relationship between loads, area and pressure.
■ Possible stresses induced in structural elements are: compressive, tensile, torsional and shear.
■ Some materials withstand the different stresses induced better than others.
■ Concrete is very good at withstanding compressive stress, whereas steel is very good at withstanding tensile stress.

REVIEW TASKS

■ Identify the differences between *compressive*, *tensile*, *torsional* and *shear stresses*.
■ Attempt to identify where these stresses may occur in a typical framed building.
■ Visit the companion website at www.palgrave.com/engineering/riley2 to view sample outline answers to the review tasks.

1.3 | Structural behaviour of elements

Introduction

■ After studying this section you should be aware of the implications of loads applied to structural building elements.
■ You should understand the terminology associated with the structural behaviour of buildings and the elements within them.
■ In addition you should appreciate the implications of the structural performance upon the selection of materials.

Included in this section are:

  - The nature of forces acting on buildings
  - The nature of building components

Overview

The effects of the types of loadings or forces exerted on a building depend on the way in which those forces are applied. Maintaining the integrity and structural stability of a building relies on its ability to withstand inherent and applied loadings without suffering movement or deformation, although it is possible to allow for a limited amount of movement or deformation within the building design, as is common in mining areas, for example. Resistance to movement and deformation results from effective initial design of the structure as a unit and the ability of materials used for individual components to perform adequately.

Limited movement, of certain types, is inevitable in all structures and must be accommodated to prevent the occurrence of serious structural defects. The effects of thermal and moisture-induced changes in building materials can be substantial, producing cyclical variations in the size of components. This dictates the inclusion of specific movement accommodation details, particularly when dealing with elements of great size, such as solid floors of large area. Additionally, the period shortly following the erection of a building often results in the minor consolidation of the ground upon which it is located; this will generally be very minor in nature however. These forms of movement and deformation are acceptable but other forms must be avoided, their nature and extent depending upon the nature and direction of the applied forces. Three main categories of applied forces combine to give rise to all types of building movement.

The nature of forces acting on buildings

Vertical forces

Vertically applied forces, such as the dead loading of the building structure and some live loadings, act to give rise to a tendency for the structure to move in a downward direction, i.e. to sink into the ground. The extent of any such movement depends upon the ability of the building to spread the building loads over a sufficient area to ensure stability on ground of a given loadbearing capacity. The loadbearing capacities of different soil types vary considerably, and the function of foundations to buildings is to ensure that the bearing capacity of the ground is not exceeded by the loading of the structure. In most instances the bearing capacity of the ground, normally expressed in kN/m^2, is very much less than the pressures likely to be exerted by the building structure if placed directly onto the ground. The pressure is reduced by utilising foundations to increase the interface area between the building and the ground (**foundation design**), thus reducing the pressure applied to the ground (Figure 1.13).

The need to withstand such vertical loadings is not exclusive to the lower elements of the building structure, although such loadings are greater in magnitude at the lower sections due to the effects of accumulated loadings from the structure. All structural components must be of sufficient size and strength to carry loadings imposed upon them without failure or deformation. Columns and walls, often carrying the loads of floors, roofs and so on from above, must resist the tendency to buckle or to be crushed by the forces exerted (Figure 1.14). The way in which columns and walls perform under the effects of vertical loadings depends on the *slenderness ratio* of the component. In simple terms, long, slender units will tend to buckle easily, whilst short, broad units will

Foundation design utilises the same principles as stress calculation. Increased loading will increase the stress induced in the soil, but increasing the cross-sectional area will reduce this stress.

Figure 1.13
Reduction of pressure applied to the ground resulting from the use of foundations.

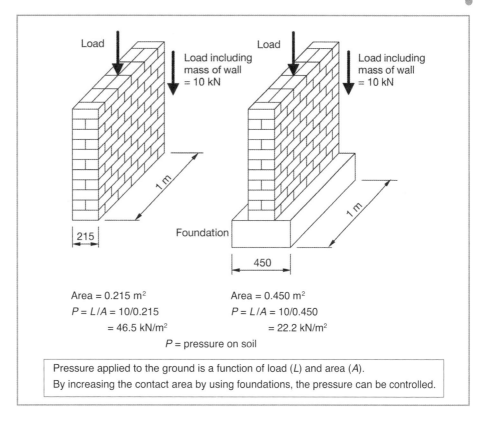

Load

Load including mass of wall = 10 kN

Load

Load including mass of wall = 10 kN

1 m

215

Foundation

1 m

450

Area = 0.215 m²

$P = L/A = 10/0.215$
  $= 46.5$ kN/m²

Area = 0.450 m²

$P = L/A = 10/0.450$
  $= 22.2$ kN/m²

P = pressure on soil

Pressure applied to the ground is a function of load (L) and area (A).
By increasing the contact area by using foundations, the pressure can be controlled.

resist such tendencies. In long, thin components the risk of buckling can be greatly reduced by incorporating bracing to prevent sideways movement; this is termed *lateral restraint*.

If overloaded significantly, even short, broad sections may be subject to failure; in such instances the mode of failure tends to be crushing of the unit, although this is comparatively rare.

Horizontal components, such as floors and beams, must also be capable of performing effectively while withstanding vertically applied loadings (Figure 1.15). This is ensured by the use of materials of sufficient strength, designed in an appropriate manner, with sufficient support to maintain stability. Heavy loadings on such components may give rise to **deflection** resulting from the establishment of moments, or in extreme cases puncturing of the component resulting from excessive shear at a specific point. When subjected to deflection, beam and floor sections are forced into compression at the upper regions and tension at the lower regions. This may limit the design feasibility of some materials, such as concrete, which performs well in compression but not in tension. Hence the use of composite units is common, such as concrete reinforced in the tension zones with steel.

Vertical forces applied to buildings may also be in an upward direction. These must be resisted, usually by making the best use of the mass of the building. Upward loadings may be generated from the ground, as in zones of shrinkable clay or soils that are prone to expansion due to frost, for example. The upward force exerted by the ground in such cases is termed *heave*.

Deflection of structural elements is allowable within given limits. Excessive deflection can lead to failure, but a more common problem is deflection to a level that produces unsightly cracking.

Figure 1.14
Effects of vertical
loading on columns
and walls.

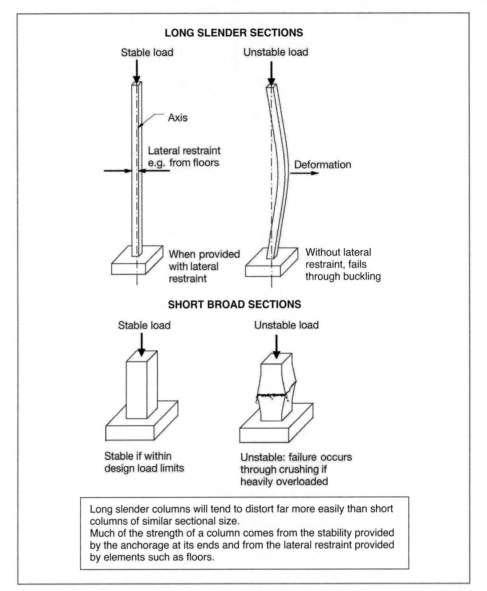

Horizontal forces acting on buildings derive from many sources and it is difficult to generalise about their origins and effects. Typically, however, such loadings may be exerted by
sub-soil pressure, as in the case of basement walls, or by wind or physical loading on the
building. The effects of such forces are normally manifested in one of two ways:

Horizontal forces

Horizontal forces acting on buildings derive from many sources and it is difficult to generalise about their origins and effects. Typically, however, such loadings may be exerted by
sub-soil pressure, as in the case of basement walls, or by wind or physical loading on the
building. The effects of such forces are normally manifested in one of two ways:

- Overturning, or rotation of the building or its components
- Horizontal movement, or sliding of the structure.

These forms of movement are highly undesirable and must be avoided by careful
building design. The nature of the foundations and the level of lateral restraint or

Figure 1.15
Vertical forces on
horizontal building
elements.

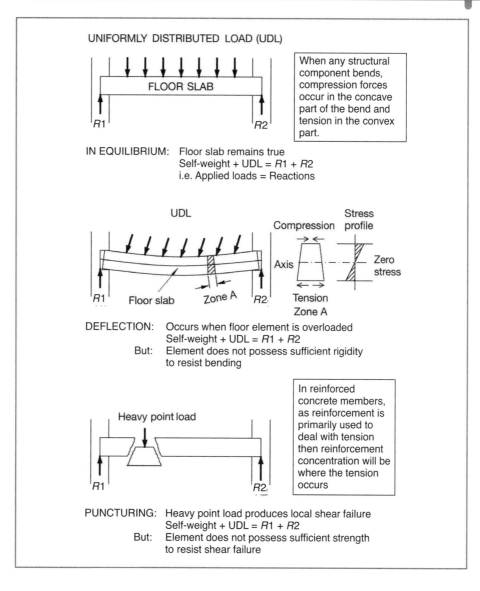

buttressing incorporated into the building design are fundamental to the prevention of such modes of failure. Additionally, particularly in framed structures, the use of bracing to prevent progressive deformation or collapse is essential (Figure 1.16); this could be described as resistance of the 'domino effect'.

Oblique forces

In some areas of building structures the application of forces at an inclination is common (Figure 1.17). This is generally the case where pitched roofs are supported on walls. The effects of such forces produce a combination of vertically and horizontally applied load-

Figure 1.16
Resistance to the
effects of horizontal
forces on buildings.

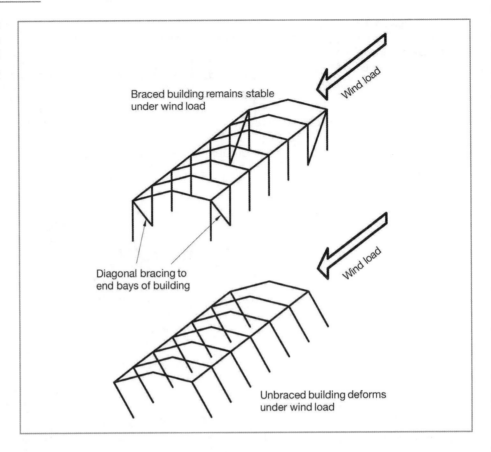

Braced building remains stable
under wind load

Wind load

Diagonal bracing to
end bays of building

Wind load

Unbraced building deforms
under wind load

ings at the point of support. These effects are resisted by the incorporation of buttressing and/or lateral restraint details. The horizontal effects of these loadings are sometimes ignored, with disastrous effects.

The nature of building components

The most common materials used for the construction of the supporting structure for multi-storey or large-span industrial and commercial buildings are:

- Reinforced concrete
- Structural steel
- Timber (this is a relatively recent development).

The properties of each of these systems, with regard to withstanding tensile and compressive stresses, have been discussed previously and can be summarised as follows:

- Structural steel as a material is excellent at withstanding tensile stresses, but may need to be overdesigned to withstand compressive stresses.
- Timber is an excellent material at withstanding tensile and compressive stresses but, because it is a natural material and defects can be 'hidden' within a section, the section sizes used may need to be increased.

Figure 1.17
The effect of oblique
forces on buildings.

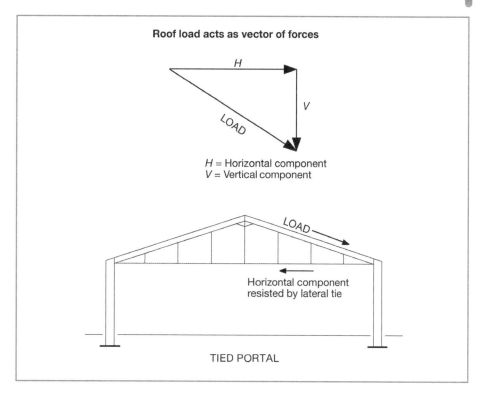

■ Reinforced concrete combines the compressive strength of concrete and the tensile strength of steel to allow for efficient and cost-effective frame design.

Although reinforced concrete is very efficient, the section sizes required can be very large, especially for massive structures. There are methods, however, that allow for smaller section sizes whilst retaining strength. These are known as pre-tensioning and post-tensioning of concrete.

The development of the use of this type of concrete started in the 1940s because of the shortage of steel that occurred during redevelopment after the Second World War. Using these methods, one tonne of prestressing can result in a structural element taking up to 15 times the load that one tonne of structural steel can take.

Principles

Stressed concrete can be defined as compressed concrete. A compressive stress is put into a concrete member before it is incorporated into a building structure and positioned where tensile stresses will develop under loading conditions. The compressive stresses, introduced into areas where tensile stresses usually develop under load, will resist these tensile stresses – for example at the bottom of a beam section, as has been discussed previously. The concrete will then behave as if it had a high tensile strength of its own. The most commonly used method for the pre-compression of concrete is through the use of tensioned steel tendons, incorporated permanently into a member. The tendons are usually in the form of high-strength wires or bars, used singly or made into cables known as tendons. The two basic methods used are pre-tensioning and post-tensioning.

Pre-tensioning

In pre-tensioning (Figure 1.18), the steel tendons are tensioned usually through a 'pulling' mechanism and the concrete placed in formwork around them. When the concrete has reached sufficient compressive strength, the steel is released. The force of the release is then transferred to the concrete. Pre-tensioning is usually carried out off-site in factory conditions. The formwork is manufactured to the correct size, and the tendons are positioned and then threaded through stop ends and anchor plates before being fixed to a jack. The correct amount of stress is then induced, the tendons anchored off and the jack released. The concrete is then poured. The bond between the concrete and the steel is vitally important and the steel must be kept perfectly clean to ensure the quality of this bond. When the concrete has cured, any temporary supports are removed and replaced with jacks that are slowly released. As the tensioned steel tries to return to

Figure 1.18
Pre-tensioning of concrete beams.

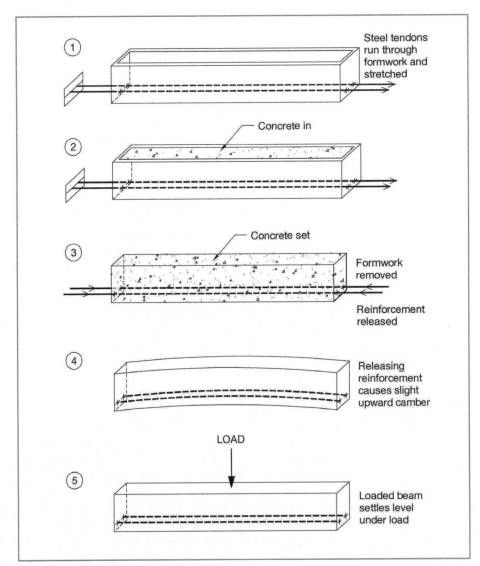

its original shape the bond between the concrete and steel will resist this and the concrete is placed in compression.

Post-tensioning

In post-tensioning (Figure 1.19), the concrete is cast into the formwork and allowed to harden before the stress is applied. The steel tendons are positioned in the correct position in the formwork, in a sheath to prevent the concrete and steel from bonding. The concrete is then placed and allowed to cure. The tendons are then tensioned by anchoring one end of the tendon and jacking against the face of the fixed steel at the other end, or alternatively by jacking or pulling the steel from both ends. When the desired load has been reached the tendon can be anchored off and the jacks released. When all the tendons have been stressed the ducts are filled with a cement grout under

Figure 1.19
Post-tensioning of concrete beams.

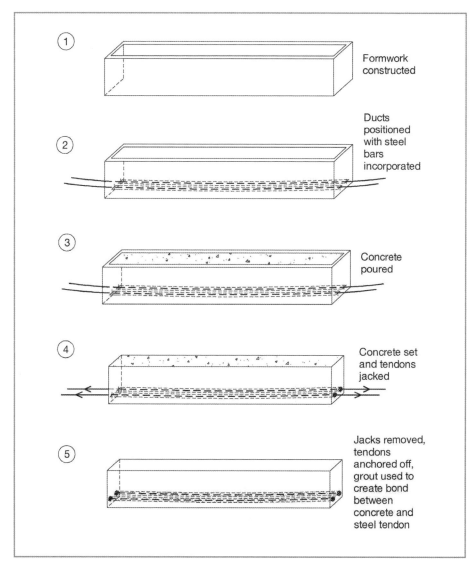

Figure 1.20
Steel tendons awaiting post-tensioning.

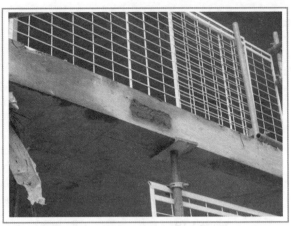

Figure 1.21
The edge of a post-tensioned concrete slab.

pressure. This grout prevents steel corrosion and creates a bond between the concrete and the tendon.

The curved profile of the steel permits the effective distribution of prestress within the member at positions in the member where the greatest tensile stresses are expected. Post-tensioning may be used in factory production both on- and off-site, but there are quality assurance issues relating to the production of post-stressed concrete on-site. Figure 1.20 shows reinforcing tendons in place in a concreted slab awaiting stretching.

In Figure 1.21 the tendons have been tensioned and the concrete made good. Propping is required until the concrete is fully cured.

As can be seen, the process of post-tensioning is undertaken on-site. As with all on-site activity, ensuring quality of processes is of paramount importance.

From a sustainability point of view, pre- or post-tensioning of concrete can be deemed to be positive actions. Tensioning the steel effectively means it is stronger and therefore less of it is needed to create a structural element that will carry the same loads. Therefore the amounts of new materials that are needed are also less. Also, if pre-tensioning techniques are used, they are done under factory conditions. This inevitably leads not only to higher quality but also to reduced waste.

REFLECTIVE SUMMARY
When considering the structural behaviour of building elements, remember:

- The direction of the applied loading is important in dictating the effect upon the building element.
- Failure of building elements can occur in a variety of ways such as:
    - buckling of slender columns
    - bending of beams and slabs
    - shear at support points
    - crushing of localised areas.
- Individual structural elements can act together to create a stronger form.

- Some deflection of elements is essential and acceptable, but if an element is allowed to reach its elastic limit due to overloading, the structural element may fail.
- Prestressing of concrete can increase its strength by up to a factor of fifteen.

REVIEW TASKS
- Briefly explain how *vertical*, *horizontal* and *oblique* forces evolve, and how the effects of these forces can be minimised in industrial and commercial building forms.
- Explain the principles of pre- and post-tensioning of concrete, and the advantages of these techniques.
- Visit the companion website at www.palgrave.com/engineering/riley2 to view sample outline answers to the review tasks.

PART 1

2 The building process

AIMS

After studying this chapter you should be able to:

- Demonstrate a knowledge of how methods of building for industrial and commercial forms have evolved
- Understand the term *system building* and be able to demonstrate an understanding of open and closed building systems
- Be able to outline the sequence of construction works required for given scenarios
- Understand to what extent expenditure on building work occurs
- Appreciate the costs required to manage buildings after completion and be able to demonstrate a knowledge of how life cycle costs are achieved

This chapter contains the following sections:

2.1   Methods of building
2.2   Building sequence
2.3   Expenditure on building
2.4   Costs in use and life cycle costing
2.5   Building Information Modelling (BIM)

INFO POINT

- BS 5964: Building setting out and measurement (1989)
- BRE Digest 53: Project network analysis (1976)
- BRE Report 367: Study on whole life costing (2010)
- *Achieving excellence in construction: whole-life costing and cost management*. Procurement Guide 07, OGC (2007)
- Bew, M. and Richards, M. (2008) *Bew–Richards BIM Maturity Model*
- Building Cost Information Service
- *CIOB occupancy costs*, CIOB construction paper 3 (1992)
- *Obsolescence in buildings: data for life cycle costing*, CIOB paper 74 (1997)
- *Planning and programming in construction: a guide to good practice*, CIOB (1991)
- *RICS Surveyors Construction Handbook* (2001)
- Smith, J. and Jaggar, D. (2007) *Building cost planning for the design team*, 6th edn, Butterworth–Heinemann [978–0–750–68016–5]

2.1 | Methods of building

Introduction

- After studying this section you should be able to discuss the issues surrounding the shift from traditional construction processes to more system-based approaches.
- You should also be able to appreciate why system building has not been as widely embraced in the UK as in other countries, and the reasons why system building can fail.
- You should also understand the need for accuracy in construction work, and will have developed an understanding of tolerances that are allowable.

Overview

Buildings are formed by the assembly of large numbers of individual elements and components, of varying size and complexity. Historically the production and assembly of these components would have taken place on-site. However, vernacular architecture has developed as a result of the ability to fabricate components from locally available building materials. This was dictated by limitations in the ability to transport materials and fabricated components over even modest distances. With advances in transport networks and technology, notably during the Industrial Revolution, it became possible to transport materials and components over large distances. These advances introduced not only the possibility of using non-local materials, but also of producing even sizeable components away from the site. The possibility to mass-produce components in a factory environment initiated a change in approach to the whole building process, with the beginnings of industrialised building.

Traditional building

Building form has historically been dictated by the availability of building materials, the local climate and the lifestyle of the population. The need of nomadic peoples to re-site their dwellings regularly imposes very different constraints on building form from those of people with a more static lifestyle. These forms of vernacular architecture could be considered to be the results of truly traditional building production. The nature of traditional building as it exists at present derives from this type of construction and the principles involved in it.

Traditional labour-intensive building crafts might include stone masonry, decorative leadwork, the building of brickwork arches, and architectural plasterwork.

The use of **traditional labour-intensive building crafts** in the production of buildings is now restricted to the building of individually designed, 'one-off' structures and to the area of specialised building refurbishment. The cost implications of such a building method are considerable, with highly trained craftspeople fabricating components on-site, although a number of prefabricated components would inevitably be used in all but very specialised cases. The nature of this method, in which most sections of the buildings are formed from a number of small parts, made to fit on-site, is inherently slow. Hence its use for large buildings or large numbers of buildings in today's economic climate, where time and cost are of the essence, is impractical.

There are a number of advantages in the adoption of this method, in that it allows tremendous flexibility, with the potential to ensure that all parts can be adapted to ensure a good standard of fit, since all parts are 'made to fit'. This allows for flexibility both during the construction stage and throughout the life of the building. Elements of this mode of construction are still to be found in some modern building, such as house building using traditional rafter and purlin roofs. In this case the roof structure is fabricated by carpenters, on-site, from straight lengths of timber. Even this type of traditional building has become rare, with increased levels of off-site fabrication. One of the main disadvantages of traditional building using on-site fabrication and the formation of components *in situ* is the difficulty of manufacturing components in a hostile location. The vagaries of the climate and the conditions which prevail on-site restrict the ability to work to fine tolerances and to fabricate elements with consistency. Although building is essentially a manufacturing process, the conditions in which it is undertaken are far removed from those in a factory environment. The disadvantages of cost, time and uniformity of standards of quality, together with difficulties in obtaining suitably skilled labour, have resulted in the adoption of what has become known as 'conventional' or 'post-traditional' building.

Post-traditional building

The evolutionary nature of the building process, with the periodic introduction of new materials and new techniques, has ensured that even traditional building has developed and progressed significantly in the past. Examples of such developments are the introduction of ordinary Portland cement, allowing the production of large, complex building sections by casting of concrete; and the development of reinforcing techniques using steel, allowing very strong sections to be produced. Such advances in building technology, allied with the need to minimise construction time and cost, have resulted in the adoption of the post-traditional method of building. It must also be noted that this would not have been possible without the advent of developments in transport mechanisms, allowing non-local materials to be transported to site and components to be produced some distance away and transported to site for use.

This form of building is a combination of traditional labour-intensive craft-based methods of construction with newer techniques, utilising modern plant and materials. The use of mechanical plant is one area where post-traditional building differs greatly from traditional building. Post-traditional construction is often adopted for the erection of buildings of large scale. In such instances, the craft-based techniques of traditional building, such as plastering and joinery, are not dismissed from the construction process, but rather are aided by the use of mechanised plant. Machinery used for earth moving, lifting of elements, and mixing of concrete and plaster is now essential on most building sites as a result of the magnitude of the operations being undertaken. Additionally, increased use is made of prefabricated components, manufactured in large numbers in factory conditions. Hence an element of industrialisation was introduced into the construction process. However, the nature of building design has not evolved to the extent of **system building**, where a series of standard parts may be used to produce the end result. Instead, the mass-produced components are placed and finished in a typically traditional way, using traditional crafts such as joinery and plastering. An example of this approach is the use of prefabricated trussed rafters for roof construction, in favour of the

The shortage of skilled labour and advances in factory-type production of components assisted in driving the adoption of **system building** in the middle of the 20th century.

traditional site-fabricated rafter and purlin form of construction. It is now very unusual to find the on-site fabrication of components such as windows and doors, which can be produced more cheaply and to higher quality standards in a factory environment. Although the traditional building activities have, in essence, changed little, the scale of post-traditional building, together with the demands of cost and time effectiveness, have placed great importance on the active planning of building operations. It is this planning to ensure efficiency which is one of the main trademarks of post-traditional construction.

Rationalised and industrialised building

Rationalised building, as considered today, is the undertaking of the construction of buildings adopting the organisational practices of manufacturing industries, as much as this is possible, with the previously described limitations of the construction industry. Such an approach to the construction process does not necessarily imply the adoption of industrialised building techniques, but is more usually based upon the organisation and planning of commonly used existing techniques. The key to the effective use of rationalised building is the ensuring of continuity of all production involved in all stages of the overall construction process. This continuity relies upon the evolution of building designs to allow full integration of design and production at all stages. The aim of this process of building is to ensure cost-effective construction of often large and complex buildings, within given time parameters, whilst maintaining acceptable standards of quality. This requires the construction process to be, as near as possible, continuous, hence necessitating the efficient provision of all resources, in the form of labour, plant, materials and information. This can, to some extent, be enhanced by the use of **standardised prefabricated components**, and the effective use of mechanical plant, thus separating fabrication from assembly and reducing on-site labour costs. It will be seen that this approach is a logical evolution from post-traditional construction.

The current shortage of skilled operatives in the construction industry could lead to further increases in the use of **standardised prefabricated components**.

In the later part of the 20th century, notably in the 1960s, great demands were placed upon builders to construct buildings quickly and cheaply. To cope with these demands, systems of building were developed based upon the use of standard prefabricated components to be assembled on-site, largely removing reliance upon traditional building techniques. Such systems attempted to introduce industrial assembly techniques to the building site. This approach is often termed *system building*. Within this description two basic approaches exist: *open system* and *closed system* building.

Open systems of building (often referred to as 'component building') utilise a variety of factory-produced standard components, often sourced from a variety of manufacturers, to create a building of the desired type. The construction of buildings adopting this approach makes little or no use of the traditional 'cut and fit' techniques of traditional and post-traditional building. An example of such an approach is the construction of lightweight industrial buildings, which are based upon designs that utilise a selection of mass-produced components which are not exclusive to the specific building. Hence flexibility of design is maintained.

In contrast, closed systems adopt an approach that utilises a dedicated series of components, specific to the individual building and not interchangeable with components made by other manufacturers. Such a method is beneficial to the speedy and efficient erection of buildings when aided by the use of large-scale mechanical plant. These systems, however, do not allow adaption of the design on- or off-site – this can be very restricting, particularly during the later life of the building, when changing user needs may require flexibility in the design. Such systems have also been subject in the past to many problems associated with on-site quality control and lack of durability of materials. These problems arose, in part, from a lack of familiarity of the workforce with new building techniques, the use of untried materials and the need to construct quickly, thus encouraging the short-cutting of some site practices. The occurrence of such problems and the inherent lack of flexibility in buildings of this type have resulted in the general rejection of closed systems in favour of open systems. There are also social issues that surround the use of closed system building. In certain cultures, as in the UK, uniformity is not acceptable and leads to an increase in unsociable behaviour, such as damage through graffiti. This was evidenced in multi-storey tower blocks and council estates where all houses were the same. Uniformity of design is much more accepted in mainland Europe and this needs to be taken into consideration alongside improved quality of work and reduced cost. If it produces good-quality and cheap buildings, then that is fine; but if such buildings will be quickly destroyed because they are socially unacceptable, then they are neither good quality nor cheap in the long term.

System building: some examples from the past

In order to examine the benefits and limitations of system building it is useful to examine some examples of the techniques that have been adopted in recent years. As previously noted, in the past the construction industry in the UK has been reluctant to embrace the principles of industrialised construction. However, there is a renewed interest and growing confidence in system building as systems are developed, having learnt valuable lessons from the past. The systems and techniques examined here are based on three well-known examples of 'non-traditional' construction. They are as follows:

- CLASP
- No-fines concrete construction
- Cross-wall construction.

CLASP

In the 1960s, eight Local Authorities agreed to combine their efforts to form a building consortium which became known as CLASP (Consortium of Local Authorities Special Programme). They attempted to act as a group of public authorities pooling their resources to develop and control a system of construction for their own use. The primary aim was to exploit the advantages of economy, quality and time which are inherent in industrialised building.

The Consortium organisation was financed by means of a contribution from each member, and additional income was obtained from licensing arrangements by which CLASP was made available commercially both in Britain and abroad.

One of the primary uses for CLASP buildings was the extensive school building programme that took place in response to the 'baby boom' of the 1960s. Many of these buildings were constructed on sites that were known to suffer from poor ground conditions, and many were subject to mining subsidence. The CLASP buildings that were developed aimed to cope with this by providing a low-weight building form that was able to cope with

Figure 2.1
CLASP construction.

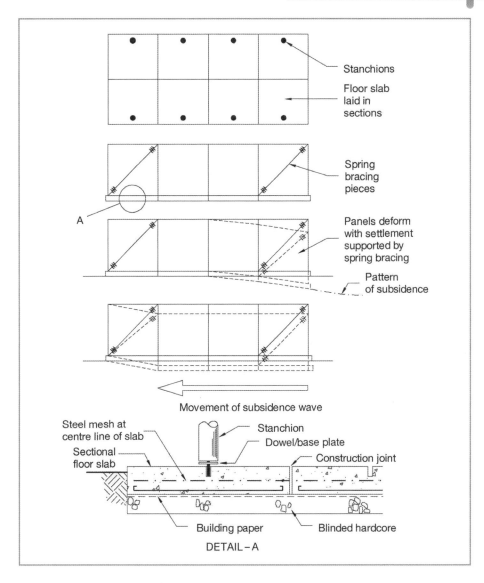

ground movement. This was achieved by creating a flexible sectional floor structure made up of reinforced concrete panels, with the reinforcement placed at the mid-point to act as a flexible link rather than to cope with tension in the structure (Figure 2.1). A lightweight superstructure was formed using a framed structure located to the floor using pin joints to allow movement and incorporating sprung bracing to ensure that the superstructure remained vertical and structurally stable whilst responding to ground movement.

No-fines concrete construction

No-fines concrete is made up from large aggregate bound together with a cementitious slurry. The resulting voids in the material assist in providing enhanced insulation properties over dense concrete.

The main principle of this form of construction is the casting of external walls formed of **no-fines concrete**, complete with internal plaster coat and external rendering in one operation. This is made possible by the use of shuttering incorporating movable steel sheets to divide it into three separate compartments. As the mixes are poured these sheets are raised to bring the three materials into direct contact with each other. This process was used extensively for the construction of houses, which also incorporated traditional elements such as timber upper floors and roofs. Other details were bespoke to the system and are quite different from traditional domestic construction. Typically a reinforced concrete raft foundation would be utilised; alternatively, a dense concrete upstand would be cast to form the section of wall up to damp-proof course (DPC) level (Figure 2.2). Above this level the walls would be cast *in situ* using no-fines concrete with 'clinker' aggregate. These would typically be in the region of 200 mm thick and would be poured at the same time as the external render coat and the internal plaster undercoat. Steel reinforcement bars were incorporated to provide support around openings and dense concrete lintels were used for larger openings. A traditional timber roof would then be formed to complete the enclosure, with internal partitions formed using no-

Figure 2.2
No-fines concrete construction.

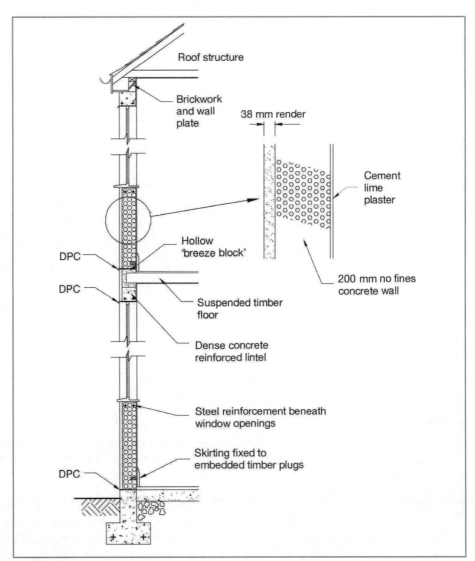

fines concrete of 100 mm thickness. The ground floor would generally be of traditional solid construction.

Cross-wall construction

The principle of cross-wall construction relies on the use of precast reinforced concrete panels to act as structural cross-walls to the building. These would be craned into position and located using threaded steel dowels to create the main structural elements of the building. These systems were used extensively to construct flats and retail units and were generally clad with a brick outer skin.

The process of construction is indicated in Figure 2.3. The concrete raft is formed first, and this provides the base for the location of precast cross-wall panels. The panels would be cast in a factory and would incorporate the cavity wall ties necessary to tie the brick

Figure 2.3
Cross-wall construction.

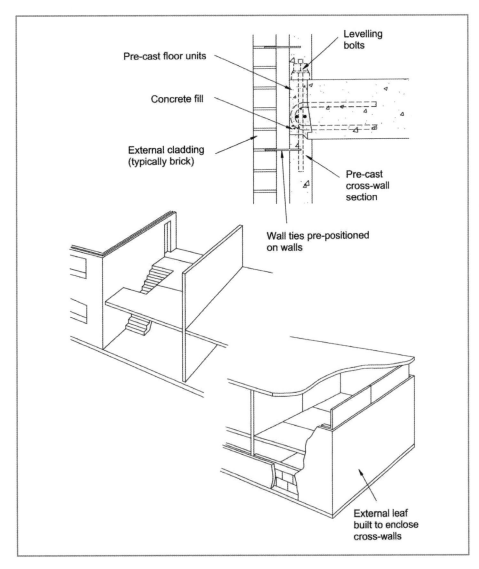

cladding to the structure. Having located the ground floor panels, precast concrete floor units are placed, supported off the cross-walls. The upper cross-walls are then positioned and the gap at the exposed edge of the floor units is filled with concrete grout *in situ*. The flat roof panels are then located in the same fashion as the upper floor and the front, rear and side elevations are enclosed with brick cladding or traditional cavity wall construction.

Modern methods of construction

The term 'modern methods of construction' (MMC) does not yet have a definition that is accepted, universally, by the construction industry. However, it is generally accepted as a term used to refer to a range of construction methods that seek to introduce benefits in terms of production efficiency, quality and sustainability. The methods that are currently being introduced into the building industry in the UK and elsewhere differ from 'traditional' approaches significantly. Many of the approaches considered as MMC involve prefabrication and off-site manufacture. There is an ongoing debate regarding the ways in which MMC could, and should, be classified. There are strong arguments for adopting a classification system based on the performance attributes of the building and the building process, such as quality and efficiency of production. An example of a system that classifies MMC buildings by construction form can be summarised as follows:

- Off-site manufactured – Volumetric
- Off-site manufactured – Panelised
- Off-site manufactured – Hybrid
- Off-site manufactured – Sub-assemblies and components
- Non-off-site manufactured modern methods of construction.

These classifications will be used for the purposes of this text although many other terms that can be used to refer to MMC are in common usage. Commonly used terms include:

- Modular building
- Industrialised building
- System building
- Prefabrication.

The very nature of these types of construction should reduce waste which can improve the sustainability performance during construction. Additionally, some forms of construction which can be badged as MMC can also lead to energy efficiency improvements via the structural form. An excellent example of this is the use of Structured Insulated Panels (SIPs). SIPs consist of two facing boards of 11 mm OSB (oriented strand board) bonded by pressure injection to CFC-free/ODP-zero polyurethane closed cell foam. The facings and core act as a composite structure. The board comprises softwood flakes/strands (from sustainable sources) bonded together with resin, binder and waxes. The boards can provide an exceptionally low U value in the external wall, meet all the requirements of the Building Regulations with regard to fire resistance and provide as durable a building as traditional systems.

The advantages of SIPs from a sustainability perspective are:

- They are environmentally friendly as the timber flakes derive from sustainable forest resources
- They have excellent thermal performance – through the insulation, lack of thermal bridging and airtightness, and thus reduce energy use
- They save construction time and cost and therefore energy used on-site
- They are a lightweight form of construction and therefore the size of foundations can be reduced.

Figure 2.4 shows a steel frame with the SIPs in the process of being fixed.

Figure 2.4
Structured insulated
panel construction.

Other examples of MMC will be discussed in Part 3.

Background to the development of MMC

Skills shortage

In recent years the nature of the construction industry, and those entering it, have changed. The industry has suffered from problems of participation in training, and the 'boom or bust' image of construction has discouraged entry to some. Although the industry has made significant efforts to increase participation in training and many initiatives have been developed, this has tended to benefit the individuals that are employed permanently within organisations. In addition, the increased use of contract labour means that many workers do not benefit from these training programmes. The problem is exacerbated by increased activity in the industry at all levels. Hence, the problem of skills shortages has become a major issue in terms of both construction operatives and supervisory staff. Clearly this has implications on quality and pace of building.

More recently there has been a migration towards partnering and other forms of procurement that positively encourage stable employment and participation in training. However, this alone cannot address the skills shortage problem. The use of MMC techniques, where much of the process leading to the construction of the building is undertaken in a factory production environment, reduces the need for skills on-site. Repetitive manufacture of components and building sections in a controlled factory environment assists quality assurance and reduces the 'risks' associated with quality control on-site. Quality control is still extremely important on the construction site, however it can be controlled more readily if the extent of site-based work is reduced and simplified.

Quality enhancement

The previous section relating to skills shortage made reference to the issue of quality assurance. The reputation of the construction industry in terms of its ability to maintain quality is mixed. Over the years much has been written about declining quality although the true situation is difficult to measure effectively. Without question there have been periods during which the industry has suffered from visible quality assurance problems; the aftermath of the 1950s and 1960s system building programme is a good example. In reality the issue of quality that is currently the concern for the industry hinges on enhancement rather than assurance.

The industry has sought to improve quality through advances in technology and training, while at the same time Government agenda for improved productivity, sustainability and 'right first time' principles have changed the context of the construction sector. Added to this, the increases in customer expectation, particularly in the housing sector, have forced a continuous programme of quality enhancement. The vagaries of the construction site environment and the fluidity of the contract labour market result in limitations upon the ability to enhance and maintain quality using traditional construction processes. Hence, the use of MMC using controlled factory manufacture, with site processes effectively limited to assembly of pre-manufactured buildings, provides an effective solution to the quality issue.

Developments in Building Regulations

The Building Regulations and their evolution were discussed extensively in Chapter 2 of *Construction Technology 1*, and it is not the intention to repeat that material here (but see section 3.2). However, the evolution of the Building Regulations and other similar frameworks outside England and Wales has had a significant impact upon the potential for the adoption of MMC in building construction. The requirements for threshold performance levels in terms of thermal and acoustic insulation and overall energy conservation demand a predictable and assessable construction approach. The potential for performance assessment of building post-construction introduces a much greater drive for repeatable, reliable construction detailing than was previously necessary. Clearly the use of factory manufacturing techniques lends itself to achieving reliability far more than does traditional site-based techniques.

Sustainability and environmental performance

It is generally accepted that around 10 per cent of all building materials that are ordered and delivered to site are unused and go to waste. The implications of such a wasteful construction process in terms of economics of production and sustainability are great. The increased focus on the environmental implications of construction in terms of process and whole-life performance places great emphasis upon the sustainability of building and approaches to building. The adoption of MMC techniques using factory manufacture, efficient materials and supply chain management provides for a much less wasteful construction process. In addition, the quality control of MMC can result in higher levels of airtightness, thus increasing thermal performance of the building in use. A further environmental benefit of factory production is that it removes much of the local environmental disturbance from the immediate site environs. Hence, sites adopting MMC will have lower levels of noise, dust and general nuisance than do equivalent 'traditional' sites.

REVIEW TASK

■ Consider a typical commercial building development that you are familiar with. How might the use of MMC affect the design, production and costs of the development?

■ Visit the companion website at www.palgrave.com/engineering/riley2 to view sample outline answers to the review task.

Accuracy in building

In traditional construction, accuracy in building was to some extent ensured by the ability to make components to fit specific spaces. This ability to be flexible in the production of elements of the building has been removed to a great extent, as many components are fabricated away from the site. The industrialised manufacture of components dictates an increased emphasis on the accuracy of component sizes in order to ensure that mismatches are reduced to a minimum on-site.

The use of factory-produced building materials and components in even small-scale post-traditional construction has resulted in some standardisation of building dimensions. For example, the widespread use of plasterboard, which is manufactured in a range of sizes that are based on multiples (or 'modules') of 600 mm is made more efficient if minimal cutting of panels is required. Hence room sizes based on 600 mm modules are common, thus reducing fixing time and wastage of materials on-site. Such an approach is also evident in the manufacture of components which are designed to fit into openings in brick walls, such as windows, which are manufactured in a range of sizes which correspond to multiples of whole brick sizes. This is termed a 'modular approach' to component size. The approach is based on simple logic and is of increased importance when related to the construction of larger, more complex buildings. When dealing with larger buildings, the size and number of components which must be assembled increase substantially. The degree of accuracy with which they are assembled must be adequate to ensure that the building is erected without undue difficulty and without the risk of compromising its performance. The modularisation of such buildings is immensely beneficial in maintaining an acceptable degree of building accuracy, and this is sometimes effected by the use of 'dimensional coordination'. Dimensional coordination relies on the establishment of a notional three-dimensional grid within which the building components are assembled. The grid allows for some variation of component size, providing a zone within which the maximum and minimum allowable sizes of a given component will fit. This variation in the sizes of elements of buildings is inevitable for a variety of reasons, including the following:

■ Some inaccuracy in manufacture is unavoidable, due to manufacturing technique. The production of a component in concrete is subject to size variation as a result of drying shrinkage following casting, together with great limitations in the manufacture of very accurate formwork.

■ The cost of producing components with great accuracy may be substantial and considered unnecessary in a given situation; hence a degree of variation in size may be accepted.

■ The accuracy of location of the component, resulting from fixing variations, also has some effect.

Tolerances for the accuracy of building work are set out in the British Standards. Wet trades tend to have higher acceptable tolerances than dry trades, and this can prove problematical when a dimensionally perfect element such as a metal window is to be fitted into a brickwork wall, where minor inaccuracies in dimension are acceptable.

Hence building components are not designed to fit exactly into a given space or position of a given dimension, and instead allowances for jointing and component linking, taking into account possible variations, are made. For these and other reasons (including the need to allow for thermal and moisture-induced variation in size following construction), a degree of allowable variation or '**tolerances**' in component size is an essential feature of modern building. The use of modular design and the allowance for tolerance whilst maintaining acceptable accuracy are of particular importance in the design and construction of system-built structures. In such buildings, ease, and consequently speed, of site assembly are of paramount importance. In these situations, the ease of assembly of the parts of the building depends upon a number of factors including:

■ The degree of accuracy with which components have been manufactured
■ The degree of setting out accuracy on-site
■ The nature of construction and assembly methods on-site
■ The nature of jointing of components and the degree of tolerance which is acceptable in a given situation.

It must always be remembered that the production of components to very fine degrees of accuracy has a direct cost implication, which may not be justifiable in a given situation. For this reason and those noted above, it is common to refer to component sizes of an acceptable range, rather than an exact dimension. This may be given in the form of a nominal dimension and an acceptable degree of variation, e.g. nominal size 1200 ±10 mm. The degree of required accuracy depends on the exact situation of the components, but may be of particular importance where structure and services interrelate, since services are generally engineered with finer degrees of tolerance than is the building fabric.

REFLECTIVE SUMMARY

■ The building process has evolved over many years and is constantly evolving with each and every new innovation adding a new dimension.
■ System building leads to better quality buildings only if the systems are fully understood by the constructors.
■ Extensive system building can only be facilitated if there are enough suppliers of standardised components.
■ System building may be socially unacceptable because of uniformity of design.
■ Open system building allows for more design flexibility than closed system building.
■ Closed system building is the most effective way of reducing costs of building work and duration of construction, and of improving quality.
■ Accuracy of construction work is essential, and it is far easier to achieve a high level of accuracy using 'dry systems' of construction than 'wet systems'.
■ A great deal has been learnt about using system building methods from the failures of past buildings.
■ Advances in industrial manufacturing processes and approaches to the building process itself have resulted in the increased use of modern methods of construction with benefits in terms of time, cost and quality.

REVIEW TASK
- Think about the traditional construction of a commercial building. How many of the elements could feasibly be constructed using system building techniques, and what form could these systems take?
- Visit the companion website at www.palgrave.com/engineering/riley2 to view sample outline answers to the review task.

2.2 | Building sequence

Introduction

- After studying this section you should understand the link between knowledge of construction technology and effective sequencing of construction activities.
- You should also be able to put together simple logic diagrams for basic building forms and understand some of the terminology that is used in construction programming and the techniques that can be used to present a construction programme.
- An appreciation of the use of method statements to identify the quickest, cheapest and most safety-conscious methods of working should also be developed.

Overview

Most construction programmes are expressed in weeks, with the major time-scale in months. However, where the project is large and takes place over a long period, major time-scales may be in years with the minor time-scales expressed in months. For short-term programming a more detailed study is often necessary, and this will probably be expressed in weeks on the major scale and days on the minor scale. The reason for this is that the study needs to be at a 'fine' enough level to enable resources to be assessed so as to ensure efficient use of labour and key items of plant. This type of planning is sometimes called a 'sequence study', which, simply put, is a detailed analysis of a construction activity at resource level. This is often carried out on projects where there are alternative methods of construction available which need to be assessed in detail for overall economy and efficiency.

Alternative forms of constructing the same element of a building

In some instances there may be a variety of different ways of constructing the same type of building. This will depend upon the nature and location of the site, the amount of work to be completed inside the building that may require a watertight enclosure to be completed quickly, the skill and past experience of the site management team, and the quality requirement of specific elements of the building.

An example of this might be the ground floor slab construction for a new factory. The contractor may have intended to construct the slab in six bays using the 'long-bay' technique at tender stage and accordingly based the tender bid on this method. However, once the contract is won, a large-scale 'single' pour might have been suggested as an alternative.

Thus the two methods would need to be analysed in detail to see which offered the best solution in terms of direct and indirect cost. For instance, the original solution might be cheaper but take longer, whilst using a single pour with a laser screed might be more expensive in terms of direct costs but quicker and therefore less expensive in terms of preliminaries.

The precise economics of the situation would need to be established through a detailed analysis of the sequencing, resources and overall time-scale of each method so that the project manager can make an informed choice before committing the company one way or the other.

Making sure that there is continuity of work, therefore, is an important aspect of the planning process, irrespective of whether the main contractor is doing the work or whether it is sublet, or both. This is another reason for doing a 'sequence study'.

Excavation work is an aspect of the programme which may need a sequence study approach, especially where there are potential complexities associated with the work in hand. It is one thing having a large amount of bulk excavation to do, but quite another when this has to be done in stages working around other activities such as temporary works, piling, concreting or drainage.

Where there is a lot of 'muck' to get at it is fairly straightforward to size the fleet of wagons or dump trucks to the output of the excavators and thereby get an efficient dig–load–haul–dump–return 'cycle'. Where there are complexities, however, it is not so easy and a detailed sequence study of the activities concerned may be required. This again may involve 'balancing' resources and compromising on the 'ideal' gang size and working cycle normally anticipated. In this situation it is sometimes easier to undertake a *method statement* approach to evaluating which would be the most effective method to use. Method statements are a means of describing the manner in which various parts or elements of work are to be undertaken. They describe the planned sequence of events for a particular site operation and may also include how the health and safety issues involved in carrying out the work are to be addressed.

R.L.: Reduced level.

Banksman: The person that directs the machine driver.

Hydraulic power shovels: Commonly known as bulldozers.

Hydraulic excavators: Commonly known as JCBs, although this is a trade name.

Worked example

The example below shows how for one activity a method statement can be used to assess the best method possible.

Reduce Level Excavation is to be carried out on a town centre site to be used for a low-rise industrial building; area = 300 × 300 m.

Average depth of excavation = 0.6 m, to be taken away to local tip.

Table 2.1 compares the available methods.

Notes to Table 2.1 (*opposite*)

- You only need include variables – for example the cost of tipping is the same for each method.
- The choice of methods must be realistic. For example, in Method 1 if you used 12 machines then the duration would be reduced to 16 days, but this would create serious health and safety problems and be very expensive. It will be difficult even to procure 12 machines for such a short duration.
- Do not overcrowd the site with plant – see above.
- Consider other activities which could occur at the same time as the activity in question.
- You *must* consider health and safety issues when preparing the statement.

Table 2.1 Comparison of excavation methods.

Operation	Quantity	Method	Plant and labour	Output	Duration
Method 1 Excavate to R.L.	54,000 m³	Use hydraulic excavators fitted with backactor arms. Excavators fitted with 1 m³ buckets Load direct onto lorries *Costings*: Hydraulic excavator hourly rate £30.00 = 4 × (30.00 × 8 × 49) = £47,040 Banksman hourly = £7.20 = 4 × 7.20 × 8 × 49 = £11,289 Total cost = £58,329	4 hydraulic excavators 4 banksmen	35 m³/hour × 4 = 140 m³/hour	54,000/140 = 385 hours 385/8 = 49 days
Method 2 Excavate to R.L.	54,000 m³	Use 3 No. bulldozers to push excavated material to spoil heaps adjacent to road. Use 3 No. hydraulic excavators to load onto lorries from spoil heaps *Costings*: Bulldozer hourly rate £40 = 3 × (40 × 8 × 19) = £18,240 Excavator hourly rate £30 = 3 × (30 × 8 × 56) = £40,320 Banksman hourly rate = £7.20 = 2 × 7.20 × 8 × 75 = £8,640 Total cost = £67,200	3 bulldozers 2 banksmen 3 excavators 2 banksmen	120 m³/hour × 3 = 360 m³/hour 40 m³/hour × 3 = 120 m³/hour	54,000/360 = 150 hours 150/8 = 19 days 54,000/120 = 450 hours 450/8 = 56 days

Summary

Method	Cost	Duration	Comments
1	£58,329	49 days	This is a feasible method and avoids double handling of the spoil.
2	£67,200	19 days to clear site so that foundation work can start Further 56 days to remove spoil	This method will allow foundation work to start after 19 days, which is a saving of nearly 30 days. Using Method 1 some foundation work could start before the reduced level excavation is complete, but work would still be delayed. It will cost £8,871 more using this method, but the time saving would more than compensate for this. This method is however only suitable if there is enough room on the site for the spoil heaps not to encroach on any other works.

For notes to table, and a marginal note, see opposite.

Line/logic diagrams

Once the method of work has been decided on and the duration for each activity calculated, then a line or sequence of work diagrams needs to be drawn. This is a simple diagram that shows the relationships between activities. There are different types of activity that can be shown:

■ *Sequential activities* are those that follow on one from the other – one activity must be complete before the next activity starts. In Figure 2.5, activity A must be complete before activity B can start, which in turn must be complete before activity C can start.

Figure 2.5
Sequential activities.

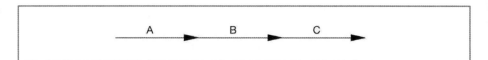

■ *Parallel activities* are those activities that can be carried out independently of others at the same time. In Figure 2.6, on completion of activity A, activities B and D can commence. Activity B must be complete before activity C can commence and activity D must be complete before activity E can commence. There is no relationship between activities C and E.

Figure 2.6
Parallel activities.

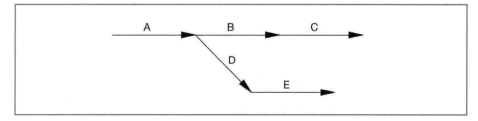

■ *Restraining activities* – all activities do not have to wait for the preceding activities to be finished before they can be started. This is shown in Figure 2.7. The formwork can start before all the reinforcement is complete, and the concrete can start before all the reinforcement is fixed.
■ *Dummy activities* are activities that have no duration and are represented with a broken line. They are not actual site activities, but show interrelationships between activities, therefore giving a logical sequence. Figure 2.7 shows that activities B and D must be completed before activity C can start, but only activity D needs to be complete before activity E can start.

Methods of presentation

Once the diagram has been drawn, it can then be reproduced as:

■ A bar chart
■ A critical path analysis diagram, and then a bar chart
■ A precedence diagram, and then a bar chart.

Figure 2.7
Restraining activities
and dummy activities.

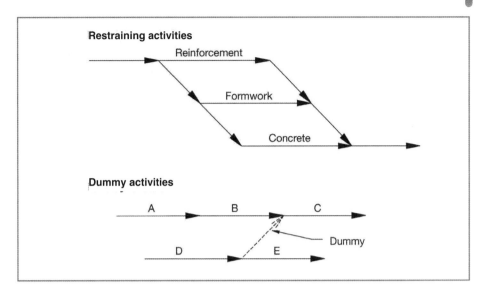

Precedence diagrams are probably the most popular tool, as they enable more complex relationships to be shown. The relationships between activities may be:

- Start to Start
- Finish to Start
- Finish to Finish
- Start to Finish.

These relationships are shown in precedence diagram format in Figure 2.8.

Figure 2.8
Precedence
relationships.

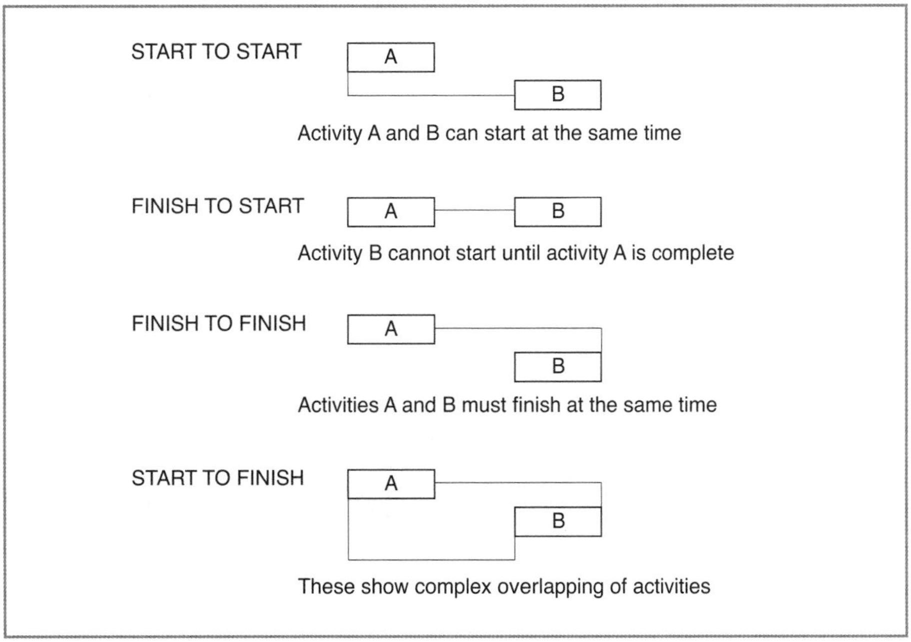

START TO START
Activity A and B can start at the same time

FINISH TO START
Activity B cannot start until activity A is complete

FINISH TO FINISH
Activities A and B must finish at the same time

START TO FINISH
These show complex overlapping of activities

All of these relationships can include a duration known as either *lead in* or *lag times*. These can show delays in the relationships. For example if the Start to Start relationship example showed a number 2 on the link line, this would mean that activity B can start 2 days after activity A has started.

Once the diagram is complete and the durations added, calculations can be undertaken to identify the *critical path*. This comprises the activities that have to start and finish on prescribed dates in order for the project to be completed on time. When the critical path has been identified, all other activities will show an element of *float*, which is the amount of time by which an activity can run over its predicted duration and the project will still be completed on time.

These days, manual calculation is not required as there are numerous computer software packages that will do this for you and produce a bar chart. However, you have to be aware that the computer will only produce a good bar chart if the information that is input is of good quality. Only a sound knowledge of construction technology and the sequencing of construction work will enable a good quality programme to be established.

Figure 2.9 shows an example of a basic bar chart produced for the construction of a floor slab. Two methods are shown that enable a comparison of different methods from a time perspective to be made. However in order to make a decision as to which method would be the best to use, a cost analysis would need to be undertaken and potential saving in duration balanced against any extra costs associated with the quicker process.

Figure 2.9
Bar charts for two different methods of large-scale floor construction.

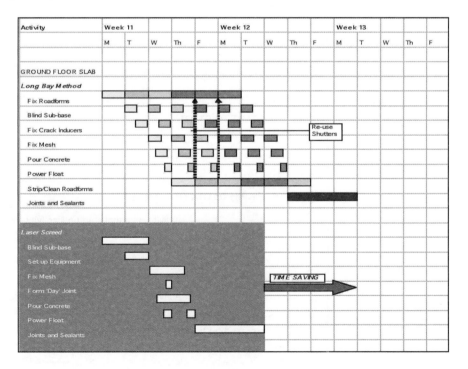

CASE STUDY: CONSTRUCTION PROGRAMMING

This Case study relates to a very simple, single-storey steel frame large-span building, as seen below and also in Figure 2.10.

The sequence of works can be dictated by something as simple as the finish to the concrete in the ground floor slab.

If the finish is specified as tamp finished with a sand/cement screed topping, then the concrete in the slab could be poured before the roof is erected. However, if the finish is specified as power floated, the roof will need to be erected before the slab is concreted, as a power float finish will be destroyed if rain falls on it. The roof would protect against this.

The sequence of activities could be as given in Table 2.2.

Terminology

A construction programme does not use the same terminology as a Bill of Quantities. For example, in a Bill of Quantities, a ground floor slab will have all the elements of it priced individually. They would be:

1. Hardcore
2. Sand blinding
3. Damp-proof membrane
4. Reinforcement
5. Reinforcement spacers
6. Expansion joints
7. Formwork
8. Concrete

On a construction programme these would be grouped together and called 'Slab Build Up and Concrete'. The durations required for each activity will be calculated, any potential overlaps in the work identified and an overall duration used in the programme.

Larger elements will have activities that are just related to one element; for example 'Erection of Steel Frame'.

(continued)

CASE STUDY: CONSTRUCTION PROGRAMMING (*continued*)

Figure 2.10
Large-span building outline diagram.

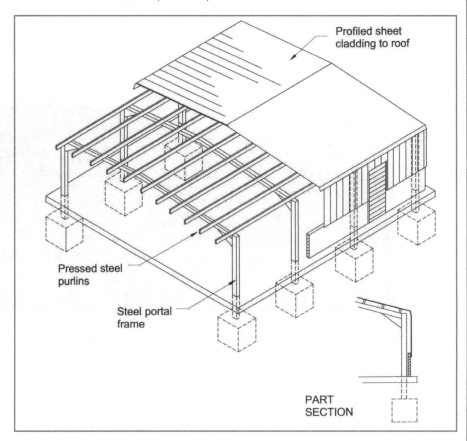

Profiled sheet cladding to roof

Pressed steel purlins

Steel portal frame

PART SECTION

A **power float finish** can produce a slab that is level to ± 3 mm tolerance. The concrete is allowed to cure until it can be walked on, and the power floater is then used to finish the concrete to a smooth finish.

A **tamp finish** will produce a rough surface that will allow for a bond to an applied screed. The tamping is generally undertaken using timber tamping rails, whilst the concrete is still wet.

Table 2.2 Alternative scenarios.

Scenario 1: Power float finish	Scenario 2: Tamp concrete and screed finish
Excavate foundations	Excavate foundations
Fix holding down bolts and concrete foundations	Fix holding down bolts and concrete foundations
Erect steel frame	Internal drainage and slab build up
Fix roofing	Concrete slab
Internal drainage and slab build up	Erect steel frame*
Concrete slab*	Fix roofing
*indicates when external brickwork can commence	The screed will be placed when the internal walls are complete.

Procurement and programming of projects using MMC

The implications of MMC that utilise off-site manufacture of significant elements of the building affect the design, procurement and programming of projects. Unlike traditional projects, the pre-construction phase will encompass major activity in the creation of elements of the building.

The procurement and construction programme associated with traditional forms of building is essentially a linear sequence of activities. The completion of earlier stages is generally a prerequisite for the commencement of subsequent activities. A simplified sequence of activity is illustrated in Figure 2.11.

Figure 2.11
Traditional procurement
and construction.

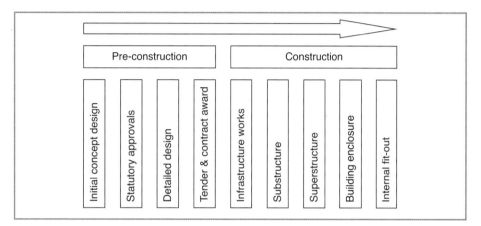

This model of procurement and programming restricts the construction activity to the period following detailed design and commencement on-site. If MMC techniques are adopted, much greater flexibility in the sequencing can be introduced since sections of the building superstructure and fit-out can be manufactured at the same time as substructure work is taking place on-site. The effect of this is to reduce the time required on-site for the construction phase. The greater the level of manufacture off-site, the lower the requirement should be for time on-site. However, increasing levels of pre-manufacture of building sections place much greater emphasis on the pre-construction phase in terms of programming. This is important in respect of the final agreement of the detailed design. Once the order has been placed for the manufacture of the panels, frames or volumetric units there is a requirement for 'design freeze'. The ability to amend the design after this point is extremely limited and will come with significant cost implications. Issues that need to be considered when procuring MMC buildings include:

- Design freeze: timing and extent
- Delivery timetable for units (storage on-site is undesirable but manufacturers will not want to store at the factory either)
- Responsibility for measuring on-site and agreement of tolerances
- Sanctions for late delivery and quality failures
- Protocols for accepting completed units and quality checking
- Defects liability periods and scope
- Manufacturer's role and responsibility during erection/assembly of units on-site.

Typical programmes of activity for MMC alternatives are illustrated in Figure 2.12.

Volumetric construction

Volumetric construction is sometimes referred to as 'modular construction'. This is not to be confused with the commonly used reference to temporary, portable buildings, such as site huts and so on, as 'modular buildings'. Modern volumetric construction is a well-developed and sophisticated process in which three-dimensional units or blocks are assembled and fitted out in the factory. They are then transported to site where they are assembled or 'stacked' onto pre-prepared foundations to provide the completed building.

The use of this technology is particularly suited to buildings that are essentially cellular in form. Hence, there is limited flexibility in the arrangement of the building during its design life. Many materials are used to create the modules or volumetric units, including timber, lightweight, cold-rolled steel and concrete. One of the most visible, and commonly utilised, forms of volumetric construction is the provision of bathroom 'pods' within existing or new buildings such as hotels, student accommodation and so on. The most effective use of volumetric construction is in buildings which have large numbers

Figure 2.12
Some activity programmes for MMC.

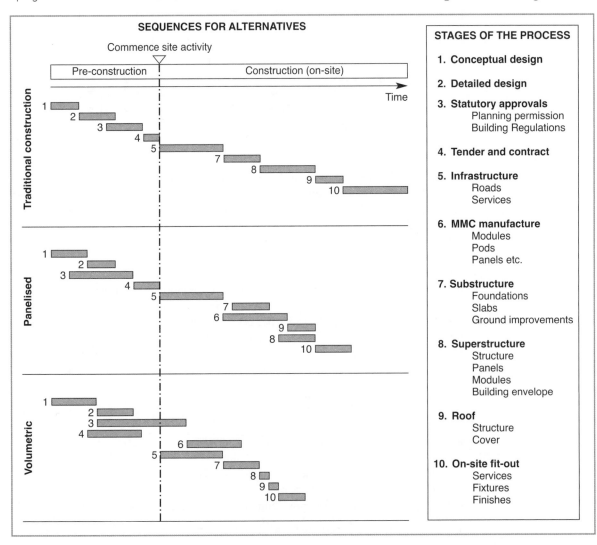

of identical room units, where repetition of the units can be achieved. Hence, their use is increasing in the development of flats, student residences and hotels.

Panelised construction

Panelised construction relies on the use of pre-manufactured flat panels that are assembled to create the building on-site. The panels are manufactured using a variety of materials. The more advanced systems adopt panelised construction for intermediate floors and roofs as well as the wall units.

Panelised construction can be effected using a number of different approaches or types of panel which may be loadbearing or non-loadbearing depending on their position and function within the overall structure. The most common forms used in construction are as follows:

Open panel
The skeleton frames, typically in timber or lightweight steel, of these panels are manufactured in the factory. They are then transported to site where the insulation, vapour barriers, linings and claddings are fitted along with the services and components such as windows.

Closed panel
These panels are fully assembled and enclosed in the factory. They are likely to comprise the same components as described above for open panels but the entire assembly and sealing of the panels is based on off-site manufacture.

Concrete panel
These often provide a structural wall for the building and can incorporate windows, services, insulation and externally applied cladding materials such as brick 'slips' to provide a complete walling assembly. These panels are sometimes referred to as 'cross-wall' systems.

Structural Insulated Panel (SIP)
Unlike open and closed panel forms, these panels do not have a skeleton structure to provide structural performance. Instead they comprise a central core of rigid insulation material that has external cladding and internal lining panels bonded to each side. The resulting, rigid panel can be used for forming wall and roof sections of the building.

Sub-assemblies and components
The use of prefabricated components within the traditional construction process is now widely accepted. The use of trussed rafter units for roofs could be taken as an example of this. However, as MMC systems have developed the use of larger components or building elements that have been manufactured off-site has become more sophisticated and more widespread. The following components/elements may be found in modern MMC construction:

- *Modular/prefabricated foundation systems*: Foundation systems based on the use of precast reinforced concrete beams supported by piles or piers allow fast assembly of foundations for dwellings.

- *Floor and roof cassettes*: Prefabricated panelised units for the floor and roof assemblies can be used to assemble floors and roof sections with reduced labour input, faster completion times and, importantly for roofs, accelerated weather-tightness of the building envelope.
- *Roof segments*: Entire sections of pitched roofs can be factory assembled or assembled at ground level on-site. These are then craned into position to create the finished roof structure and fabric. The use of these technologies reduces the health and safety risks associated with working at height.
- *Wiring 'looms'*: Adopting technologies that have been long-established in other industries, such as car manufacturing, the speed and quality of electrical installations can be enhanced using pre-manufactured 'plug-in' wiring looms. Cables systems are assembled in the factory to allow easy installation and the locations of sockets, lighting points and so on are terminated with plugs so that the finishing elements can simply be plugged in to complete the installation.

REFLECTIVE SUMMARY

- Sequence studies enable building programmes to be produced that are effective as tools of control during construction works, and provide the most efficient method of working.
- Method statements allow for the most economical way of working to be identified. They can be prose or tabular and can include required health and safety provisions.
- A good construction sequence will allow for as many overlapping activities as possible, without causing safety problems.
- 'Modern Methods of Construction' refers to construction methods that benefit production efficiency, quality and sustainability.
- MMC involves prefabrication and off-site manufacture.
- Commonly, forms of MMC for construction include:
  - Volumetric
  - Panelised
  - Hybrid
  - Sub-assemblies and components.
- Reasons for developing MMC for construction relate to:
  - Skills shortage
  - Quality enhancement
  - Developments in Building Regulations
  - Sustainability and environmental performance.
- MMC techniques allow flexibility in sequencing of site operations so reducing the time required for the on-site construction phase.
- Pre-manufacture of building sections places emphasis on the pre-construction programming and requires 'design freeze'.

REVIEW TASK

- For the building details given in the Case study on pages 47 and 48, produce a simple line diagram for the construction works that would be required to construct the shell, thinking carefully about which activities can take place at the same time as others.
- Visit the companion website at www.palgrave.com/engineering/riley2 to view sample outline answers to the review task.

CASE STUDY: MODERN METHODS OF CONSTRUCTION (MMC)

This is a Case study of a single-storey commercial building, constructed using a number of MMC techniques.

This photograph shows a concrete panel with insulation incorporated, fixed to a concrete foundation. The panels were delivered to site, fixed in place and then propped to ensure stability until the structure was stable.

All the outer wall panels are now in place and propped.

Any profile of the structure can be produced using off-site manufactured precast concrete. The excellent quality of the finished concrete can be seen in this photograph.

Conduits have been embedded in the panels before delivery to site, and chases for socket boxes formed in the factory.

Roof trusses are shown here being delivered to site after manufacture in the factory. These are easily and quickly lifted into place using a mobile crane.

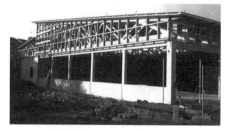

External and internal panel wall units are now fixed in place, roof trusses secured and are ready for roof coverings. The site staff had no previous experience of this type of construction but were amazed at the speed of construction, the quality of the finished product and the lack of labour required to reach this stage. The Case study demonstrates all the benefits of MMC. However the design is very basic and unimaginative, which is often mooted as the downside for using MMC.

2.3 | Expenditure on building

Introduction

- After studying this section you should gain an appreciation of the link between expenditure and time while constructing a framed building.
- You should appreciate the various elements that incur costs during, and prior to, the building process.
- In addition, you should recognise the relative implications of selecting prefabricated components and site-assembled elements upon the expenditure profile.
- The concept of a *'lazy S' curve* profile as a way to represent expenditure commitments during the construction process will also be considered.

Overview

The nature of the property industry is such that the vast majority of industrial and commercial buildings are built for profit by developers. It is quite rare for the construction of buildings to be commissioned by individuals who are not driven by the Developer's equation that was discussed in *Construction Technology 1*. Most building is speculative and the Developer's equation is key to the viability of the project; hence the control of building costs is essential. It must also be recognised, however, that the quality of the end product reflects greatly upon the potential for income generation. Modern office buildings must be constructed with a quality of end user in mind. Office space which is of inappropriate quality will command much lower rents than high-quality, well-equipped office space. Hence the level of investment at the time of construction will have a direct implication on the financial return over the building's life. This differs rather from the situation in house building, where the construction process is often driven by the wish to build down to a cost.

The expenditure profile of any construction contract is directly related to the sequence of building operations. In the case of most commercial and industrial developments the contract will generally be subdivided into work packages or phases with a range of specialist contractors each undertaking their own area of work. There are a number of implications arising from this kind of project relating to the fact that completion of individual packages or phases of the contract is achieved prior to completion of the entire project. It is normal for the subcontractors and main contractors to seek interim or stage payments for the work undertaken, so the client will be required to pay sums at regular (normally monthly) intervals. Indeed, it is also quite normal for parts of the development to be completed and handed over to the client while other parts are in the early stages of construction. The reason for this is often related to the Developer's equation and the need to generate income on the project as well as controlling expenditure.

Building costs

The total cost of a building project is made up of a number of individual elements in addition to the cost of the materials, labour and plant used for the building itself. Aspects of setting up the site (as identified elsewhere), providing insurance and so on, are all elements that have a cost implication. These are often referred to as *preliminary items* and are normally included within the overall costings on the basis of a percentage addition to each costed item or as specific items identified within the project costings. It is not the purpose of this book to consider the nature of construction contracts and tendering; hence, we will not consider the basis of pricing in detail. However, it is important to understand the nature of the total project costs and the individual elements that it comprises. Since every construction project is individual it is not possible to set out a definitive list of items of expenditure. Table 2.3 sets out the typical list of items included within the total cost of a building project from the point at which work commences on-site, but this should not be considered as definitive.

Expenditure profile

As previously stated, the majority of industrial or commercial building projects include several phases. In order to understand the totality of the project we must first consider the expenditure profile relating to an individual phase.

The previous section considered the process of building and identified some of the key operations in the construction process. Each of these operations has a cost attributed to it that will be made up of three components: materials, plant and labour. The *materials* are the raw materials (such as steel framing, sand and cement) and the manufactured components (such as windows, lintels and elements of services) that are required to facilitate the construction of the building. *Plant* is the term used for mechanical tools and vehicles such as lifting equipment and dumper trucks used to assist the construction process. *Labour* is the human resource in the form of skilled and unskilled tradespeople taking part in the construction process. Each of these has a direct cost that can be calculated for any element of the project.

By applying these costs to each activity within the construction sequence we are able to build up an expenditure profile that illustrates the cost of the project at any given point. When expressed in terms of a cumulative cost profile, i.e. the total cost of all elements up to any given point, this produces a curve which is referred to as a 'lazy S' (Figure 2.13).

The profile of the curve arises as a consequence of the fact that the expenditure accelerates at the beginning, achieves a consistent level in the middle and decelerates at the end of the project. This is because there is a period at the beginning of the project during which the site is established and preparation works are underway. This results in modest outlay initially, before expenditure on materials and assembly begins. Once the preparatory period has been completed, expenditure on plant, labour and materials remains fairly constant until the later stages. At this point items of finishing and tidying are undertaken. These take time but are not costly in terms of materials etc. Although each project is individual, some generic models of expenditure have been developed. One of the most common is the quarter:third mode. In this the expenditure is anticipated on the basis that one quarter of the cost will be attributable to the first third of the project period, one quarter to the final third and the remainder spent uniformly in the period between.

The nature of the building design will affect the shape of this curve. For example, the selection of a steel-framed building solution will result in a quite different expenditure

Table 2.3 Costs associated with construction.

Item	Description	Phasing
Statutory and other fees	Costs associated with inspection of the works in progress and statutory approvals, including energy performance certification	Various stages within the project
Insurances	Public and employer's liability insurance, together with insurances for equipment, plant and buildings paid by the contractor and subcontractors	Normally paid at the outset of the project
Utilities connections	Costs of connecting to gas, electric, water, drainage and other utility services	During the early stages of the project
Temporary works	Protective hoardings, scaffolding, fencing, temporary supplies of power, water etc.	At the early stages of the project
Accommodation and facilities	Site accommodation for offices, canteen facilities, toilets, secure storage of plant and equipment etc.	At the outset of the project
Plant hire	Hire of specific items of plant and equipment such as excavators and hoists	Throughout the project
Infrastructure	Roads, access and services installation	At the early stages of the project
Labour	Costs of skilled and unskilled labour to undertake the work, including supervision costs	Throughout the project
Plant	Costs of purchase, hire or depreciation of existing plant involved in the works	Throughout the project
Materials	Costs of materials and components used in the building process	Throughout the project
Making good	Costs associated with removal of temporary works, making good damage to any areas affected by the works outside the boundaries of the site	At the end of the project
Possession/access charges	Costs associated with temporary access or possession of land/space to allow for the works to proceed, e.g. pavements, air space for traversing of cranes, road closures	At the start and throughout the project

Figure 2.13
Plotted project
expenditure – the S
curve.

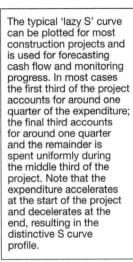

The typical 'lazy S' curve can be plotted for most construction projects and is used for forecasting cash flow and monitoring progress. In most cases the first third of the project accounts for around one quarter of the expenditure; the final third accounts for around one quarter and the remainder is spent uniformly during the middle third of the project. Note that the expenditure accelerates at the start of the project and decelerates at the end, resulting in the distinctive S curve profile.

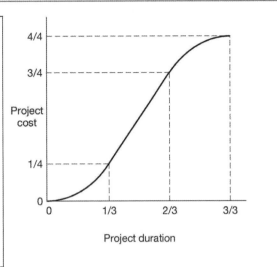

PART 1

profile from the selection of a concrete frame cast *in situ*. This is because the steel frame will be ordered as a single prefabricated item demanding a considerable outlay at the point of ordering or delivery. Clearly in this situation the curve will start off with a steep section due to the high-cost item at the early stage of the contract. The *in-situ* frame, in contrast, will show a shallower profile, since expenditure will be more measured and distributed more evenly throughout the contract period.

This is one of the reasons why some design decisions are made on the basis of the client's cash flow rather than physical performance criteria.

REFLECTIVE SUMMARY

When considering expenditure on building, remember:

■ The expenditure profile for a project is directly linked to the sequence of building operations.
■ Contracts will often be subdivided into specialist work packages.
■ Contractors will seek interim payments at regular intervals.
■ Building cost will be made up of a variety of components in addition to materials and labour.
■ The expenditure profile can be plotted over time to manage cash flow and budgets.
■ The S curve profile is typical for most contracts, but the shape of the curve will vary.

REVIEW TASK

■ With reference to a simple building with which you are familiar, try to think about the various items of expenditure during the period of construction. Plot these on a bar chart and attempt to generate a simple S curve based on your findings.
■ Visit the companion website at www.palgrave.com/engineering/riley2 to view sample outline answers to the review task.

2.4 | Costs in use and life cycle costing

Introduction

- After studying this section you should have developed an appreciation of the principles of life cycle costing related to buildings.
- You should understand the building life cycle and the various elements that require expenditure during the life of a building.
- In addition, you should have an overview of the various methods adopted to calculate life cycle costs and you should appreciate the implications for building use and design.

Overview

In order to understand the principles of life cycle costing we must first gain an understanding of the concept of the 'building life cycle'. When a building is designed and constructed there is always an expectation that at some point in its future it will become obsolete or will have served its intended purpose and will be disposed of. These two points, construction and disposal, mark the start and end points of the building life cycle. Different buildings will inevitably be designed with very different intentions for their expected period of occupation and use – this is termed the 'design life' of the building. Components and details are selected with this design life in mind – clearly the specification of high-quality long-life components for a building which itself is to have a short functional existence is uneconomic. For most commercial buildings a design life in the region of 60 years is common, whilst a rather shorter life might be envisaged for industrial buildings. A glance around any town or city will reveal that buildings often survive in functional occupation long beyond these projected lifespans with the aid of major refurbishment and careful maintenance. However, some intended time-scale must be used at the outset of the design process in order to allow for the correct specification of materials and components.

Building costs in use

Most buildings are designed with a finite **lifespan** in mind. The functional and technical performance of the building will deteriorate as it ages, triggering a response to refurbish or upgrade. When the building no longer satisfies the user's requirements, even though it may be in a physically sound condition, it may be considered to be obsolete.

During its **lifespan** the building will be subject to repairs, maintenance, alteration and upgrading to allow it to maintain an acceptable level of performance and to maintain user satisfaction. This demands a degree of regular expenditure throughout the building's life which must be considered along with the initial building cost to fully assess the total cost of the building during its life cycle. This is the principle of life cycle costing.

During the life of a building it may undergo a number of changes, driven by changing user needs, advances in IT, evolution of working practices and so on. In addition to the basic costs of maintaining the building, cleaning and redecorating there will be costs associated with essential upgrading to avoid obsolescence. This is particularly so in the case of building services, which are likely to require major upgrading and alteration several times during the life of a building. The structure and fabric of the building are likely to remain largely unchanged, other than by maintenance and redecoration for a relatively long period, possibly even for the entire building life (60 years or more). The interior layout may change much more often – typically less than 10 years. This is due

Figure 2.14
Life cycle cost graph.

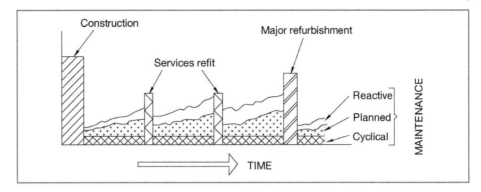

to changing user needs or even changing users; it must be remembered that commercial buildings tend to be occupied on a leasehold basis and occupiers change regularly. These changes are facilitated by the use of loose fit building designs and the adoption of demountable partitioning etc. Building services will also be upgraded or renewed more often (typically every 10 to 20 years) than the structure and fabric as a result of the drivers for change identified earlier.

These elements of expenditure, combined with the initial building costs and the ongoing maintenance and repair costs, make up a package of expenditure which reflects the whole-life cost of a building. Using this information designers can evaluate different design solutions not simply on the basis of initial cost, but on the basis of life cycle cost (Figure 2.14).

Approaches to life cycle costing

When attempting to consider the life cycle or whole-life costs of a building we must take into account all of the elements of expenditure that are likely to be incurred during its life. This is sometimes referred to as a 'cradle to grave' approach and relies on a realistic vision of the probable life expectancy of the building before it becomes obsolete. Thus the life cycle as illustrated in Figure 2.15 is generally considered, with expenditure at each stage taken into account. The totality of life cycle cost will include, among other things:

- Costs associated with inception, design and planning
- Initial costs of construction
- Running costs, including maintenance and repairs
- Operating costs, including energy costs
- Costs of alteration, upgrading and refurbishment
- Costs associated with renewal of services
- Demolition and disposal costs.

Since the frequency of, and hence costs associated with, refitting and alterations due to changes of user or user needs are very difficult to predict with any accuracy at the outset of the process, they are often removed from the assessment of costs. The other costs can be assessed with the aid of a series of assumptions relating to operation of the building and they provide a useful aid to developers in selecting between available options for new buildings.

Three main methods are commonly used to evaluate life cycle costs. These are based on simple aggregation of costs, assessment of net present value and assessment of annual equivalent expenditure. Each of these methods can be applied taking inflation into account or basing consideration on a static approach. Each has its own merits and may prove to be more appropriate in a given set of circumstances.

The simple aggregation of costs, taking the initial build cost and adding known expenditure items throughout the building's life, will give a single, gross figure for life cycle expenditure. This approach tends to favour schemes that allow a higher initial building cost in order to reduce long-term annual costs. However, this approach to financial assessment is rather simplistic.

The calculation of net present value allows for the aggregation of total costs taking into account an assumed interest rate for the period over which expenditure is envisaged. The assumption is that money invested now will yield interest until it is needed for expenditure on the building, and thus the amount needed now will be lower than the actual amount spent at a later date. Thus expenditure in the future has less of an impact than expenditure now when considered in terms of its net present value.

The annual equivalent cost method takes a fundamentally different approach to the foregoing. The total cost for the building construction and operation is converted to an equivalent level of annual expenditure throughout the life of the building.

As is the case in most aspects of construction, there is not a single correct answer. However, these tools provide a useful mechanism for assessing the true cost associated with the construction and operation of a building throughout its life. Figure 2.16 summarises the components of a building's whole-life cost.

REFLECTIVE SUMMARY

When considering building life cycle costs remember:

- The total cost of a building during its life is made up of initial build cost, maintenance and operating costs.
- Buildings are designed with a specific lifespan in mind, but they often exceed this.
- During their lives buildings may be refurbished to extend lifespan or to satisfy changing user needs.

REVIEW TASKS

- Generate a list of typical costs associated with buildings during their life.
- Identify examples of buildings that you know which have been subject to extension of their lifespan by refurbishment.
- Visit the companion website at www.palgrave.com/engineering/riley2 to view sample outline answers to the review tasks.

2.5 | Building Information Modelling (BIM)

Introduction

- After studying this section you should have developed a basic understanding of BIM and the benefits that BIM can bring to improving all aspects of the building design and construction process.

Figure 2.15
The building life cycle.

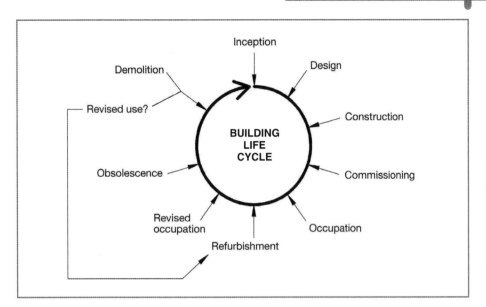

Figure 2.16
Building whole-life cost.

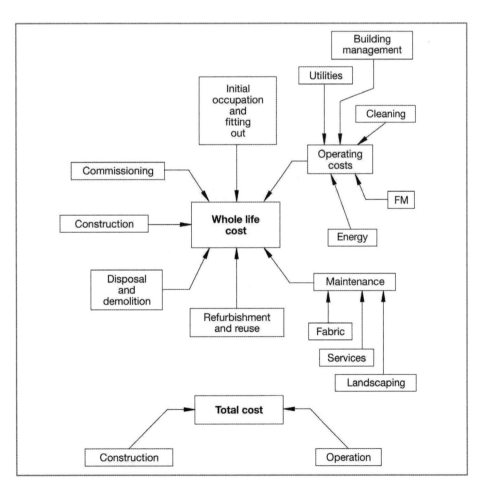

- You will have gained an insight into why the development of BIM is deemed to be so important to the construction industry and the imperative from the Government to fully implement it.
- You will also gain an appreciation of why it is so important for all built environment professionals to work collaboratively on the development of BIM.

Overview

Building Information Modelling (BIM) is about the collection, management and manipulation of data created by the members of the project team during the project's life cycle, from inception to decommissioning. All information is stored on an integrated database and is shared in a collaborative manner by members of the project team. The data is displayed as a 3D model demonstrating the design of the project with its building form and construction. The 3D model also offers the potential to display schedules, quantities and costs, life cycle maintenance and energy consumption alongside health and safety information, as all information is interconnected and changes made by one member of the project team will be observed by other members of the team.

The development of BIM

BIM is the current industry buzzword. It has been hailed as one of the measures to be adopted to increase efficiency and reduce costs in procurement and to improve the overall level of information delivered to the end user. However BIM is not a totally new concept. In the 1960s the introduction of Computer Aided Design (CAD) allowed design to become digital which in turn produced 2D drawings used throughout the construction process. In 1986 the first 3D model software, Archicad, became available which allowed designers to produce a full dimensional view of their design as opposed to the previous 2D version. In recent years this 3D model has been taken into real time and enabled changes to elements to be updated in all other views within this 3D model. The impetus to its application in the UK has stemmed from the Government Construction Strategy 2011. The UK Government has stipulated that BIM should be adopted on all public projects by 2016. This provides a massive incentive to contractors and construction professionals alike to ensure that their organisations train their staff and develop and utilise BIM tools and techniques from 2016.

The BIM process

BIM is seen by many as simply a model but in order for it to realise its maximum potential it is necessary to understand that it is the process underlining the model that is the important factor. BIM should therefore be considered as the process of managing project information. In order to do this we need to consider the labelling or grouping of the data, the method of exchange or manipulation of the data and the method of exchanging and issuing of the data (Figure 2.17).

Figure 2.17
Process of data/information management in a BIM-enabled project.

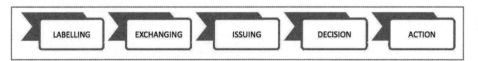

The BIM process will increase the speed and accuracy by which decisions can be made. In construction there are many decisions to be made through the life cycle of a building and many different software packages being utilised to support and inform those decisions. Fully integrated BIM will allow for the amalgamation of all these softwares into a single unified database from which collaborative, better informed decisions can be made and effective action taken in real time.

The Government's BIM strategy

The Government directive is that all Government-funded projects shall be at BIM Level 2 by 2016. Level 2 maturity is a managed 3D environment incorporating separate discipline BIM tools with data attached. This level of BIM may utilise 4D construction sequencing and/or 5D cost information. The Government by committing to maturity Level 2 has committed construction project teams to provide their own distinct outputs by using a BIM tool and managing it via a series of self-contained models using proprietary information connections between each of the project team's own distinct systems. The next stage, Level 3 BIM, is where the data is fully integrated and interoperable, with all the data provided by the individual project teams being stored in a fully integrated web services BIM hub (Figure 2.18). This is regarded as iBIM or integrated BIM potentially employing concurrent engineering processes.

BIM – the current situation

Two years after the publication of the UK Construction Strategy 2011, much of the construction industry is unaware of the different levels of BIM. The industry is still unclear as to what BIM actually is, with the majority of the industry being less than confident in their knowledge and skills. However the adoption of BIM is on the increase, with those that have utilised BIM claiming greater cost efficiencies, better coordination of construction documents, improved productivity arising from the easy retrieval of information and improved visualisations. The recent publication of Construction 2025 (July 2013) by the UK Government further promotes the take-up and use of BIM. Targets to reduce the initial cost of construction and the whole-life cost of built assets by 33 per cent along with a reduction in overall time from inception to completion of 50 per cent

Figure 2.18
BIM development and collaboration, based on Bew and Richards (2008).

LEVEL 0	LEVEL 1	LEVEL 2	LEVEL 3
• CAD	• 2D, 3D	• BIM	INTEGRATED BIM,
• PAPER DRAWINGS	• MODELS, OBJECTS	• MODELS, OBJECTS,	LIFE CYCLE MANAGEMENT
	• COLLABORATION	COLLABORATION	INTEGRATED,
		• LIBRARY MANAGEMENT	INTEROPERABLE DATA, BIM HUB

INCREASED TECHNOLOGY – GREATER COLLABORATION, LESS WASTE, GREATER EFFICIENCIES

Figure 2.19
3D model school
building (source
Cadassist, 2013).

have been set for 2025, with BIM and smart technologies being the key drivers behind the achievements of these targets.

There is little evidence to suggest that BIM is currently being used other than to produce designs, drawings and schedules from the model. Designers are in the main promoting the use of BIM and are leading the field in its take-up. There has been limited use of 4D and 5D BIM, possibly due to the inappropriate level of detailed information stored within the model supporting it being of little value in relation to manipulating data for 4D programming, 5D costs and 6D facilities management. Contractors are slow to recognise the advantages that BIM can bring to their work process and the efficiencies it can help create, particularly in relation to programming. Many contractors are using it more as a document management system than as a collaborative tool.

BIM has the potential to impact on every aspect of the project team. It has applications for those involved with building construction, property, facilities management, building surveying, quantity surveying and civil engineering.

The benefits of BIM

Improved visualisation and collaboration

BIM promotes collaboration at an early stage. Better visualisation tools early on in the design allow a virtual walk-through of the 3D model, providing improved project understanding and resulting in better coordination of design and deliverables between the project team. This in turn facilitates an improved understanding of design changes and their implications on the cost, programme and maintenance of the project. Changes as to the type of frame of the building can be considered and its impacts assessed quickly on all other elements of the project by the adoption of a single, shared 3D model, which provides focus on achieving best value, from project inception to eventual decommissioning (Figure 2.19).

Regulatory compliance

The 3D model allows for simulation and analysis for regulatory compliance, and has the ability to simulate and optimise energy and wider sustainability performance. Changes made to the specification can immediately be assessed as to the impact on the thermal dynamics of a building.

Reduction in waste

The model allows for the project to be built before it goes to site. Complex technical solutions and difficult construction arrangements can be tried 'virtually' (the components can be assembled in the model) before reaching site. This has the potential to reduce reworking on-site and ultimately project costs as all problems are discovered and solutions found 'virtually'. One of the most significant benefits highlighted to date includes savings made by identifying clashes, particularly in relation to building services and structural elements, and so once again avoiding reworking on-site. In addition, the model generates exact quantity take-offs, thus reducing overordering of materials and reducing waste.

Faster project delivery

Project durations have been reduced as a result of improved collaboration at an early stage. Decisions made early on in the project do not need to be reconsidered because the virtual construction has considered all clashes and resolved complex construction virtually.

Better informed site management

Just-in-time delivery of materials and equipment are optimised by the production of precise programme schedules from the 3D model. The use of BIM for automated fabrication of equipment and components enables more efficient material handling and waste recovery. The information can be produced and used to schedule materials in addition to indicating which elements belong to particular work packages.

Better informed facilities management

The model allows for the creation of a facilities management database in real time. The ability to update the model with real time information offers the potential to undertake costing in relation to operation and management of the built asset during the project's development.

REFLECTIVE SUMMARY

- Building Information Modelling (BIM) is about the collection, management and manipulation of data created by the members of the project team during the project's life cycle, from inception to decommissioning.
- BIM should be considered as the process of managing project information.
- The Government directive is that all Government-funded projects shall be utilising BIM Level 2 by 2016.
- Currently there is little evidence to suggest that BIM is being used other than to produce designs, drawings and schedules from a 3D model.
- The benefits of BIM include improved visualisation and collaboration, regulatory compliance, easy identification of clashes within the design, reduction in waste, faster project delivery, better informed site management and facilities management.

REVIEW TASK

- Integration of services is one of the most difficult things to achieve during the design and subsequent construction process. After you have read Chapter 13, reflect on the chapter and determine how BIM could be used to make these tasks easier.
- Visit the companion website at www.palgrave.com/engineering/riley2 to view sample outline answers to the review task.

3 Preparing to build

66

AIMS

After studying this chapter you should be able to:

- Understand the reasons why site investigation is important
- Describe the processes involved in site investigation
- Appreciate the various elements of statutory control that affect the building process
- Recognise the health and safety issues associated with design and construction
- Appreciate the nature of utility services and their relationship to buildings
- Understand the reasons for the need to control ground water during construction and the main methods of doing so
- Describe the various stages and processes of setting up a construction site

This chapter contains the following sections:

3.1   Site investigation and associated issues
3.2   Overview of statutory control of building
3.3   Ground water control

INFO POINT

- Building Regulations Approved Document C: Site preparation and resistance to contaminants and moisture (2010)
- Building Regulations, Part L (2010)
- BS 5930: Code of practice for site investigations (1999)
- BS 5964: Building setting out and measurement (1996)
- BS 10175: Investigation of potentially contaminated sites – Code of practice (2011)
- BS EN ISO 22475-1: Geotechnical investigation and testing – sampling methods and groundwater measurements. Technical principles for execution (2006)
- BRE Digests related to site investigation: 318 (1987), 322 (1987), 348 (1989), 381 (1993), 383 (1993), 411 (1996), 412 (1996), 427 (1998)
- *Introduction to sustainability*, CIBSE (2007)
- *Role and responsibility in site investigation*. Special Publication SP 73, CIRIA (1991)
- *Site safety handbook*, 4th edn. Publication C669, CIRIA (2008)
- The Construction (Design and Management) Regulations 2007
- The Control of Substances Hazardous to Health Regulations 2002. Statutory instruments SI 2002/2677 Legislation [formerly SI 1999/437]
- The Health and Safety at Work Act 1974
- The Technical Building Standards, Part J (1999)

3.1 | Site investigation and associated issues

Introduction

- After studying this section you should have developed an understanding of all the factors that could influence the choice of a particular site for the construction of a given building.
- These issues will need to be addressed at the design stage by the main designers, and at **pre-tender stage** by the contractor for different reasons.
- Some issues will affect cost and others buildability, and you should be able to appreciate the implications of each issue for all aspects of the work.
- You will be able to assess which data can be collected at a desktop and which will need a site visit.

The **pre-tender stage** of a contract is the time between the drawings and specification/BOQ being sent to a contractor and the date for submission of a tender price.

Overview

The process of site investigation occurs early in the development and even before the purchase of the land. The process reveals a huge amount of information, which often has a major influence on the design chosen. Possibly the main reason for an early investigation is the type of ground on the site. Very poor ground will require expensive foundations and this could make the building prohibitively expensive. In this section we will look at the elements of site investigation that are required at the pre-design and pre-tender stages.

Pre-design stage investigations

- *Information required*: Ground conditions, including type of ground, water table level and extent of temporary supports required, and whether or not there is any ground contamination.
  Reason: Selection of suitable foundation system. If the ground is contaminated, a system of remediation or containment will be required.
- *Information required*: Location of tipping facilities, distance to tip, tipping fee.
  Reason: If the nearest tip is a long way and tipping costs are very expensive, the designer could reduce tipping by employing the 'cut and fill' idea to the site (Figure 3.1). Rather than excavating to a reduced level and tipping all the excavated material, the reduced level could actually be raised to reduce tipping to a minimum.
- *Information required*: Are there any existing buildings that need to be demolished?
  Reason: The foundation type may be affected by the existence of footings to other properties, some of the demolished material could be reused in the new building to promote a more sustainable form of construction, and there may be elements of the existing buildings that could be salvaged and sold to reduce the cost of the new building for the client.
- *Information required*: Site access.
  Reason: Certain types of construction may be unbuildable if a site is very restricted. For example, the construction of a precast concrete frame would require a lot of very

PART 1

large deliveries to a site. If access was limited to only a few hours per day, then a different choice of frame would be more feasible. Local roads may have to be upgraded as part of the contract if they are unsuitable for large deliveries, and this will have to be communicated to the client.

■ *Information required*: Restrictions due to site location.
Reason: An example of how this could affect design is restrictions that could be applied to the positioning of scaffolding. If encroaching the highway with scaffolding is unacceptable to the Local Authority, then the building line will have to be moved back to allow for the scaffolding.

■ *Information required*: Labour issues, such as availability of skilled labour.
Reason: For example, if it is known that there is a shortage of bricklayers in the local area, then it may be advisable for the designer to use another system for external wall construction, such as prefabricated panels that look like brickwork.

■ *Information required*: Issues related to main services. Proximity of water, electricity and telephone mains. Location of foul and surface water facilities.
Reason: Any new building will require these services, and if the mains are a long way from the site this will increase the cost of the statutory body connections. This information will have to be relayed to the client.

■ *Information required*: Vicinity of the site and surrounding area.
Reason: Site security requirements will differ depending on the area in which the construction work is proposed, and this can affect the cost of construction work. Also, the design should attempt to blend in with other buildings in the area. If there is a problem with graffiti or damage to property in the area, then the design should be robust enough to withstand these problems.

■ *Information required*: The **topography** of the site.
Reason: Tipping costs money, and tipping on landfill sites damages the environment. This means that the tipping of soil from greenfield sites needs to be avoided as it fills up the sites, is not contaminated in any way and costs a great deal, especially if the nearest tip is a long way from the site. This can be reduced by reviewing carefully the excavation requirements on sites when undertaking the design. If the designer does not do this then the contractor could easily suggest it as a way of saving money and behaving in a more sustainable way. Rather than excavating the whole site to the lowest point before excavating for foundations and then tipping all the excavated materials, the 'cut and fill' idea can be used. In this system excavated soil from the higher levels on the site can be used to raise the levels on the lower parts of the site. If this exercise is carried out correctly then tipping should be reduced to a minimum. Figure 3.1 illustrates the principle.

Topography is the term used to refer to the surface features, contours and general profile of a site.

Figure 3.1
Principle of 'cut and fill'.

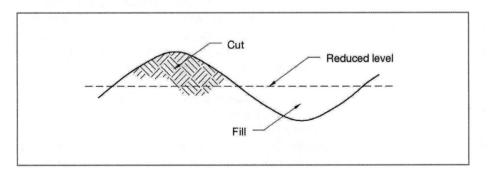

■ *Information required*: Other factors, including the location and nature of the site in relation to the movement of plant, extent of overhead or underground obstructions, possibility of the existence of mine shafts or other workings, possibility of old foundations and old watercourses, and history of freak floods.
Reason: These factors will influence the design, but if their extent is not known then the client needs to be consulted and the cost of encountering these problems explained.

COMPARATIVE STUDY: EXTERNAL WALLS

The issues detailed will increase or decrease in relevancy depending on the type of site. Concentrating on new build work, Table 3.1 shows a comparative study of how the issues are likely to impact on city centre sites as opposed to greenfield sites.

  D indicates that the information can be obtained by a desk study and research. S indicates that a site visit is the best way of gathering the data. DS indicates that a combination of both will be required to assess likely implications.

Table 3.1

	Survey type	City centre site	Greenfield site
Ground conditions, including type of ground, if there is any ground contamination	DS	Will need to be assessed in detail. Likely to be contamination, more so if on the site of an old industrial unit	Will need to be assessed in detail. Unlikely to be contamination
Location of tipping facilities, distance to tip, tipping fee	D	Should be relatively near	Could be a long way to nearest facility
Are there any existing buildings that need to be demolished?	DS	Highly likely	Very unlikely
Access	S	Difficult	Easy
Location of the site	DS	Deliveries should not have to be delivered from a long way away	Suppliers may be far away
Labour availability	D	Should not be too much of a problem	Could be a big problem
Mains services	DS	Will be close to the site	Could be a long way from the site
Vicinity of the site	S	Probably requires greater security	May require less security, but theft can still occur from more isolated sites due to the lack of people out of working hours

> **REFLECTIVE SUMMARY**
>
> With reference to site investigation, remember:
>
> - As a process this needs to happen at a very early stage (project feasibility) by the designer.
> - It also needs to be undertaken by the contractor at the pre-tender stage.
> - Some of the issues will cause the tender price to increase.
> - Some of the issues will require greater coordination skills by the site management team, such as managing a site with access problems.
> - Probably the most important element that needs to be investigated is the ground conditions. Foundation design will depend on the results of the findings.
> - Even if a thorough investigation of ground conditions is undertaken there can still be unknown ground conditions that are not apparent until excavation starts.

> **REVIEW TASK**
>
> - Identify the most important aspects that would need to be included in a site investigation report being undertaken by the building designer.
> - Visit the companion website at www.palgrave.com/engineering/riley2 to view sample outline answers to the review task.

3.2 | Overview of statutory control of building

Introduction

- After studying this section you should be able to appreciate the evolutionary development of building control through public health and other associated legislation.
- You should appreciate that building control is effected by using Building Regulations (in England and Wales) and Building Technical Standards (for Scotland).
- You should be able to outline the basic processes required to obtain formal approval to build and how control extends to the building process.
- You should be able to develop an understanding of the future changes to the regulations that may occur and the implications for future buildings.
- You should develop an understanding of the legal responsibilities of all the parties involved in the construction process with regard to the health, safety and welfare of site workers and the public during construction work and also develop an understanding of relevant regulations.

Overview

The application of controls over the building process is a relatively new concept. Prior to the first set of Building Regulations in 1965, control over the building process was limited. The first real major realisation of the need for control probably came in the aftermath of the Great Fire of London in 1666, and in 1667 using timber for the external structure and fabric of buildings was outlawed in London. A main feature of the development of control in the rest of England was the various Public Health Acts (PHA), and probably the most notable of these was the PHA 1875. This had three main focuses: structural stability, dampness and sanitation. Following this Act a set of Model Bye-Laws was issued in 1877 as a guide for Local Authorities, who were delegated the responsibility for setting and enforcing minimum standards of construction.

Following limited progress in the 19th century, legislation concerning construction was inadequate and mainly in the area of public health. By the 1950s many Local Authorities were issuing bye-laws peculiar to their locality, and this made it difficult because laws varied from area to area. Because of the need for consistency, National Bye-Laws were established in 1952 and, following these, items of significance include the Public Health Act 1961 and the first set of national Building Regulations in 1965. The aims of the Building Regulations were to set out the minimum standards that are acceptable for building works.

A major reference for the safety of persons associated with the building process is the Health and Safety at Work Act 1974, which applies to all industries and aspects of construction work. The Act is an 'umbrella' for a huge set of regulations that are specific to individual industries and aspects of those industries. Some of the most recent and important regulations affecting construction work will be discussed in this section.

Building Regulations

Building Control legislation is empowered by the Building Act 1984, which is the basis for the Building Regulations 2000 (as amended)

The Building Regulations contain various sections dealing with definitions, procedures, and what is expected in terms of the technical performance of building work. For example, they:

- Define what types of building, plumbing and heating projects amount to 'Building Work', and make these subject to control under the Building Regulations
- Specify what types of buildings are exempt from control under the Building Regulations
- Set out the notification procedures to follow when starting, carrying out and completing building work
- Set out the 'requirements' with which the individual aspects of building design and construction must comply in the interests of the health and safety of building users, of energy conservation and of access to and use of buildings.

The following list shows which part of the Regulations deals with which aspect of buildings:

A   Structures
B   Fire safety
C   Site preparation and resistance to contaminants and moisture
D   Toxic substances

E   Resistance to the passage of sound
F   Ventilation
G   Hygiene
H   Drainage and waste disposal
J   Combustion appliances and fuel storage systems
K   Protection from falling, collision and impact
L   Conservation of fuel and power
M   Access to and use of buildings
N   Glazing – safety in relation to impact, opening and cleaning
P   Electrical safety.

For each part there is the Regulation, plus the Approved Document giving detailed guidance on how to show compliance. It is up to the applicant to demonstrate compliance by means of drawings and specifications. The most difficult parts to deal with are – A: Structures; B: Fire safety; and L: Conservation of fuel and power.

Part L of the Regulations is intended to improve the energy efficiency of all new construction work, including extensions and refurbishment work, with an emphasis on higher standards of insulation for the building fabric. The Regulations allow for traditional forms of construction to be used, but require more insulating products to be incorporated, and a reduction in cold bridges and air leakage. The equivalent Building Technical Standard for Scotland is Part J.

Design and construction issues related to Part L

Airtightness could become the Achilles' heel, as designers will be required to ensure that a building meets the air leakage target of $10\,m^3$ per hour from the building fabric at an applied pressure of 50 Pa. Mobile testing units can be used by building control bodies, and if buildings 'fail' they will have to be adapted to ensure compliance. Robust standard details will have to be produced, but they may require improved standards of workmanship. Current industry operatives may need retraining. Building control bodies will have to monitor compliance and may need extra resources.

The revisions discourage reliance on mechanical cooling systems and encourage the reduction of overheating through shading, orientation, thermal mass, night cooling etc. There can be a trade-off between the elemental **U values** and the efficiency of the boilers, but this could lead to constantly changing boilers that are increasingly more efficient and could lead to reliance on boilers to achieve compliance. Every building will need some low-energy lighting systems. Refurbishment projects may not be able to comply with the regulations, and Design and Build as a procurement tool may drop in popularity because of the amount of work required during the pre-tender design stage by the contractor, which could prove financially uneconomic. However, the changes are excellent for the sustainability-minded designer and Design and Build contractors.

U values are measured in W/m^2K. A U value of 0.25 means that one quarter of a watt of heat is lost through each square metre of the element when a one-degree temperature difference exists between the inside and the outside of the element.

Facilities management issues

All buildings will have to have energy meters and building log books, which are essential if building owners and operators are to monitor energy consumption against a benchmark figure.

Health, safety and welfare on construction sites

The Health and Safety at Work Act 1974

This is an umbrella for a multitude of regulations regarding the health and safety of employees in the workplace, and covering every industry.

The purpose of the Health and Safety at Work Act is to provide the legislative framework to promote, stimulate and encourage high standards of health and safety at work. The Act is implemented by the Health and Safety Executive and the Health and Safety Commission:

- **Health and Safety Commission**: Responsible for analysing data and publishing results regarding accidents, and for proposing new regulations to help solve problems.
- **Health and Safety Executive**: Responsible for ensuring that regulations are complied with.

The latter is the body that can issue improvement and prohibition notices, prosecute, and seize, render harmless or destroy any substance which is considered dangerous.

The construction industry has a poor record with regard to health and safety, and many regulations have been introduced which deal specifically with the industry, such as lifting appliances and excavations. High-risk trades include steel erection, demolition, painting, scaffolding, excavations, falsework, maintenance, roofwork and site transport.

The HSE produces checklists for contractors to ensure that all precautions that can be taken to prevent accidents are implemented. Some areas of construction work have to be checked weekly and the findings entered in a register. These are scaffolding, excavations, lifting appliances and cranes. Although there is a large body of law covering many aspects of health and safety at work, we could define the main statutes that are particularly important in construction as:

- The Health and Safety at Work Act 1974
- The Management of Health and Safety at Work Regulations 1999
- The Construction (Design and Management) Regulations 2007 (CDM)
- The Work at Height Regulations 2005
- The Manual Handling Operations Regulations 1992
- The Provision and Use of Work Equipment Regulations 1998 (PUWER)
- The Lifting Operations and Lifting Equipment Regulations 1998 (LOLER)
- The Control of Substances Hazardous to Health Regulations 2002 (COSHH)
- The Personal Protective Equipment at Work Regulations 1992
- Control of Asbestos at Work Regulations 2006
- The Confined Spaces Regulations 1997
- Control of Lead at Work Regulations 1998
- Noise at Work Regulations 1989
- Corporate Manslaughter and Corporate Homicide Act 2007.

Regulations of major importance

Although all the regulations are of great importance, there are five which stand out as being most able to influence the overall health, safety and welfare of site workers:

1. The Management of Health and Safety at Work Regulations 1999
2. CDM – The Construction (Design and Management) Regulations 2007
3. The Work at Height Regulations 2005
4. COSHH – The Control of Substances Hazardous to Health Regulations 2002
5. Corporate Manslaughter and Corporate Homicide Act 2007.

The Management of Health and Safety at Work Regulations 1999

The Management Regulations deal with the assessment of risk and arrangements for and competence in the measures needed to protect individuals and prevent accidents at work etc.

The impact of this legislation is to ensure that issues identified by risk assessments are dealt with by effective planning, organisation and control, and that procedures to monitor and review such arrangements are put in place.

The Management Regulations are also important because they impose obligations on employers to provide adequate health and safety training for employees and to communicate health and safety risks and the measures planned to deal with them.

The COSHH Regulations 2002

These Regulations were designed to address some of the long-term health problems that can arise when working with materials. The material that is commonly quoted as being the most dangerous material used in construction is asbestos; however, there are many other commonly used materials that can cause severe health problems in the short and long term if adequate protection during use is not observed.

To comply with the COSHH Regulations, designers and contractors must:

■ Assess health risks arising from hazardous substances
■ Decide what precautions are needed
■ Prevent or control exposure
■ Ensure control measures are used and maintained
■ Monitor the exposure of employees
■ Carry out appropriate health surveillance
■ Ensure employees are properly informed, trained and supervised.

COSHH applies to all substances except:

■ Asbestos and lead, which have their own regulations
■ Radioactive materials, which have their own regulations
■ Biological agents if they are not directly connected with the work and are outside the employer's control (e.g. catching a cold from a workmate)
■ Commercial chemicals that do not carry a warning label (e.g. washing-up liquid used at work is not COSHH relevant, but bleach would be).

CDM Regulations 2007

The CDM Regulations spread the responsibility for health and safety on construction sites between the following parties:

■ The client
■ The designers

- The CDM coordinator
- The principal contractor
- Subcontractors.

CDM is a set of management regulations dealing with the responsibilities of the parties identified above and with the documentation necessary to enable construction operations to be carried out safely. Within CDM are detailed regulations aimed at controlling the risks arising from particular construction tasks such as demolitions, excavations, vehicles and traffic movement etc. The Regulations also provide specific rules regarding site welfare facilities.

The Regulations state that where CDM applies, pre-construction information must be provided by the client and by designers and must be collected by the CDM coordinator in order to ensure that contractors are able to identify and manage site-specific risks. Before starting work on-site, this information must be developed by the principal contractor into a construction phase (health and safety) plan. After being appointed, the principal contractor needs to develop the construction phase plan and keep it up to date. The Health and Safety File, which is prepared by the CDM coordinator in conjunction with the principal contractor, is a further statutory document required under CDM. This is a record of information for the client or end user, and includes details of any work which may have to be managed during maintenance, repair or renovation. This must be handed to the client on completion of all the works.

Health and Safety legislation is part of the UK criminal code and breaches can lead to heavy fines and, in serious cases, custodial sentences for offenders. Additionally, companies and organisations can be prosecuted for health and safety management failures under the Corporate Manslaughter and Corporate Homicide Act 2007 where there is a gross breach of duty of care.

Safe systems of work

In order to ensure that construction work is undertaken in as safe a manner as is feasibly possible, safe systems of work strategies should be adopted. A safe system of work strategy should include:

- *A safety policy*
  A statement of intent setting down appropriate standards and procedures for establishing safe systems of work in an organisation.
- *Risk assessments*
  Identify hazards and associated risks to the workforce and others where a safety method statement will be required to control them.
- *Safety method statements*
  Provide details of how individual safe systems of work may be devised for particular tasks.
- *Permits to work*
  Restrict access to places of work or restrict work activities which are considered 'high risk'.
- *Safety inductions*
  General safety information provided by principal contractors to workers, contractors and visitors when first coming onto site.

- *Site rules*
  Set out the minimum standards of safety and behaviour expected on-site.
- *Tool box (task) talks*
  Provide particular safety information and a forum for exchange of views regarding specific operations or activities about to commence on-site.
- *Safety audits*
  A procedure for reviewing and appraising safety standards on-site, both generally and for specific activities where appropriate.

Designing for health and safety

'Design' is a complex process of many interrelating and interacting factors and demands, including:

- Form and appearance
- Structural stability
- Heating and ventilation
- Sound and insulation
- Access, circulation and means of escape
- Environmental impact
- Cost.

Good design requires a balance of these and other factors, but as design is an iterative process, different design decisions have to be made as each stage develops. As the design process unfolds, from the concept design stage to the detail design stage, so designers of buildings, whether new build or refurbishment, are constantly seeking technical solutions to design problems.

For instance, at the **concept design** stage, decisions are made concerning:

- Plan shape
- Plan size
- Number of storeys
- Floor to ceiling height
- The nature of the building 'envelope'.

Design choices here will not only be influenced by the size and shape of the site and any planning restrictions, but also by adjacent buildings, the need to incorporate services (in ceiling voids, for instance) and by the tone and general impact of the design required by the client. However, once decided, these decisions will have an irrevocable influence on the project which the design team and the client have to live and work with thereafter. These decisions are particularly influential on the economics of the design and they will determine the cost 'bracket' within which the building naturally falls.

During **scheme design**, designers' considerations will turn to the major components of the building such as:

- The substructure
- The type of frame

- Choice of external cladding
- Building services
- The design of the roof structure.

Consequently, designers will need to consider alternative foundation choices in the context of the prevailing geological conditions on the site and methods for the remediation of contamination. The economics of alternative structural forms (e.g. steel frame vs. *in-situ* concrete) and whether to choose curtain walling or precast concrete panel cladding solutions etc. will also be considered at this stage of the design.

When the **detail design** stage is reached, design questions will include:

- Staircase construction (steel, timber or precast concrete)
- Lifts and access
- The quality and specification of items such as doors and windows
- Choice of finishes (walls, floors and ceilings)
- Internal fittings (fixed seating, reception desks, special features etc.).

In making decisions throughout the design process, the designer will be constantly balancing issues of **appearance**, **function** and **cost** consistent with the client's brief and budget. As the design develops, more information becomes available, and this may give rise to technical problems or design choices that might have an impact on the financial viability of the project. A further consideration for designers is that their decisions might give rise to hazards in the design, and this will also influence their final choice.

An example of this approach might be the choice of façade for a multi-storey building. The aesthetics of the design would clearly be an important consideration in this respect, but a design which required operatives to work at height fixing curtain walling panels from an external scaffold would not be attractive from a health and safety perspective. However, a design solution whereby cladding units could be craned into position and fixed by operatives working inside the building attached by appropriate harnesses to fixed points or running lines would be a much more proactive design solution.

This objective approach to health and safety in design is eminently sensible because specific solutions to particular problems have to be found, rather than adopting 'generic' designs which might not be project-specific or address the risks involved on a particular site.

Of course, no designer sets out to design unsafely on purpose, but this is what happens unconsciously in practice. Designers therefore need to give health and safety proper weighting in their design considerations by applying the principles of **preventative design**.

However, all design considerations must also be viewed in the context of any prevailing statutory legislation. This will include:

- Permission to build (planning)
- Building standards (building regulations or codes)
- Heat loss and insulation
- The rights of third parties (such as adjoining properties)
- Statutory requirements relating to the health and safety of those who may be involved in the construction or maintenance of the building or who may be affected by the building process, such as visitors to the site or passers-by.

PART 1

Of course, different legal jurisdictions around the world have their own approaches to legislating the criminal code relating to the issue of health and safety in the design and construction of buildings, and there is no common standard.

In the USA, for instance, health and safety legislation is far more prescriptive than in the UK and Europe, where the basis for such legislation is a 'goal-setting' or objective standard. Thus, in the USA, legislation comprises detailed sets of rules which must be followed to ensure compliance, whereas in the UK the less prescriptive approach relies on designers and others working out their own 'reasonably practicable' solutions to particular problems within a general duty of care.

However, this objective or 'goal-setting' approach to health and safety legislation creates problems for building designers. They must not only find solutions to their design problems, but they must do so by identifying the hazards emanating from their design, by eliminating or reducing the ensuing risks to health and safety, and by considering appropriate control measures to protect those who may be affected. This is a much more difficult standard to achieve than following detailed sets of rules where design solutions are prescribed or set down in statutes.

For instance, under the UK Construction (Design and Management) Regulations 2007 (CDM), designers have a statutory duty to prepare designs so as to avoid the foreseeable risks which might have to be faced by those with the task of carrying out the construction work or cleaning the completed structure or using it as a workplace. Allied to this statutory duty is the higher duty imposed by the Management of Health and Safety at Work Regulations 1999 to assess the risks involved and to follow the general principles of prevention, that is to:

- **Avoid** the risk
- **Evaluate** risks that cannot be avoided
- **Combat** risks at source
- **Adapt** work to take account of technical progress.

The best way to achieve this is by considering the health and safety impact of design decisions right from the outset of a design and throughout its development. This will mean thinking about the risks attached to both concept-stage decisions and decisions made during scheme and detail design. The issues will be different at each stage, but the thought processes will be the same. Thought processes, alternatives considered and decisions made should be recorded to provide an audit trail of the designer's thinking.

When designers are unable to eliminate risk from their designs, UK legislation requires that this is stated in the pre-construction information which informs the contractor of residual design risks, thereby enabling him to develop safe working methods during construction. A further duty is to consider the possible impact of the design on maintenance activities to be carried out when the building is completed and commissioned by the client, such as window cleaning and cleaning out roof gutters.

Table 3.2 takes the re-roofing of an existing 1950s factory building as an example.

In the UK, this objective thinking does not stop when the building work is complete. The CDM Regulations require completed structures to have a **health and safety file** which is passed on to the building client or subsequent owner of the property. This file contains information which would be indispensable to those who might be involved in the maintenance, adaption, alteration or demolition of the structure at some time during its life cycle.

Table 3.2 Re-roofing a 1950s factory building.

Step	Think about	Examples
1	The design element concerned	1. Removing existing roofing 2. Removing existing rooflights 3. Installing new roof covering 4. Installing new rooflights 5. Installing new guttering
2	The hazards which could be present	1. Operatives working at height 2. Fragile rooflights 3. Asbestos cement roof sheeting 4. Roofing materials stacked at height 5. Weather/wind
3	The persons in danger	1. Roof workers 2. People working below 3. Passers-by 4. Maintenance crews
4	The likelihood of an occurrence and the severity if it happened	Judge the risk by considering the chance of falling and the extent of injury likely. This is a common occurrence where fragile roofs are concerned. Consider the chance of materials falling on people below. A simple calculation could be used to give a measure of risk or a judgement could be made as to whether the risk is high, medium or low
5	The control measures required	1. Can the design be changed to avoid the risk? 2. Could permanent edge protection be included in the design, e.g. parapet wall or permanent protected walkway? 3. Specify loadbearing liner sheets for new roof for use as a working platform 4. Specify non-fragile rooflights to replace existing 5. Specify a permanent running line system for the new roof for use by construction and maintenance workers
6	Any residual risks that need to be managed	These will be design issues that cannot be resolved, leaving people at risk. They should be raised in the health and safety plan. This could mean that the contractor will have to provide edge protection and/or collective fall arrest measures such as safety nets

PART 1

Included in the health and safety file might be:

- Information on how existing hazards have been dealt with in the design (e.g. contaminated sites, asbestos in buildings, underground services)
- As-built drawings including the location of key services, pipework and wiring runs
- The structural principles underpinning the design (this could be important for future demolition)
- Any hazardous substances remaining (e.g. asbestos or lead-based paints in existing buildings).

REFLECTIVE SUMMARY

With reference to statutory control remember:

- This is a process only recently applied in the history of building.
- The first set of national Building Regulations was published in 1965.
- The environmental impact is immense during both the construction and occupancy phases.
- Part L of the Building Regulations deals with the energy efficiency of buildings, and future revisions are likely to be increasingly stringent.
- Many companies are now developing Environmental Management Systems in order to improve their environmental performance.
- Health and Safety needs to be considered in great detail when building forms and systems materials are specified.
- Major references for health and safety include:
  - The Health and Safety at Work Act 1974
  - Corporate Manslaughter and Corporate Homicide Act 2007
  - The Management of Health and Safety at Work Regulations 1999
  - The Construction (Design and Management) Regulations 2007 (CDM)
  - The Work at Height Regulations 2005 (amended 2007)
  - The Control of Substances Hazardous to Health Regulations 2002 (COSHH).

REVIEW TASKS

- What are the main implications of the requirements of Part L of the Building Regulations for multi-storey and large-span commercial and industrial buildings?
- What are the requirements of a client under the CDM Regulations?
- What are the implications of the CDM Regulations for designers?
- Visit the companion website at www.palgrave.com/engineering/riley2 to view sample outline answers to the review tasks.

3.3 | Ground water control

Introduction

- After studying this section you should be aware of why water is found in the ground, why it needs to be controlled and the methods of controlling ground water either permanently or temporarily.

■ Included in this section are:

- The origins of water in the ground
- Methods of permanent exclusion of ground water
- Methods of temporary exclusion of ground water.

Overview

Surface water, usually due to rainfall, lies on the surface of the ground and tends to flow into excavations as they proceed, but is usually only small in quantity except in times of flood. Ground water is present when rainfall soaks into the ground and is held temporarily in the soil above the water table. Below the water table level is the subsoil water, which is the result of the natural absorption by the subsoils of the ground water until they become saturated. In coastal areas the water in the ground will rise and fall with the tide, and even areas which are presumed to be inland can be affected by tides as river courses have changed over the centuries, but the water passage still exists. For example, in London the water level in the ground increases up to two miles from the Thames when the tide comes in.

The presence of ground water can cause the following problems:

■ A high water table, which can cause flooding during wet periods
■ Difficulties with excavation works due to the tendency of ground water to flow into the void created by the removal of soil during the excavation period
■ It can give rise to unacceptable humidity levels around the finished building.

In the summer, as a general rule, evaporation exceeds saturation, whilst in the winter months saturation exceeds evaporation.

When any excavation is carried out, control of surface, ground and subsoil water is very important for both safety and economic conditions and ground water control can take one of two forms, which are referred to as 'permanent' or 'temporary' exclusion (Figures 3.2 and 3.3).

Figure 3.2
Permanent exclusion.

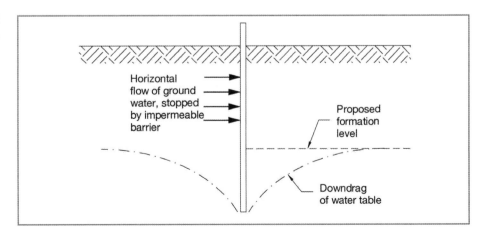

Horizontal flow of ground water, stopped by impermeable barrier

Proposed formation level

Downdrag of water table

Figure 3.3
Temporary exclusion.

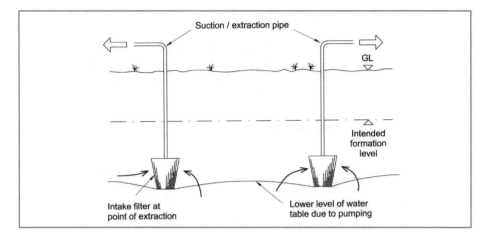

Figure 3.3
Temporary exclusion.

Methods for the permanent exclusion of ground water

Figure 3.4 summarises the various ways of excluding ground water.

Figure 3.4
Summary of methods
of ground water
exclusion.

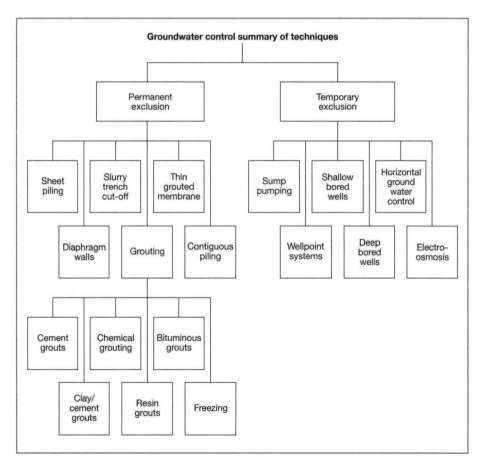

Steel sheet piling

Steel sheet piling can be used in all types of soils except boulder clay. The sheets are used to form a barrier to prevent the flow of water (Figures 3.5 and 3.6). Sheet piling is permanent where used as a **retaining wall**. However it can be used to form a temporary barrier in excavations. The disadvantage with using sheet piles is that they have to be 'hammered' into the ground which causes noise and vibration. If the steel sheets are to be used in a built-up area, then the noise generated may prove unacceptable to neighbouring building occupiers. If the steel sheets are to be used next to existing properties then excessive vibration may cause damage to these structures. In addition the process of 'hammering' is quite slow.

A **retaining wall** is any wall that supports material and prevents collapse.

Figure 3.5
Steel sheet piles and support.

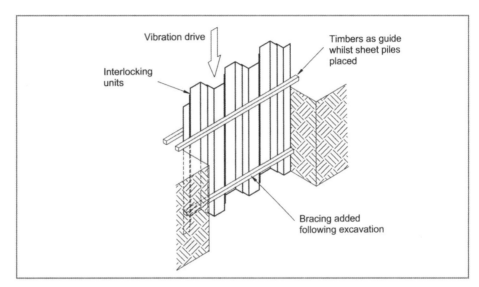

Contiguous piling

Contiguous pile walls are formed using a series of interlocking *in-situ* concrete piles. This system is especially useful if a basement is to be formed as it becomes the external wall, and is covered in more detail in Chapter 5.

Grouting methods

Grouting methods are used to form a wall in soils where pumping methods are unviable. **Grouts** are injected into the soil at high pressure and, once cured, form the wall. The walls are non-structural and therefore further support will be required during the excavation work. They are quick and relatively cheap to install, do not cause any vibration and minimal noise is created. There are five main types of grout used for this purpose:

Grouts are thin fluid mixtures that can be manufactured from a variety of different materials, dependent on the properties required of the dried grout mixture.

- Cement grouts: these are a mixture of cement and water
- Chemical grouts
- Resin grouts
- Bituminous grouts
- Clay/cement grouts.

Figure 3.6
Larsen steel sheet piles
and steel sheet
trenching.

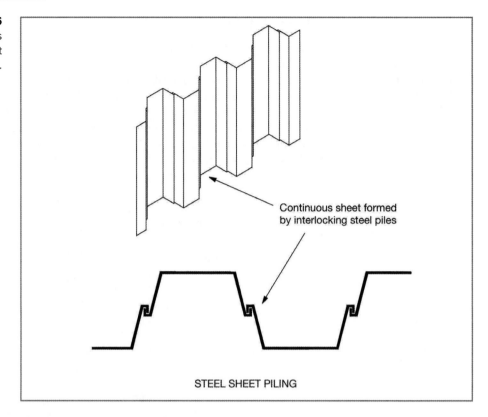

Continuous sheet formed
by interlocking steel piles

STEEL SHEET PILING

Thin grouted membranes

These walls are permanent and non-structural, used mainly in sandy and silty soils. Column sections with grout introduction pipes fixed to the web are 'hammered' into the ground (Figure 3.7). The sections are removed once the grout has cured. The major advantage of using this system is the speed of construction, but there are depth limitations.

Diaphragm walls

Often used for constructing basements and other underground structures such as car parks, diaphragm walls can be used in all types of soils. They are formed using *in-situ* concrete. An excavation is formed and this is filled with Bentonite slurry. The slurry is then replaced with the concrete (Figure 3.8). In contrast with sheet piling this method does not cause excessive noise and vibration. However, if the wall does not form part of the new structure, it will be an expensive choice.

Precast diaphragm walls

These walls are made up of a series of precast panels. The panel units are installed in a trench filled with slurry which then cures to achieve the stability of the wall. The fact that the panels are already precast reduces any quality issues which can occur with *in-situ* concrete, but it does mean that design flexibility is limited.

Figure 3.7
Thin grouted
membranes.

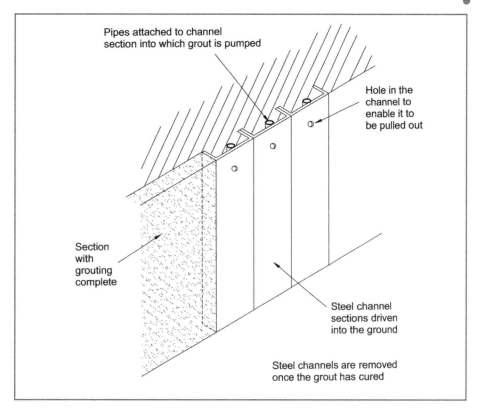

Pipes attached to channel
section into which grout is pumped

Hole in the
channel to
enable it to
be pulled out

Section
with
grouting
complete

Steel channel
sections driven
into the ground

Steel channels are removed
once the grout has cured

Figure 3.8
Typical cast *in-situ*
diaphragm walls.

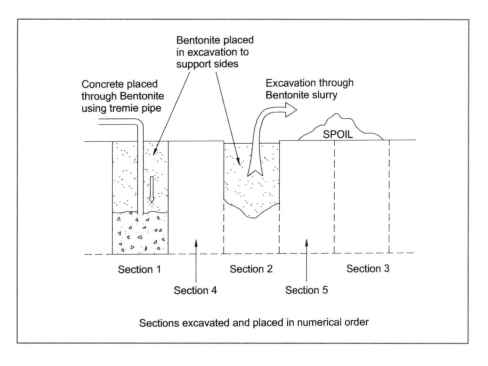

Bentonite placed
in excavation to
support sides

Concrete placed
through Bentonite
using tremie pipe

Excavation through
Bentonite slurry

SPOIL

Section 1   Section 2   Section 3

Section 4   Section 5

Sections excavated and placed in numerical order

Slurry trench cut-off

Slurry trench cut-off walls are used if there is adequate space to house a wall separate from the proposed building structure. They are non-structural, do not include reinforcement and are generally used when the mass of soil is light – such as sandy soil. The benefits of using these wall types are that they are relatively quick, easy and cheap to construct.

Ground freezing methods

Ground freezing is suitable for any soil with a high moisture content or water table. The idea is that the water in the ground is frozen and this frozen layer or section forms a barrier to prevent water ingress. Tubes are inserted into the ground and freezing brine circulated in the tubes. As the ground is frozen it acts as an impermeable frozen wall (Figure 3.9). This is an expensive and specialised system and is only used when alternative systems are unviable.

Methods for the temporary exclusion of ground water

See Figure 3.4 for a summary of the available methods.

Figure 3.9
Ground freezing
methods.

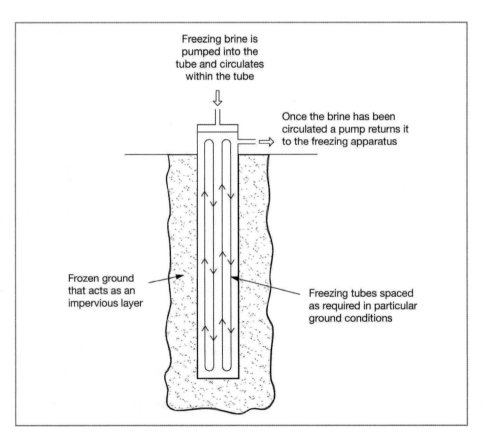

Open sump pumping

In sump pumping, an excavation is started and usually at the outer perimeters a trench or 'hole' is dug deeper than the formation level of the excavation. The sump pump is placed in this deeper excavation and the water is electrically pumped to the surface (Figure 3.10) and discharged, usually to a butt.

Figure 3.10
Sump pumping.

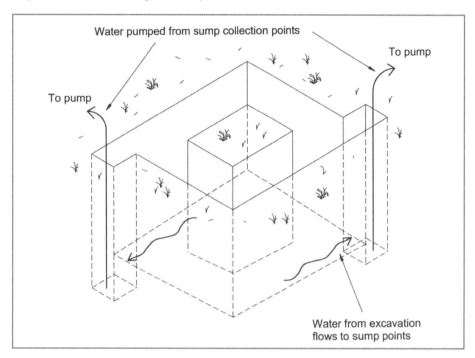

Jetted sump pumping

Jetted sump pumping is similar to open sump pumping except that with this system you do not need to undertake excavation first. A metal tube is jetted into the ground using water pressure and then the soil is removed. A disposable wellpoint is then lowered into the tube. The wellpoint is covered with sand that acts as a water filter and the tube removed. Water that collects around the wellpoint is then sucked out using a mechanical pump. The pump in the ground is left *in situ* and remains in place after construction has finished although the tube is removed.

Wellpoint systems

The principle behind wellpoint dewatering is to lower the level of the existing water table so that there is no water in the ground that needs to be prevented from entering an excavation. Systems can be single stage or multi stage if the water table needs to be more significantly reduced. The method of installation is similar to jetted sump pumping but a series of wellpoints are installed, rather than just one or two. These are connected to a common pipe which is connected to a mechanical pump and so water is removed from the soil (Figure 3.11).

Figure 3.11
Wellpoint
arrangements.

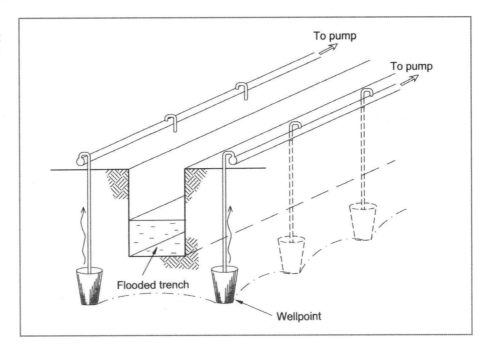

Figure 3.11 Wellpoint arrangements.

Other available methods

There are other methods available to facilitate temporary dewatering, but these methods are only used in exceptional circumstances. These other methods include:

- Shallow bored wells
- Deep bored wells
- Horizontal ground water control
- Electro-osmosis.

REFLECTIVE SUMMARY

- Ground water can cause problems during the construction work and/or after the building is complete.
- Ground water tends to create more problems in the winter. Therefore construction work should be planned to ensure that the minimum of ground working is undertaken during the winter months.
- Coastal areas and land on flood plains will generally suffer the most from ground water problems.
- The decision to choose permanent or temporary dewatering systems will depend on the nature of the site, the building to be constructed and the extent of ground water.
- Ground dewatering systems can be very expensive, and the cost of such a system needs to be seriously considered before the decision to build is made.
- Virtually all basement construction will require some ground water control, and the system will be most cost-effective if the ground dewatering system is built into the basement design.

- All the systems defined, with the exception of sump pumping, will be recommended by the engineer.

REVIEW TASKS
- After studying the text and the diagrams, which do you believe is the most cost-effective: having a shallow-sided excavation that requires more material to be excavated, or a steep-sided excavation that requires less excavation but allows more ground water to flow into the excavation (see Figure 3.12)?

Figure 3.12
Shallow and steep-sided excavations.

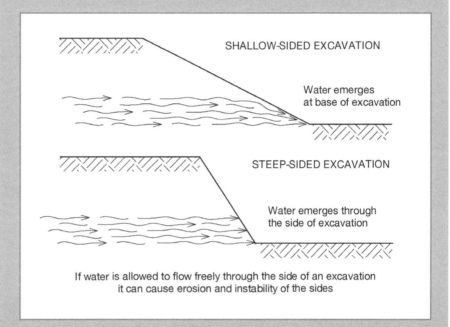

SHALLOW-SIDED EXCAVATION

Water emerges
at base of excavation

STEEP-SIDED EXCAVATION

Water emerges through
the side of excavation

If water is allowed to flow freely through the side of an excavation
it can cause erosion and instability of the sides

- Why might the steep-sided excavation with a form of ground water control be the only suitable option?
- For each of the systems discussed, identify the positive and negative points from environmental impact and health and safety perspectives.
- Visit the companion website at www.palgrave.com/engineering/riley2 to view sample outline answers to the review tasks.

Building substructure

4 Foundations

AIMS

After studying this chapter you should be able to:

- Demonstrate a knowledge of the various alternatives available for the foundations of industrial and commercial buildings
- Understand the relationship between foundations, ground conditions and the design of the building superstructure
- Outline the sequence of construction works required for given deep and shallow foundation forms
- Understand the implications of the existence of basements when considering foundation design

This chapter contains the following sections:

INFO POINT

- BS 8004: Code of practice for foundations (1986)
- BRE Report 440: BRE building elements: foundations, basements and external works (2002)
- BRE Report 473: Geotechnics for building professionals (2005)
- *Design and construction of deep basements including cut-and-cover structures*, IStructE (2004)
- *Ground engineering spoil: good management practice*. Report R179, CIRIA (1997)
- *Integrity testing in piling practice*. Report R144, CIRIA (1997)
- *Piling handbook*, 7th edn, Parts 1–12, BSSPC (1997)
- Reynolds, T., Lowres, F. and Butcher, T. (2010) *Sustainability in Foundations: A review*. BRE Information paper, IP 11/10, April [978–1–84806–134–7]
- *Steel bearing piles*. Guidance Publication 156, SCI (1997)

4.1 | Soils and their characteristics

Introduction

- After studying this section you should have an understanding of the range of soils and ground conditions in which foundations are required to function.
- You should appreciate the limitations on performance associated with each of the major soil types and the implications for selection of an appropriate foundation option.
- You should also have an overview of the possible nature of contamination of soils on building sites and the ways in which these are recognised and dealt with.

Overview

The stability and integrity of any structure depend upon its ability to transfer loads to the ground which supports it. The function of foundations is to ensure the effective and safe transfer of such loadings, acting upon the supporting ground whilst preventing over-stressing of the soil. The nature of the structure, its foundations and the soil onto which they bear dictate the ways in which this function is achieved. It is inevitable that in the period shortly after the construction of a building some consolidation of the soil takes place. Minor initial settlement is, therefore, to be expected; more serious movement must be avoided, in particular that which is uneven and which may result in differential settlement, causing cracking and deformation of the building. The specific design of a foundation depends upon the structure, the way in which its loads are delivered to the foundation and the loadbearing characteristics of the soil type. It is logical, therefore, in considering the performance requirements and design of foundations, to first consider the various soil types and their individual properties.

Soil types

In practice there is an infinite variety of soil compositions. However, these can be broadly categorised into five generic types, along with solid rock, although they can be further subdivided into a large number of specific descriptions. Engineers will tend to think in terms of classifications that distinguish soils and rock, with arbitrary divisions of type based on notional soil strength. Since the properties of soils depend on the size of the particles from which they are comprised, a system of classification that is often used is based upon particle type and size. This is appropriate, since the strength of soils is linked strongly to the characteristics that arise as a result of this. In addition to solid rock, the five generally recognised categories of soil are *gravels*, *sands*, *silts*, *clays* and *peats*.

Rock

Solid rock provides a sound base upon which to build, as a result of its very high load-bearing characteristics. However, the notion that rock is a non-compressible base for building is incorrect. The bearing capacity of rock will be related to the way in which the rock was formed and the characteristics arising from its subsequent deformation due to movement of the earth etc. Hence we often refer to geological distinction and rocks may be classified as one of three types:

- *Igneous rocks*, such as granites and basalts, which were formed as a result of solidification of molten material
- *Sedimentary rocks*, such as sandstone, chalk and limestone, which were formed by deposition of granular material or sediment
- *Metamorphic rocks*, such as slate, which formed as a result of heat or pressure acting on igneous or sedimentary rocks and changing the crystalline structure.

Typically, the safe loadbearing capacity of sedimentary rock, such as sandstone and limestone, is approximately 10 times that of a clay soil, with igneous rocks, such as granite, having capacities 20–30 times that of clays. These high loadbearing capabilities suggest that solid rock is an excellent base upon which to build. However, there are also great disadvantages, resulting from the difficulty in excavating and levelling the sub-strata. This presents a considerable cost implication in building on such sub-strata. An additional problem, manifested in some sedimentary rocks, is that posed by the presence of inherent weak zones, created as a result of fault lines in the rock, together with the presence of slip planes between adjacent layers of differing composition. The loadbearing capacity of the rock is greatly altered if there is substantial presence of flaws or faults.

Gravels and sands

Gravels and sands are often grouped together and considered under the broad heading of coarse-grained cohesionless soils. The particles forming these soil types, ranging in size from 0.06 mm upwards, show little or no cohesion, and hence they tend to move independently of each other when loaded, if unrestrained. This property can result in many problems if not addressed, however, if treated properly these forms of soil can provide an adequate building base. When initially loaded some consolidation is likely to occur, which may pose problems in relation to the connection of services and so on. The voids present between individual particles are relatively large, but represent a low proportion of the total volume. The presence of such voids results in the soils being permeable, and hence water is not held in the ground for long periods; neither is the volume of the ground affected by the variation of moisture content.

Silts and clays

These types of soil are generally considered within the broad category of fine-grained cohesive soils. The small size of the particles making up these forms of soil, with a high proportion of small voids between particles, results in the tendency of clays and silts to display considerable variation in volume when subject to changing moisture content. This results in drying shrinkage in dry weather and swelling and heave in wet weather. The nature of the soil is such that moisture is retained for considerable periods, due to low permeability. Such soil types may also be compressible when loaded, especially at shallow depths; however, at suitable depths, below the region which is likely to be affected by moisture variation (normally 1 m or deeper), satisfactory bearing is normally found. The effects of trees, particularly those which have been felled recently, can be significant,

Table 4.1 Possible contamination risks on building sites.

Contaminant	Signs	Action
Metals	Affected vegetation Surface materials	Extraction of contaminated matter or cover technology
Organic compounds	Affected vegetation Surface materials	Biological remediation, extraction of affected matter or cover technology
Oil/tar	Surface materials	Extraction of affected matter
Asbestos/fibres	Surface materials	Extraction of affected matter or cover technology
Combustible materials	Surface materials Fumes/odours	Cover technology with venting facility or extraction of affected matter
Gases (methane/CO_2)	Fumes/odours	Barrier or cover technology
Refuse/waste	Surface materials Fumes/odours	Extraction of affected matter or cover technology

as they affect the amount of water which is removed from the ground. A further problem that is sometimes present in areas of clay soil is that posed by the aggressive actions of soluble sulphates in the soil upon ordinary Portland cement. This may necessitate the use of sulphate-resisting cements in some areas.

Peat and organic soils

Soils which contain large amounts of organic material are generally subject to considerable compression when loaded and to changes in volume resulting from decay of the organic matter. Hence they are generally unsuitable for building. For these reasons top soil is removed prior to building, and excavation will normally take place to remove soil with a considerable quantity of organic matter present.

Contaminated ground

The risk of the presence of certain forms of contamination in the ground can be of great significance to the process of construction. Although it is not possible to consider this complex issue in depth within this book, it is useful to consider the topic in overview. Contaminants that are likely to be found in the ground can pose problems in the form of potential risk to health or as aggressive substances to the materials used in construction. The Building Regulations define a contaminant as:

any substance which is or could become toxic, corrosive, explosive, flammable or radioactive and likely to be a danger to health & safety

Site investigation will be necessary to identify the nature and extent of contamination. This may have direct consequences for the viability of the proposed building project and the requirements for remediation may be onerous. The Environmental Protection Act and the Environment Act impose a responsibility for remedial action to negate the potential risks associated with land contamination. This can have very great consequences in terms of cost and time.

The main options available for treatment of the problem of contaminated ground are:

- Removal of the contaminated material to another location, normally a licensed site
- Provision of an impervious layer between the contamination and the ground surface, often termed 'cover technology'
- Biological or physical remediation to neutralise or remove the contaminant.

The types of contaminant typically found on sites that may be intended for building projects are set out in Table 4.1.

REFLECTIVE SUMMARY

When considering soil types and their relationship with building design, remember:

- The characteristics of different soils are strongly related to the size of the particles that make up the soil.
- Different soil types will perform in different ways with regard to loadbearing capacity, moisture retention and drainage, and stability during excavation.
- Generally, soils with larger particles will be free draining, whilst those with smaller particles will retain moisture.
- Soils can be subject to a range of contaminants that may need to be dealt with before building work starts.

REVIEW TASKS

- Place these soil types in ascending order of strength: clay, sand, chalk, sandstone, silt and peat.
- With reference to a site with which you are familiar, try to make a list of possible contaminants that might affect the construction process.
- Identify the soil types in the place where you live and list the general characteristics of such soils.
- Visit the companion website at www.palgrave.com/engineering/riley2 to view sample outline answers to the review tasks.

PART 2

4.2 | Functions of foundations and selection criteria

Introduction

- After studying this section you should have a broad appreciation of the functional requirements of foundations to framed buildings.

- In addition, you should be able to recognise key aspects of their performance which may be used to apply selection criteria to given scenarios with a view to adopting the most appropriate design solution.
- The nature of loads from framed buildings should be understood and the implications of these loads on foundation design and selection should be appreciated.

Overview

The foundations of a building form the interface between the structure and the ground which supports it. The primary function of the foundations is to safely transmit the building loads to the supporting strata, spreading them over a sufficient area to ensure that the safe loadbearing capacity of the soil is not exceeded. In the case of framed buildings this is often difficult to achieve, due to the localised nature of the loadings generated. Unlike low-rise traditional loadbearing structures, framed buildings concentrate the loads from the building structure through columns which exert highly concentrated loads upon the foundations. In effect, this results in the load from the building frame being converted to a series of point loads at the base of the columns or stanchions. For this reason it is often the case that foundation solutions are deep in order to transfer loads to the ground where the bearing capacity is greater. There is an added complexity in that many framed buildings incorporate basements, which are subject to considerable lateral pressure from the ground. In such instances the foundations and the basement wall structure are sometimes combined; this will be explored in more detail later. In addition, there is a potential difficulty in ensuring that the pressure on the ground at all points below the building foundations is balanced such that there is no risk of differential movement or rotation. In high-rise buildings this can sometimes be a problem because of differing building configuration and restrictions of the site. It is normal, therefore, in dealing with larger buildings to ensure that an appropriately calculated design solution is used in each individual instance. This differs somewhat from the situation in construction of dwellings, where a more standardised empirical approach based on experience and minimum acceptable standards tends to be adopted.

In order to achieve the functions required of them, the foundations to industrial and commercial buildings must possess a number of properties, as follows:

- The foundation must possess sufficient strength and rigidity to ensure that it is capable of withstanding the loadings imposed upon it without bending, suffering shear failure or 'punching through' at the point of loading.
- It must be capable of withstanding the forces exerted by the ground, and must resist the tendency to move under such loads as exerted by volume changes in the ground caused by frost and moisture action together with the considerable lateral loadings that may arise where basements etc. exist.
- It must be inert and not react with elements within the ground, such as soluble sulphates, which may degrade the material from which the foundation is formed.

Selection criteria

The basis for the selection of foundations is quite straightforward. It is important to remember that the performance of foundations is based on an interface between the loadings from the building and the supporting ground or strata. The nature and conditions of each of these may vary, and it is primarily as a result of these variations that some foundation solutions are more appropriate in certain circumstances than others. In all cases, however, the most economical solution will be selected, provided that it satisfies the performance requirements.

Factors related to ground conditions

In most cases the ground upon which industrial and commercial buildings are constructed is reasonably stable, level and of uniform composition. In some situations it is the case that the ground close to the surface is capable of supporting the loadings resulting from the construction of a framed building. Where this is so, the use of shallow forms of foundation is generally adopted. However, in some instances this is not the case, and the foundation solutions must be selected accordingly. This is particularly the case in the construction of tall framed buildings where the loads applied from the structural frame of the building are significant and the soil strength close to the surface is unlikely to be sufficient to support them. Another important factor that must be taken into account when dealing with buildings of this type is that they tend to be constructed on **brownfield**, or previously used, sites. As a consequence of this, the ground tends to be affected by the presence of elements of previous construction on the site, such as demolition debris and buried services. Their presence can have a significant effect upon the choice of foundation. The specific factors related to the nature of the ground that affect foundation selection are as follows:

Brownfield sites are sites that have been built on in the past and which have been cleared to provide the base for a new construction operation. Many such sites are located in urban areas, where virgin sites are unavailable. The need to conserve greenfield sites has increased the emphasis on reusing existing building land in this sustainable way.

- *Bearing capacity of the ground*: In section 4.1 we considered the variability of soils and the degree to which their ability to carry loads differs. This is one of the key elements in the selection of appropriate foundations for all types of building.
- *Depth of good strata*: Although the area upon which the building is to be constructed may provide ground of appropriate bearing strength, this may be at a considerable depth below the surface. In these circumstances the use of a shallow foundation form is unlikely to be efficient or cost-effective. Instead, the adoption of a deeper foundation such as piles would be considered.
- *Composition of the ground*: Ideally the construction of buildings takes place on sites with uniform, stable ground conditions. As more and more sites have been utilised in recent years the supply of these ideal sites has diminished. Hence it is becoming increasingly the case that sites available are less than ideal. The increasing trend to build on brownfield sites, driven by the need to develop in a more sustainable way, has led to greater variability in ground composition. It is not uncommon to find buildings being constructed on sites that feature areas made up of filled ground. Sites with lower than ideal bearing capacities may also be considered.
- *Ground level and gradients*: It is quite rare for building sites to be truly flat and level. In most cases, however, the slight irregularities and changes in level of the ground do not pose significant problems. In cases where the variation in level is more extensive the design of the foundation will be affected. Several factors are significant in this scenario. Firstly, there is the need to provide a level base from which to build; this results in the need to step the foundation along the slope. Secondly, there is the potential for the entire structure to slide down the slope as a result of the action of gravity.

Factors relating to loads from the building

The form and type of the building to be erected dictate the nature of the loadings that are directed to the foundations as well as their magnitude. In the case of low-rise long-span buildings the extent of the loading is relatively modest. This is one of the reasons for the widespread use of shallow foundations for this type of structure. As loads increase, the need to provide deeper foundation solutions also increases. This is because the strata at a greater depth are more highly compacted, and generally therefore have the ability to withstand higher loads. By contrast, the strata close to the surface are generally less well compacted and less able to withstand loading.

The nature of framed buildings is such that the loads are likely to be concentrated at the point of application, i.e. the column bases. Hence the use of pads and piles tends to be most common. However, there may be situations where there are also uniformly distributed loads, such as from masonry cladding. These must also be dealt with, and a combination of foundation solutions may be applied in a given situation.

REFLECTIVE SUMMARY

When thinking about the functional requirements of foundations remember:

■ Selection of foundations depends on the nature of the ground, the characteristics of the building form, and the extent and nature of loads from the building.
■ Foundations act as the interface between the loads from a building and the ground. They must therefore be robust, durable and balanced to ensure effective load transfer to the ground.

REVIEW TASK

■ Generate a simple matrix that relates the functional requirements of foundations to different types of foundation that you are familiar with.
■ Visit the companion website at www.palgrave.com/engineering/riley2 to view sample outline answers to the review task.

4.3 | Shallow foundation forms

Introduction

■ After studying this section you should understand that the nature of the foundation which is appropriate in a particular situation depends on a number of factors.
■ You should appreciate that the extent of the loading and the way in which it is applied to the foundation dictate the design, together with the nature of the supporting strata. Point loads on level sites, for example, are treated in a very different manner from uniformly distributed loads on sloping sites.
■ You should appreciate that most forms of foundation are formed from concrete, either reinforced or unreinforced, although other forms, such as steel grillage foundations, are sometimes used in specific circumstances.

- The foundation options for framed buildings tend to be restricted to a relatively few economic solutions. It must always be remembered that each solution is individually designed for a given set of circumstances, and no two situations are identical.

Overview

As previously stated, the loads experienced in the construction of industrial and commercial buildings tend to be relatively high when compared with those of low-rise traditional buildings. This is accentuated by the nature of the application of the loads, which are point loads rather than uniformly distributed loads as exerted by a masonry wall, for example. Thus we must consider the implications of attempting to deal with concentrated loads from the building frame whilst maintaining stability of the building structure. Two broad approaches are taken in dealing with such loads. Firstly, we may attempt to distribute the load over a suitable area of ground close to the surface, i.e. shallow foundations. Alternatively, we may attempt to transfer the loads to a point deeper within the ground, where the bearing capacity is greater or where we can exploit frictional resistance against the foundation, i.e. deep foundations. It is convenient to consider the foundation options available within these two broad classifications.

Where there is sufficient loadbearing capacity close to the surface of the ground and where ground conditions are relatively uniform it is generally most economical to adopt a shallow foundation type to support the building loads. When dealing with framed buildings the loads from the main structure are isolated point loads, and will be treated as such. In many instances such as when dealing with industrial buildings, we may also be required to cater for traditional uniformly distributed loads from masonry cladding and such like. Thus there may be a combination of foundation types used within the same building. The following subsections consider each of the main options in turn.

Pad foundations

In some instances, such as when constructing framed buildings, the loads from the building are exerted upon the foundations in the form of concentrated point loads. In such cases independent pad foundations are normally used. The principle of pad foundation design is the same as that of strip foundations, i.e. the pad area will be of sufficient size to safely transfer loadings from the column to the ground without exceeding its loadbearing capacity. The pad will generally be square, with the column loading exerted centrally to avoid rotation, and is normally formed from reinforced concrete. Although other formations (such as steel) are possible, they are rare in modern construction. As a result of the concentration of loading applied to pad foundations, it is vital that they are reinforced sufficiently against shear as well as bending. This is because of the highly concentrated application of the load at the base of the column, which could cause the foundation to fail by 'punching through' the pad. In normal situations reinforced concrete is likely to be sufficiently strong to resist such tendencies. However, in some extreme situations there may be a need to provide extra steel reinforcement, possibly in the form of a framework of steel beams, to resist this. This is termed a 'steel grillage foundation'.

In most situations it is possible to ensure that the load from the column of the building frame is applied centrally to the pad, thus preventing the risk of rotation of the founda-

PART 2

Figure 4.1
Pad foundation shapes.

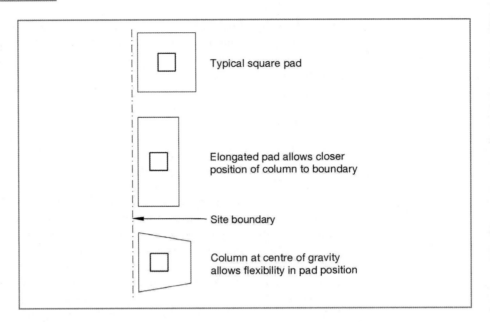

tion under load (Figure 4.1). However, in congested sites it is sometimes impossible to achieve this due to restrictions on the available area for formation of the pad. In such situations the pad may be designed to be irregular in shape, provided that the load is applied at the point of the 'centre of gravity' of the foundation. The use of elongated or trapezoidal pads is common in such situations, allowing columns to be placed closer to the perimeter of a site than would normally be possible with a square pad configuration.

Often, in practice, the foundation will actually take the form of a continuous strip resembling a deep strip foundation. This is because the economy of using a mechanical digger with a bucket to dig a simple trench outweighs the extra cost of concrete incurred by filling the spaces between the 'pads'.

The nature of the location of the loadbearing columns at the pad foundation can vary greatly in different situations. Figure 4.2 illustrates some of the more common methods.

Strip foundations and ground beams

When dealing with loadbearing walls, by far the most common form of foundation in general use is the simple mass concrete strip foundation (Figure 4.3). Used where continuous lengths of wall are to be constructed, the strip foundation consists of a strip of mass concrete of sufficient width to spread the loads imposed on it over an area of the ground, which ensures that overstressing does not take place. The distribution of the compressive loading within the concrete of the strip tends to fall within a zone defined by lines extending at 45° from the base of the wall. Hence to avoid shear failure along these lines, the depth of the foundation must be such that the effective width of the foundation base falls within this zone. The principles of strip foundations were considered in detail in *Construction Technology 1*.

As the loading exerted on the foundations increases, or the loadbearing capacity of the ground decreases, so the width required to avoid overstressing of the ground increases. In order to resist shear failure in wide foundations, they would be required to

Figure 4.2
Connections to pad
foundations.

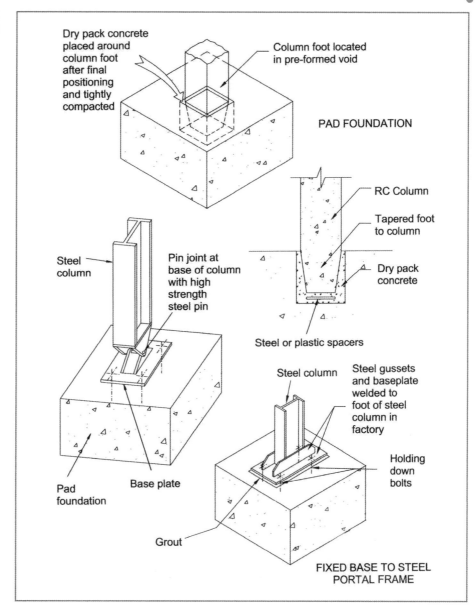

Dry pack concrete
placed around
column foot
after final
positioning
and tightly
compacted

Column foot located
in pre-formed void

PAD FOUNDATION

RC Column

Tapered foot
to column

Dry pack
concrete

Steel
column

Pin joint at
base of column
with high
strength
steel pin

Steel or plastic spacers

Steel column

Steel gussets
and baseplate
welded to
foot of steel
column in
factory

Holding
down
bolts

Pad
foundation

Base plate

Grout

FIXED BASE TO STEEL
PORTAL FRAME

PART 2

be very thick, for the reason previously described; this would normally be uneconomic. Hence to reduce the required foundation thickness, whilst still resisting failure in shear or bending, steel reinforcement is introduced to the regions subjected to tension and shear. Such a formation is termed a 'wide strip foundation'.

In many situations, however, the ground is unsuitable for carrying the uniformly distributed loads associated with, for example, masonry cladding to framed buildings. In such situations the use of piled foundations may be considered as the means of transferring the building loads to an appropriate bearing stratum. Whilst this is acceptable for the transfer of point loads acting directly on pile caps, there remains the problem of

Figure 4.3

Strip foundations.

The loads exerted on the central zone of the foundation are distributed within the concrete at a 45° angle. This results in the creation of a defined area of loading at the point of contact with the ground. The loads are resisted by the reaction force from the ground. If the width of foundation is such that this reaction is applied outside the zone of applied loading shear failure may result.

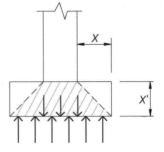

$X' \geq X$ to avoid risk of shear failure

In order to avoid the risk of shear failure the foundation is designed so that the foundation does not extend outside the zone of downward loading. The width of foundation is dictated by the strength of the soil and the extent of the applied load. (Area is adjusted to reduce pressure.) As the width increases, so the depth of the foundation material is increased so that there is no chance of shear failure.

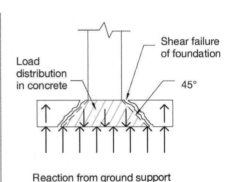

Reaction from ground support

supporting the walls between. This is overcome by the adoption of reinforced concrete ground beams (Figure 4.4) spanning between the pile caps, or in some cases pad foundations. The walls will be constructed off these beams in the same manner as they would be if using strip foundations. However, the ground beam does not transfer the load directly to the soil, but rather acts as a beam between two support points, transferring the loads to the piles or pads at each end. Thus we see a situation where the uniformly distributed loads from the cladding and the point loads from the frame are dealt with simultaneously.

Ground beams may, of course, be used for traditional loadbearing construction where there is insufficient strength close to the surface in the soil to support the required loadings. In such instances piles and ground beams would be used in combination to support the loads from the traditional wall structures.

The adoption of ground beams to support uniformly distributed loads is generally an effective and satisfactory solution. However, in some situations, such as in soils that are prone to shrinkage and swelling with changes in moisture content, there may be problems associated with uplift on the beam due to swelling of the soil beneath. This is a particular concern in clay soils. The action of uplift, or heave, on the beam could result in cracking of the structure above ground and should be avoided. This is dealt with by the use of compressible material laid beneath the beam which allows for a degree of heave without exerting upward pressure on the beam itself (Figure 4.5).

Figure 4.4
Pads/piles and ground
beams.

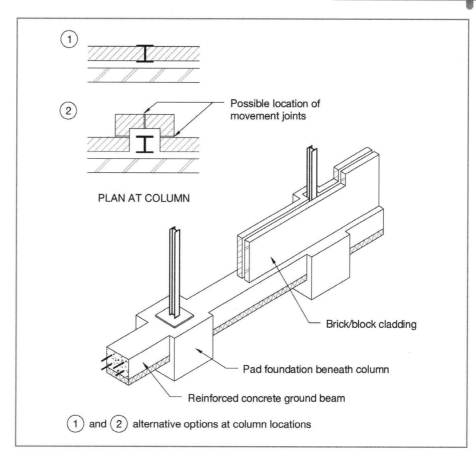

PLAN AT COLUMN

Possible location of
movement joints

Brick/block cladding

Pad foundation beneath column

Reinforced concrete ground beam

(1) and (2) alternative options at column locations

Raft foundations

Raft foundations (Figure 4.6) are constructed in the form of a continuous slab, extending beneath the whole building, formed in reinforced concrete. Hence the loadings from the building are spread over a large area, thus avoiding overstressing of the soil. The foundation acts quite literally like a raft. Inevitably, however, there tends to be a concentration of loading at the perimeter of the slab (where external walls are located) and at intermediate points across the main surface, resulting from loadings from external cladding and the structural columns of the building. Hence the raft is thickened at these locations and/or is provided with extra reinforcement (Figure 4.7). At the perimeter, this thickening is termed the 'toe' and fulfills an important secondary function in preventing erosion of the supporting soil at the raft edges; if not prevented, undermining of the raft could occur, with serious consequences. At intermediate points within the main body of the raft the thickening will take the form of a 'downstand', which is effectively a thickened rib on the underside of the raft. These will often be formed on two directions, intersecting beneath a column loading point to provide increased resistance to bending within the raft and to provide a locally strengthened area able to resist the possibility of shear failure beneath the column. As the depth of these ribs or downstands increases it becomes possible to create a series of hollow voids within the raft. This has two potential

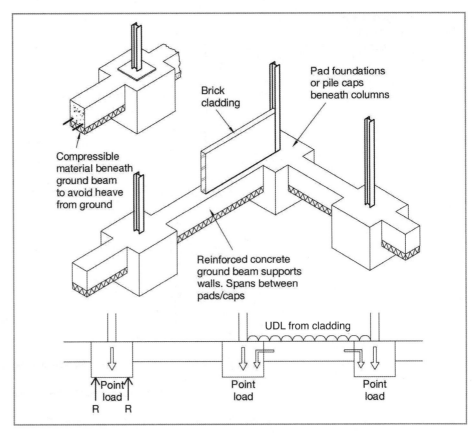

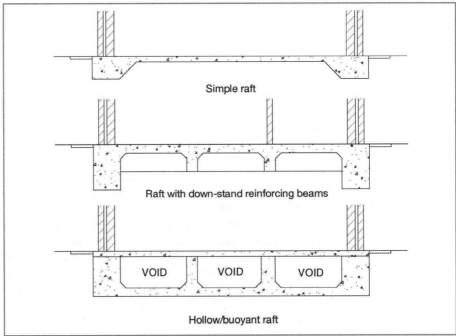

Figure 4.7
Raft foundations.

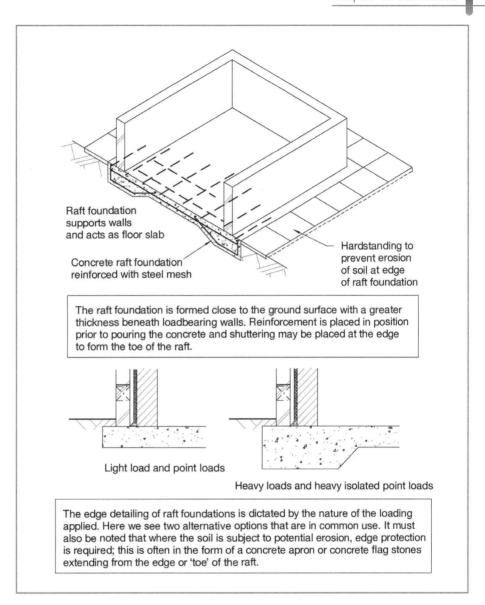

Raft foundation
supports walls
and acts as floor slab

Concrete raft foundation
reinforced with steel mesh

Hardstanding to
prevent erosion
of soil at edge
of raft foundation

The raft foundation is formed close to the ground surface with a greater
thickness beneath loadbearing walls. Reinforcement is placed in position
prior to pouring the concrete and shuttering may be placed at the edge
to form the toe of the raft.

Light load and point loads

Heavy loads and heavy isolated point loads

The edge detailing of raft foundations is dictated by the nature of the loading
applied. Here we see two alternative options that are in common use. It must
also be noted that where the soil is subject to potential erosion, edge protection
is required; this is often in the form of a concrete apron or concrete flag stones
extending from the edge or 'toe' of the raft.

benefits: firstly, if the voids are large enough they may provide usable space for the
building; and secondly, the hollow raft provides a degree of buoyancy which assists in
reducing the applied load to the supporting soil. This works on the simple principle that
if a solid mass of soil is removed from the ground and replaced with a hollow structure
of the same volume, the applied load to the ground is actually reduced. Taken to its logical
conclusion, the use of a buoyant raft could in some instances allow a building to be
constructed without applying additional load to the ground beneath. In reality this is
unlikely to be achieved totally, but there will be a reduction in applied load to some
degree.

Foundations of this type are generally utilised in the construction of relatively lightweight buildings on land with low bearing capacity. One of the advantages of this form

of foundation is that, if settlement occurs, the building moves as a whole unit. Differential movement is prevented, and hence the building retains its integrity.

REFLECTIVE SUMMARY

Shallow foundation forms for framed buildings have the following features:

- They are required to cope with point loads.
- They generally take the form of pads, strips, ground beams or raft foundations.
- Shapes of pad foundations can be varied to cope with site constraints.
- Raft foundations can be developed to become buoyant in some circumstances.
- Ground beams are used in conjunction with pads or piles to allow for point and continuous loads.

REVIEW TASKS

- Generate a list of criteria against which you might judge the performance of different shallow foundation options.
- With reference to buildings that you know, attempt to select an appropriate foundation using your list of criteria.
- Visit the companion website at www.palgrave.com/engineering/riley2 to view sample outline answers to the review tasks.

4.4 | Deep foundation forms

Introduction

- After studying this section you should have developed an appreciation for the various options available for the deep foundation solutions for framed buildings.
- You should be familiar with the different forms of piled foundation that are available and you should understand the main functional and technical differences between them.
- In addition, you should have developed an understanding of the selection criteria used to decide on an appropriate type of pile in a given situation and should be able to identify appropriate options in given scenarios.

Overview

The term 'deep foundations' is generally used to refer to foundations that are deeper than around 3 m below ground level, where their capacity to transmit loads safely to the supporting strata is unlikely to be affected by changing conditions at the surface of the ground. When considering the possible options for deep foundations to framed buildings it should be recognised that we are essentially dealing with various forms of piled foundation. Although there do exist other forms of deep foundation, such as deep pad or deep strip foundations, caissons and piers, piles are the most common form adopted in the construction of modern framed buildings. As such, we shall restrict our consideration to the various options that are available within the classification of piled foundations. Piles are often described as columns within the ground, since the basis upon which they work is similar to that of a traditional column, in that they transfer loadings from a higher level to a loadbearing medium at a lower level. They are generally categorised in two ways, either by the way in which they are installed, or by the way in which they transfer their loads to the ground. They may also be classified by the material of their manufacture and although concrete and steel are the most common materials, timber has, in the past, also been used extensively. For our purposes the following classifications are conveniently used to define pile types:

- Definition by installation method. Two common methods exist:
  - displacement piles (often subdivided into large and small displacement piles)
  - replacement piles
- Definition by load transfer mechanism. Two generic transfer mechanisms exist:
  - friction piles
  - end-bearing piles.

In addition, variants exist that are loaded in different ways, such as tensile or laterally loaded piles. Tensile piles may be used to secure retaining structures or to prevent uplift in unusual loading situations. Laterally loaded piles may be used as retaining walls or to prevent slippage in unstable rock layers, for example. Since these are specialised situations we shall not consider them in detail here. Rather, we shall restrict our considerations to the more common situations in which piles may be used. Figure 4.8 illustrates the more common families of pile used.

Although there are several variants within each type of pile, we will restrict our consideration to those most commonly used for the construction of framed buildings.

Displacement piles

Displacement piles are set into the ground by forcing or driving a solid pile or a hollow casing to the required level below ground, thus displacing the surrounding earth. In the case of solid piles, precast piles of required length may be driven into the ground using a jack or driving rig; alternatively the rig may be used to drive a series of short precast sections, which are connected as the work proceeds. The use of the second of these methods is by far the most efficient, since the length which is required may vary from pile to pile, thus the use of one-piece precast piles inevitably necessitates adjustment of length on-site. Trimming or extending of one-piece piles is a difficult task on-site.

The difficulties of providing piles of exactly the correct length, together with the danger of damage to the pile resulting from the percussive driving force, have resulted in the adoption of the use of driven shell, or casing, piles. With this method, a hollow shell or casing is jacked or driven into the ground in a number of short sections; concrete is then poured into the void as the casing is withdrawn, the steel reinforcement having already been lowered into the hole. By this method an *in-situ* pile is formed in the ground. The vibration of the pile casing as it is withdrawn from the ground results in the creation of a ridged surface to the pile sides, thus taking most advantage of any frictional support provided by the ground.

Figure 4.8

Pile 'family tree'.

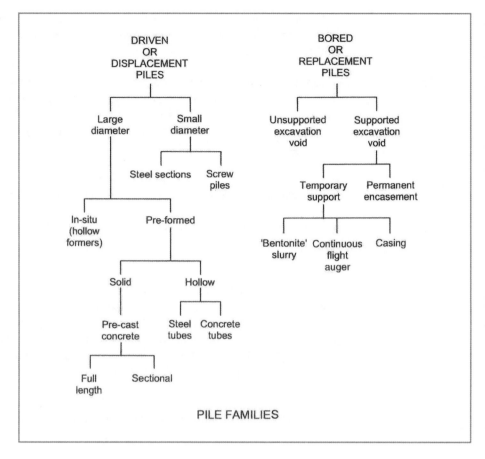

PILE FAMILIES

Methods of installation for displacement piles

There is a range of mechanisms available for installing displacement piles. Each has its own merits and disadvantages, and some are more suitable in a given situation than others. As a general rule they will be considered as either driving or jacking methods and they can be described as follows (Figure 4.9):

■ *Drop hammer*: this method is perhaps the most common method of driving displacement piles. A weight is repeatedly raised and dropped to strike the pile, forcing it into the ground. When using a hollow case, the weight will impact on some form of shoe or plug at the base of the pile case. When driving a pre-formed or sectional pile it will impact at the head of the pile direct. In some cases the hammer is driven using steam or compressed air in order to assist the driving process. It must be remembered that when using this method the pile can suffer damage from the results of the repeated impact. This can often take the form of tensile cracking due to the aftershock of the impact rather than compressive cracking. Hence concrete piles designed for this method of insertion will be suitably reinforced against tension.
■ *Diesel hammer*: this method relies on the generation of repeated impacts in quick succession driving the pile into the ground. Although the mass impacting on the end

Figure 4.9
Piling drive methods.

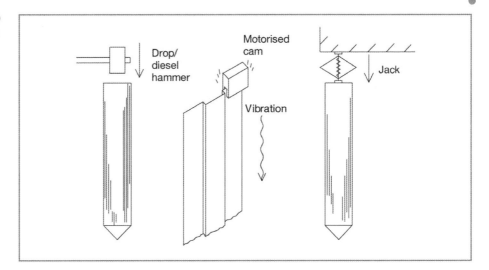

of the pile is less than in the case of the drop hammer, the increased frequency of action results in a highly effective method of pile installation.

- *Vibro-driving*: this method relies on the action of an eccentric cam, often driven by a diesel engine, inducing vibration in the body of the pile. The vibration causes mobility of the soil particles adjacent to the pile and reduces the friction at the interface between the ground and the pile. Thus the pile can be inserted with relatively little effort.
- *Jacking*: this method relies on the use of hydraulic jacks exerting pressure against the pile and braced against some solid object to gradually force the pile into position in the ground. This method tends to be restricted to remedial piling operations, such as underpinning, where access is restricted.

In the case of most of the methods described and whether in the form of a solid pile or a hollow casing, the penetration of the pile/casing is aided by the use of a driving toe or shoe, often in the form of a pointed cast-iron fitting at the base of the pile to allow easier penetration of the ground (Figure 4.10). These methods of installation have several disadvantages, in that considerable levels of noise and vibration may be generated as a result of the driving operation. Hence they are often considered unsuitable for congested sites, where adjacent buildings may be structurally affected, or areas where noise nuisance is undesirable. However, they may be used to good effect in consolidation of poor ground by compressing the earth around the piles.

Replacement piles

Unlike displacement piles, replacement (often referred to as non-displacement) piles are installed by removing a volume of soil and replacing it with a load-supporting pile. The holes are bored either using a hollow weighted grab, which is repeatedly dropped and raised, removing soil as it does so, or by using a rotary borer or auger. As the excavation progresses, the sides are prevented from collapsing by introducing a shell, or casing, normally made from steel, or by the use of a viscous liquid called Bentonite. The

Figure 4.10
Installation of
displacement piles.

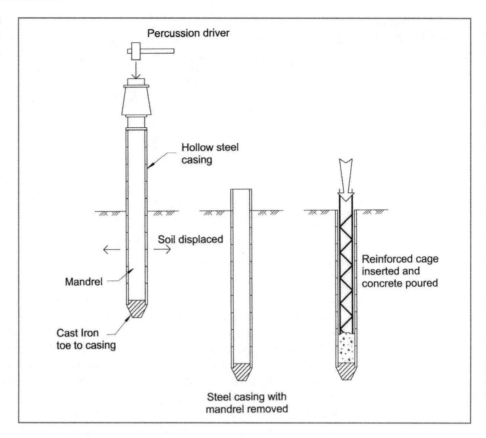

Bentonite is then displaced by concrete as it is poured into the excavation and is stored for further use. Replacement methods are quieter than the displacement method and do not result in damage to surrounding buildings. Four generic forms of replacement pile are in common usage and are described below.

Small-diameter, bored, cast *in situ* piles

These are piles of less than approximately 600 mm diameter and are generally formed with the assistance of a weighted auger which is dropped repeatedly from a portable rig. The rig is usually in the form of a simple tripod with a winch which allows the auger to be raised and dropped. Two forms of auger may be utilised depending on the nature of the soil (Figure 4.11). In cohesive soils a cylindrical auger with a sharpened cruciform section at the centre is used, which adheres to the soil when dropped. Once raised above ground level the captured soil is removed and the auger is again dropped until the required depth is reached. In non-cohesive soils the principle of operation is the same, but a hollow cylinder is used with a hinged flap at the base that allows the capture of the granular soil. In granular soils a temporary hollow casing is inserted to prevent collapse of the excavation; this may not be necessary in cohesive soils, since the excavation may be capable of self-supporting. Having formed the bored excavation the concrete is placed and a steel reinforcement cage is inserted into the wet concrete to complete the pile formation.

Figure 4.11
Auger details.

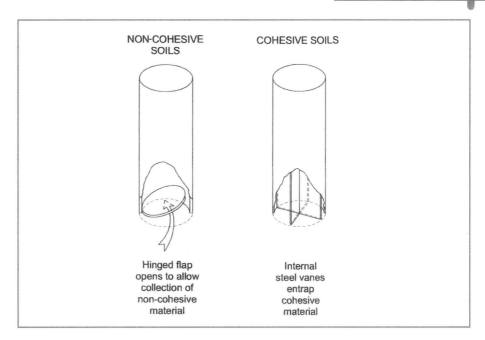

NON-COHESIVE
SOILS

COHESIVE SOILS

Hinged flap
opens to allow
collection of
non-cohesive
material

Internal
steel vanes
entrap
cohesive
material

When piles are cast *in situ* it is normal practice to cast the pile to an extended length, typically 400 mm above the required height. The excess concrete is then removed using a mechanical breaker to the required level. This ensures the integrity of the concrete throughout the entire length of the pile.

Large-diameter, bored, cast *in situ* piles

For piles in excess of 600 mm diameter a different approach is taken to formation of the bored excavation (Figure 4.12). Generally a rotary auger or drilling bucket will be used. This is often mounted on the back of a mobile unit or lorry and is operated with the aid of a diesel engine driving the auger or bucket via a square section 'kelly' bar. This technique is often assisted by the use of Bentonite slurry to support the excavation during formation and is suitable for depths well in excess of 50 m. The use of an expanding auger bit also allows the process of under-reaming, in which the base section of the pile is widened to create a greater contact area for the transmission of loads through end bearing. Having created the excavation, concrete is placed using a 'tremie' pipe, which displaces the Bentonite slurry. Reinforcement is then placed to complete the formation of the pile.

Partially pre-formed piles

This technique relies on the creation of a bored excavation into which concrete sections incorporating longitudinal voids are placed one on top of another to form a hollow column section. Steel reinforcement is then placed through the longitudinal voids and grout is pumped in to complete the pile formation. This technique is sometimes adopted where the ground is permanently waterlogged, since it does not require evacuation of water from the bore to facilitate construction.

Figure 4.12
Wide bored piles with
Bentonite.

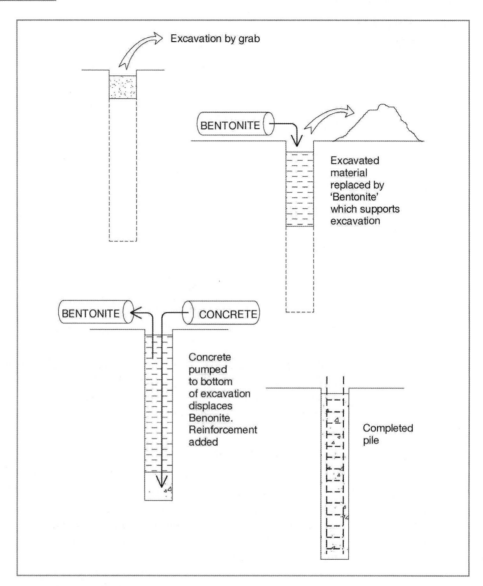

Figure 4.12
Wide bored piles with
Bentonite.

Grout injection piles

The methods of installation of replacement piles described above have largely been super-seded by the introduction of continuous flight auger (CFA), or grout injection piling (Figure 4.13). This method is highly efficient, with exceptionally fast installation possible. A continuous auger of the required length, typically up to 25–30 m, is mounted on a mobile rig and is used to excavate the pile void. The central shaft of the auger is hollow, and is connected to a concrete pumping system which allows the concrete to be placed as the auger is withdrawn. Hence there is no need to provide temporary support to the sides of the excavation. Once the auger is withdrawn, a steel reinforcement cage is forced into the concrete from above. Thus, the pile is formed quickly and efficiently with little disturbance to surrounding areas.

Figure 4.13
Installation of
continuous flight auger
(CFA) piles.

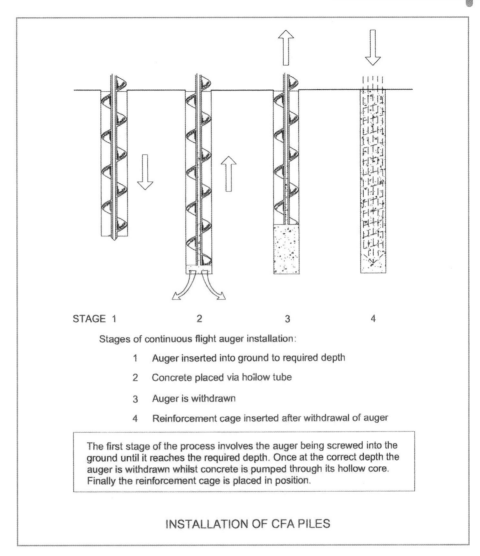

STAGE 1   2   3   4

Stages of continuous flight auger installation:

1   Auger inserted into ground to required depth

2   Concrete placed via hollow tube

3   Auger is withdrawn

4   Reinforcement cage inserted after withdrawal of auger

The first stage of the process involves the auger being screwed into the ground until it reaches the required depth. Once at the correct depth the auger is withdrawn whilst concrete is pumped through its hollow core. Finally the reinforcement cage is placed in position.

INSTALLATION OF CFA PILES

Friction piles and end-bearing piles

The method by which the pile transfers its load to the ground depends on the nature of the design of the pile and is dictated to a large extent by the nature of the ground in which it is located. Piles are utilised as a form of foundation in a variety of situations, each imposing differing demands upon pile design and creating restrictions on the ways in which the piles act. Although there are a great variety of situations which may necessitate the use of piles, the following are some of the most common:

■ Where insufficient loadbearing capacity is offered by the soil at a shallow depth, but sufficient is available at a greater depth
■ Where the nature of the soil at a shallow depth is variable and performance is unpredictable, such as in areas of filled land

- Where soils at shallow depths are subject to shrinkage or swelling due to seasonal changes
- Where buildings or elements are subjected to an uplifting force, and require anchoring to the ground.

It can be seen that in some of these instances one form of pile may be obviously more appropriate than another as a result of the way in which they act (Figure 4.14). End-bearing piles act by passing through unsuitable strata to bear directly upon soil with adequate bearing capacity, whilst friction piles are supported by the effects of frictional support from the ground to the sides of the pile throughout its length; hence the formation of ridges to the sides of piles as described earlier. In practice, all piles derive their support from a combination of these factors.

Figure 4.14
End-bearing and friction piles.

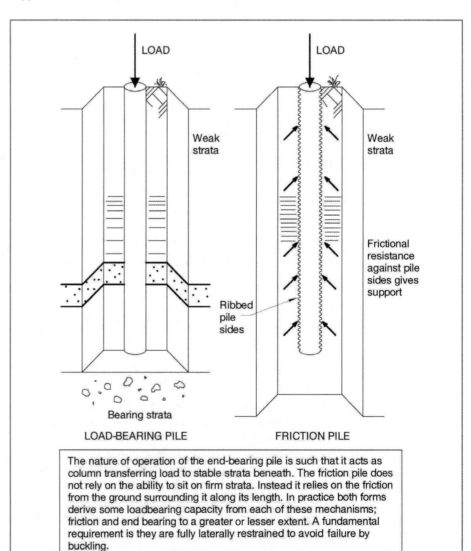

The nature of operation of the end-bearing pile is such that it acts as column transferring load to stable strata beneath. The friction pile does not rely on the ability to sit on firm strata. Instead it relies on the friction from the ground surrounding it along its length. In practice both forms derive some loadbearing capacity from each of these mechanisms; friction and end bearing to a greater or lesser extent. A fundamental requirement is they are fully laterally restrained to avoid failure by buckling.

Connections to piles

The processes of installation of piles result in the tops of the piles being far from perfectly level and true; hence a loading platform must be created to take the loads from the building and transmit them safely to the piles. This platform is known as a pile cap, which is formed in reinforced concrete and may transfer loads to a single pile or a group of piles. The caps are normally loaded by the columns of the building and may also take loads from the walls via a ground beam, as shown in Figure 4.15. In the cases of industrial and commercial buildings piles are normally formed in groups or clusters rather than as individual units. When formed in this way the pile group acts as a composite, with the piles themselves and the soil captured between them acting as a cohesive section or block. As

Figure 4.15
Pile cap configurations
and piles acting as a
block.

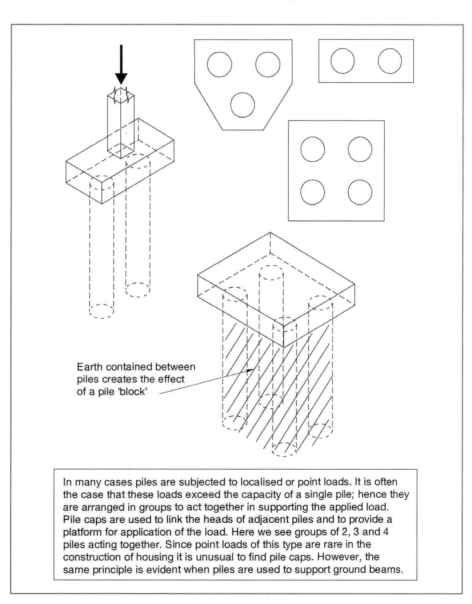

Earth contained between
piles creates the effect
of a pile 'block'

In many cases piles are subjected to localised or point loads. It is often the case that these loads exceed the capacity of a single pile; hence they are arranged in groups to act together in supporting the applied load. Pile caps are used to link the heads of adjacent piles and to provide a platform for application of the load. Here we see groups of 2, 3 and 4 piles acting together. Since point loads of this type are rare in the construction of housing it is unusual to find pile caps. However, the same principle is evident when piles are used to support ground beams.

PART 2

a consequence of the compression of the ground between the piles causing greater contact pressure along the length of each pile, the loadbearing capacity of a group may be greater than the sum of the capacities of the individual piles. The way in which the group of piles is capped will affect its performance considerably.

Selecting pile types

The choice of an appropriate pile type for a given situation will need to take into account a series of criteria that will make some forms more appropriate than others. The relative significance of each of these criteria may vary from site to site, but they should all be considered and ranked in order of importance when attempting to select the most appropriate solution. The main selection criteria are:

- *Location*: the proximity to adjacent structures must be taken into account and the possibility of damage arising from the installation of driven piles often results in the selection of bored or CFA piles for congested sites.
- *Nature of the ground*: the ground conditions restrict the type of piles that can be readily installed in some circumstances. For example, the use of driven piles in soils that contain demolition debris, rocks and boulders is not advisable. In clay soils, where heave can be problematic, driven piles may cause exacerbation of the problem and are ill advised. Similarly, the use of bored piles in cohesionless soils which have significant mobility would be considered unsuitable.
- *Cost*: the most economical design solution that effectively meets the performance needs of the foundation will generally be the favoured option.
- *Structural performance*: naturally the need to adequately support the loads from the building and the ability to cope with the nature of such loadings are an essential factor in selecting piles.
- *Durability*: the ability of the pile to survive the conditions in which it is placed is another essential factor which particularly limits the choice of materials used to form the pile.

4.5 | Enhancing the sustainable credentials of foundations

When considering the design of foundations from a sustainability perspective there are several principles that need to be adhered to. Careful scrutiny of the ground conditions should be undertaken to avoid overdesign. Overdesign means that more concrete and steel are used than are necessary. Alternatives that need fewer materials, such as screw or tapered piles, should be considered. Using piles generally means that less excavation is needed and therefore less tipping of soil is required. Using recycled materials such as pulverised fuel ash will reduce the amount of cement required.

By more careful consideration of the materials used to construct the superstructure, the weight of a building can be significantly reduced, so reducing the size of the foundations needed. Of course the most sustainable form of foundation is to reuse an existing foundation as no new materials are used and no disturbance to the area is made due to excavation. Therefore, the general principles are to use as small a foundation as possible and a system that requires the least excavation and the fewest 'new' materials.

When using wet concrete in foundations, it is very common to have a trench that is bigger than required and therefore more concrete is needed. Using a proprietary system that is produced in a factory removes this problem. Figure 4.16 shows such a system where piles have been driven into the ground and the preformed steel foundation and floor support sit on the piles. The insulation is cut to size in the factory and can then be placed correctly with no gaps. Reinforcement is placed in the inner tray and concrete is poured to form the support for the floor screed. The brickwork is then built up and supported by the deck system. This system uses minimum materials, is relatively lightweight and requires no excavation if the existing ground is at the correct reduced level to start with. Additionally the materials used in this system can all be easily recycled if or when the building is demolished.

This proprietary foundation system would not be suitable for multi-storey buildings, but it would be suitable in certain circumstances for use in large-span, single-storey commercial buildings.

PART 2

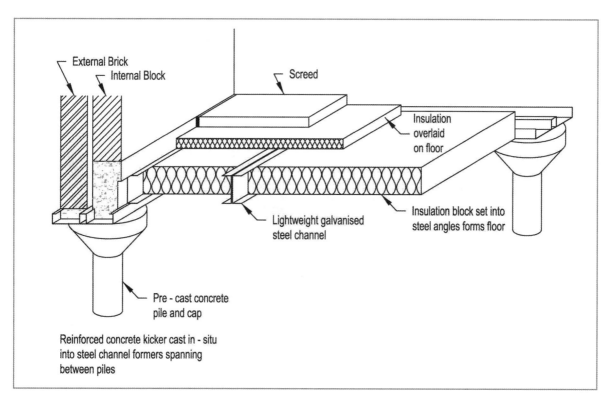

Figure 4.16
Proprietary foundation system.

CASE STUDY: FOUNDATIONS

Here we see a pile cap for a group of three piles. This forms the base for the support of columns for the building superstructure.

This photograph illustrates a series of independent pile caps supporting steel columns. Note that the column bases to the rear have been encased on concrete for protection from damage and corrosion.

Here we see the captive bolts that will secure the column bases to the foundation. Note the enlarged hole, which allows adjustment of the bolt position.

Note the gap beneath the column base plate in this photograph. This is due to the method of vertical adjustment based on metal shims placed between the concrete and the base plate. The gap will now be filled with grout.

COMPARATIVE STUDY: FOUNDATIONS

Option	Advantages	Disadvantages	When to use
Raft foundations	Economic due to combination of foundation and floor slab Shallow form requires little excavation Can cope with mixed/poor ground conditions	Require specific treatment for point loads Potential for edge erosion if not treated properly	Lightweight structures on ground of relatively poor strength Used in areas of filled ground with mixed/poor bearing capacity
Pad foundations	Shallow form of foundation needs little excavation Shape can be designed to accommodate tight sites Economic due to control of size	Can become very large if used for high point loads Limited to dealing with point loads	Ideal solution for point loads from framed buildings if bearing capacity close to the ground is sufficient The principle is essentially the same as that of a continuous deep strip foundation when applied on-site
Bored piles			
CFA piles	Speed of installation No need to support excavation during installation Quiet and relatively vibration free	Not appropriate for areas of filled ground etc. If voids are present can bleed grout into open areas Limited maximum diameter of pile restricts loadbearing	Used on sites with known ground conditions for buildings of moderate loads Often used for smaller buildings Used in sites where nuisance is a significant issue
Large-diameter bored piles	Can cope with heavy loads Large diameter allows fewer piles to be used in a group No need for permanent support by using Bentonite	Require relatively consistent soil conditions Large plant required to excavate	Large buildings with heavy loads Under-reamed piles used where loads are very high
Driven piles	Provide element of ground consolidation Can cope with variable ground conditions Made off-site and quality maintained due to factory production	Problems where dimensional stability of the ground is an issue Difficulties encountered where there is demolition debris in the ground	Sites with poor general ground conditions Often used in cohesionless soils

REFLECTIVE SUMMARY

- Piles are the only real form of deep foundation used in the construction of framed buildings.
- Piles can be categorised as displacement or replacement forms, and each may act through end bearing and/or friction.
- Piles may be preformed or cast *in situ* if made from concrete, although steel and even timber forms do exist.
- Piles act individually or as groups and will be capped to provide a sound base for connection to the frame.
- When piles are formed as a group they act as a pile block.
- Proprietary foundation systems can improve sustainability, but will generally only be suitable for large-span, single-storey buildings.

REVIEW TASKS

- Prepare a list of the criteria that you would apply to select an appropriate pile form for a framed building.
- Apply the criteria in your list to a range of buildings in the area where you live.
- Visit the companion website at www.palgrave.com/engineering/riley2 to view sample outline answers to the review tasks.

5

Walls below ground and basement construction

AIMS

After studying this chapter you should be able to:

- Demonstrate a knowledge of the main methods adopted for the construction of walls below ground and the circumstances in which each might be adopted
- Understand the alternative approaches to the construction of deep basements and shallow basements
- Be able to outline the sequence of construction works required for given scenarios
- Understand the implications of the need to exclude moisture from basements and the various ways in which this can be achieved
- Appreciate the interaction between basement structures and building superstructures

This chapter contains the following sections:

5.1   Requirements of walls below ground
5.2   Options for waterproofing of basements
5.3   Methods of basement construction

INFO POINT

- BS 8002: Code of practice for earth retaining structures (1994)
- BS 8102: Code of practice for the protection of structures against water from the ground (1990)
- *BRE Good Building Guide 72, Parts 1 and 2*: Basement construction and waterproofing (2007)
- *Building elements: foundations and basements and external works*. Report 440, BRE (2002)
- *Solid foundation – Building guarantees technical manual*. Section 3: Substructure. ZM (2006)
- *Steel intensive basements*. Publication 275, SCI (2001)
- *Water-resisting basements*. Report R139, CIRIA (1995)

5.1 Requirements of walls below ground

Introduction

- After studying this section you should be familiar with the generic requirements of walls below ground level.
- You should understand the nature and origins of loads affecting walls below ground and the ways in which these affect design.
- In addition, you should appreciate the implications of the presence of ground water and the ways in which this affects walls below ground.

Overview

It must be stated at the outset of this section that the walls that exist below ground in the context of framed buildings are generally associated with the construction of basements or large sub-ground voids. Hence they are required to perform rather differently from those associated with traditional low-rise construction forms. However, they do share common functional requirements in that they must maintain structural stability, exclude moisture and so on. The ways in which these are satisfied may differ greatly, however.

Some of the primary functional requirements of walls below ground may be summarised as follows:

- Structural stability
- Durability
- Moisture exclusion
- Buildability.

In addition, issues such as thermal and acoustic insulation may take on considerable significance depending on the form and function of the building and varying with the approach taken to conservation of fuel and power, as discussed elsewhere in this text.

The sub-ground environment in which some elements of the structure are sited can be hostile to the materials commonly used in construction. The presence of aggressive salts and the potential for high water table levels can interfere with the process of construction and can affect the durability of the building elements. These issues and others impose restrictions on the nature of construction of walls below ground level; these restrictions become particularly significant when dealing with basements, where the void below ground is to be used as an internal building space (Figure 5.1). The walls below ground are subjected to ground moisture for much of the time, although this depends upon the specific nature of the soil and the height of the water table. The presence of this moisture is often a problem during the construction process and ways of dealing with it were considered in Chapter 3. Generally, materials with low porosity are desirable for use below ground, as these are less likely to suffer deterioration as a result of the freezing of the water close to the surface of the ground in periods of cold weather.

Figure 5.1
Functional needs of
basement walls.

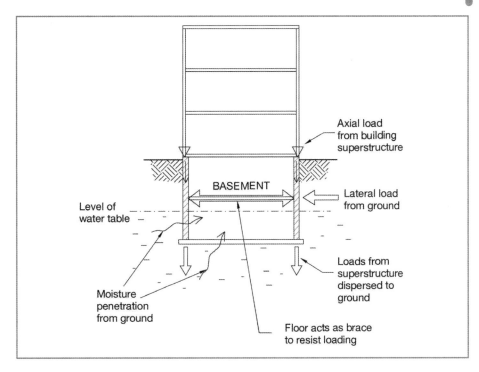

Porous materials tend to absorb ground moisture, which expands on freezing, causing spalling and friability of the material. Non-porous materials also tend to perform better in terms of moisture exclusion, since they do not promote moisture migration through capillarity. Options for excluding moisture from the building are dealt with specifically later in this chapter. The walls in these areas are subjected to high pressures, both axially (from the building above ground) and laterally (from the pressure exerted by the ground itself). The lateral force exerted by the mass of earth which surrounds the walls can have a considerable effect, particularly in the case of walls to deep basements. These lateral loads must be adequately resisted if the stability of the wall is to be maintained. This is generally achieved either by bracing the walls or by constructing walls that are sufficiently robust to cope with the stresses involved.

In order to resist the effects of this loading it is common practice to brace the walls below ground level with temporary supports or to utilise the floors of the building as permanent braces. Alternatively, the walls may be formed in such a way as to minimise ground pressure during construction and by bracing them gradually as work proceeds. This is discussed in detail later.

In most commercial buildings the utilisation of space beneath the main structure, for car parking, siting of plant rooms, storage or other uses, is necessary. The construction of basements may take many forms, but the general principles and performance requirements will generally be the same. Table 5.1 summarises the ways in which basements are graded according to BS 8102 depending on their specific performance requirements, which are linked to their intended use. From this table it can be clearly seen that the range of uses for basements and the subsequent variation in acceptable internal environment provide great scope for the quality of basement construction and the need to exclude moisture. There is a link between the grade of basement and the cost of construc-

Table 5.1 Basement grades.

Grade	Possible use	Conditions required	Moisture exclusion
Grade 1: Basic Utility	Car parking Mechanical plant rooms	>65% relative humidity 15–32°C temperature	Minor wet seepage and visible damp patches may be acceptable
Grade 2: Better Utility	Retails storage Electrical plant rooms	35–50% relative humidity Temperature depending on use: <15°C for storage, up to 42°C for plant rooms	Wet seepage unacceptable. No visible moisture patches
Grade 3: Habitable	Offices Residential use Kitchens, restaurants etc	40–60% relative humidity Temperature range 18–29°C, depending on use	Seepage and wet patches unacceptable. Possible active control of internal environment required to control temperature and humidity
Grade 4: Special	Archive storage of books, documents, art etc	35–50% relative humidity Temperature range typically 13–22°C	Environment tightly controlled by active measures. Seepage and visible dampness unacceptable

tion. Hence the end use of the basement must be defined at the design stage if excessive cost is to be avoided. There is no benefit in over-specifying the performance requirements of basements.

REFLECTIVE SUMMARY

- Walls below ground are required to act in a hostile environment and must be highly durable.
- They are required to cope with lateral and axial loads.
- The requirement to exclude moisture is highly significant due to positive hydrostatic pressure in the ground.
- The design of basement walls is related to the required grade, which may range from basic utility to special grade.

REVIEW TASK

- With reference to the grades of basement with which you are now familiar, identify a range of basement uses appropriate to each grade.
- Visit the companion website at www.palgrave.com/engineering/riley2 to view sample outline answers to the review task.

5.2 | Options for waterproofing of basements

Introduction

- After studying this section you should appreciate the need to prevent moisture penetration through walls below ground.
- You should have developed a broad understanding of the origins of water in the ground and the mechanisms by which it penetrates buildings.
- In addition, you should be familiar with the more common options for providing waterproof walls below ground.

Overview

As previously noted, the walls extending below ground level, often extending to a considerable depth, are required to resist the lateral loadings that are exerted by the surrounding earth. In this respect they may be considered as retaining walls as well as supporting the walls and structure above. As a result they share more similarity with retaining structures used in civil engineering than they do with traditional walls above ground. The nature of the construction form dictates, to some degree, the ways in which exclusion of moisture is dealt with. In addition to the grades previously outlined, basement construction is also defined by 'type' of structure.

As well as the obvious requirements to provide for structural stability and durability, the extent to which exclusion of ground water may be necessary will be a key criterion for the selection of an appropriate 'type'. Additionally, the nature of the soil and the required depth of the basement will place limitations on selection. The construction of shallow basements differs fundamentally from that of deep basements due to the ability to undertake work which is close to the surface and accessible. There is also a range of other factors that will affect the selection of an appropriate type. These relate to the likely performance and reliability of the completed structure and include aspects such as the level of ground water on the site, possible contamination of ground water, natural drainage, soil type, access to the site and so on. No two sites are the same, and it is therefore difficult to generalise; however, it is generally considered that the most significant factor when selecting structure type is the water table. Sites are often classified into three broad groups with this in mind:

- *High water table*, where the water level is likely to be constantly above the level of the basement floor
- *Low water table*, where the site is free-draining and the water level is likely to be constantly below the level of the basement floor
- *Variable water table*, where levels vary between the two situations described above.

Basement types

There are three broad types of basement wall structure, identified as types A, B and C (Figure 5.2).

PART 2

Figure 5.2
Basements wall
structure types.

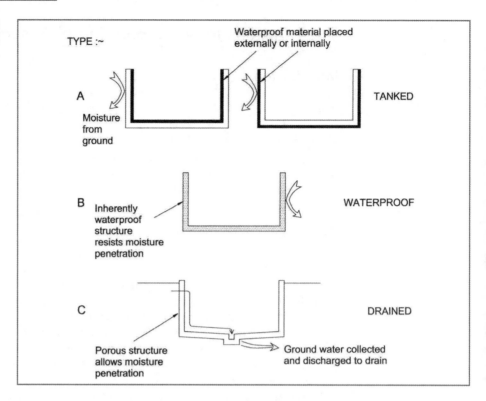

Type A structures (Figure 5.3) involve the utilisation of an impervious material applied to the structure, internally or externally, to exclude moisture. This process is termed 'tanking' and may be effected in several different ways. The selected waterproofing element must be capable of withstanding hydrostatic pressure from ground water and must be able to accommodate any other loadings required. The structural walls of the basement to which the impervious material is applied may be formed in reinforced concrete or masonry. The waterproofing material may be applied externally during the construction process, in which case it would be considered 'external tanking'. Alternatively, it may be applied to the interior face of the walls after construction, and in this case it would be considered 'internal tanking'.

Since the internal environment of the basement is totally reliant on the effectiveness of the applied waterproof membrane, any defect arising within the membrane will have a serious effect internally. The ability to gain access to the membrane to rectify any such defect is hindered by the adoption of external tanking systems. Internal systems are more readily accessible following completion of the basement, but they are at greater risk of damage due to penetration by fixings for shelving systems etc. They are also at risk of being driven off the internal face of the wall by hydrostatic pressure from ground water. For these reasons they tend to be protected by a slender internal skin of blockwork or masonry which acts to brace the membrane against the external wall and allows for fixings to the internal face without penetrating the membrane.

Type B structures (Figure 5.4) utilise waterproof materials within the structure of the walls, considered as 'structurally integrated protection'. The waterproofing is provided by the structural walls themselves, which are almost always formed from reinforced or

Figure 5.3
Type A structures:
tanking of basements.

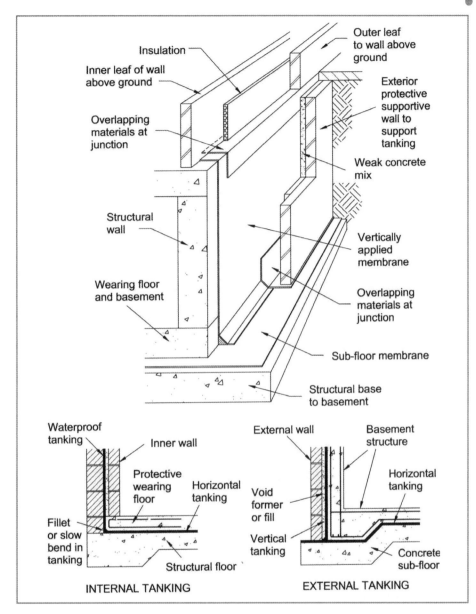

prestressed concrete designed such that it is inherently waterproof. The waterproofing of
the basement is reliant on the integrity of the external shell or enclosure. The composition
of the structural concrete and the detailing at construction joints must ensure that the
enclosure remains impervious to ground water. However, even where this is achieved, there
is still the possibility of water vapour passing through the shell. The formation of construction joints must be carefully considered, as it is in these locations that there is the greatest
risk of leakage. Joints normally incorporate water stops made from materials such as
neoprene. Alternatively, they may utilise 'hydrophilic' technology, which introduces a crys-
tallising agent that reacts in the presence of water to seal the joint in the event of leakage.

Figure 5.4
Type B structures:
waterproof concrete.

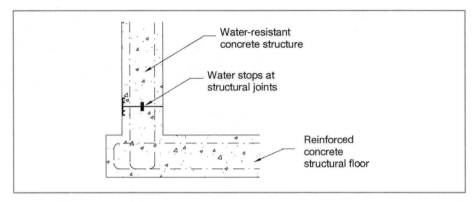

Type C structures (Figure 5.5) allow for ground water to pass through the walls, but provide a mechanism for it to be drained away. A drained cavity is formed in the basement structure, which facilitates the collection of ground water that seeps through the external basement wall. In order to be effective the drainage system must be able to cope adequately with the quantity of water that is likely to penetrate the external walls. This can present problems in situations where there is a high water table if there is any failure in the mechanical pumping systems that must be used to remove the water, or if the drainage channels become blocked for any reason, such as the accumulation of soil fines or silt.

In these cases flooding of the basement would inevitably occur. Hence type C structures tend to be selected only where there is a free-draining site with a low water table unless they are used in combination with a type B structure, in which case they are suitable for high or variable water table conditions.

Waterproofing systems

Basement constructions will fall into one of the three types defined above, but across these types there is a range of specific approaches to waterproofing systems. Most systems in use today are proprietary forms that have been developed as a complete waterproofing package. Naturally there are many variations, but they will generally fall into one of the categories in Table 5.2.

Much of the foregoing has related specifically to wall construction. However, the exclusion of moisture from basement areas requires the provision of vertical and horizontal

Figure 5.5
Type C structures:
drained cavities.

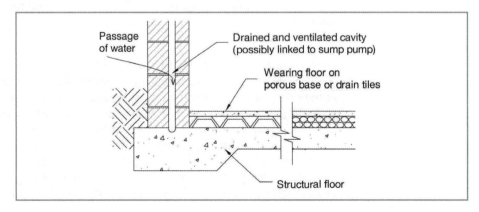

Table 5.2 Waterproofing systems.

Category	Description	Details
1	Bonded sheet membranes	Bonded (hot or cold) to structural walls externally or internally
2	Cavity drain membranes	High-density ridged or pimpled plastic sheeting generally placed internally to allow draining of moisture passing through walls
3	Bentonite clay active membranes	Sandwich of Bentonite clay between layers of cardboard. Used externally the clay swells to fill gaps when in contact with water
4	Liquid applied membranes	Epoxy, polymer or bitumen solutions applied cold to walls internally or externally. These cope well with unusual profiles and variable shapes
5	Mastic asphalt membranes	Applied hot as multi-layer application internally or externally
6	Cementitious crystallisation active systems	Applied as slurry to exterior or interior. They react with elements of concrete in the presence of moisture to seal cracks or fissures
7	Cementitious renders and toppings	Generally used internally and applied as multi-layer coatings incorporating waterproof admixtures. These cope well with variable surfaces and are sometimes used as remedial treatments for existing basements

damp-proofing elements around the entire sub-ground structure, including the floor. It is essential that the selected waterproofing system is continuous and it must retain its integrity throughout the area which is to be waterproofed. Great care must be taken, therefore, where elements of the structure, such as beams and columns, together with building services, penetrate the basement floor or walls. This is of particular importance: since the ground water acts against the basement enclosure with positive hydrostatic pressure, the damp-proofing elements must be capable of resisting such pressure, even in areas of potentially complex detailing.

REFLECTIVE SUMMARY

When considering waterproofing of basements, remember:

■ Basements may fall into one of three types: tanked structures, waterproof concrete structures and drained structures.
■ There are seven categories of waterproofing system in common use, ranging from sheet membranes to cementitious renders.

PART 2

5.3 | Methods of basement construction

Introduction

- After studying this section you should be familiar with the common methods of constructing basements to industrial and commercial buildings.
- You should understand the differences in approach to shallow and deep forms.
- In addition, you should have an appreciation of the general sequence of operations and the technical constraints that affect basement formation.

Overview

In practice there are many different approaches to the construction of basements, and the selected design solution often owes as much to economic and legal issues as it does to technical considerations. The difficulties associated with the construction process often arise due to the location of the intended building relative to adjacent structures and to limitations on access to the works. One of the key influences upon choice of construction method is the extent to which effective lateral support of the excavation can be achieved. The implications of failure to provide effective support can be cata-

Table 5.3 Issues to be considered when deciding on construction form.

Nature of the site	Access limitations to and within the site boundaries
	Topography
	Water table/drainage
	Proximity of existing structures
	Previous site use
Building form	Structural form of the building
	Depth of basement
	Intended function of basement
Legal issues	Rights of adjacent building owners
	Health and safety requirements
Economic issues	Maximisation of usable building volume
	Minimisation of costs of construction

strophic in both technical and financial terms. Basement construction also presents a very specific set of health and safety requirements that demand an exacting approach to the construction process. Table 5.3 summarises some of the key issues that must be considered in deciding on a suitable approach to the construction form.

Taking these and other issues into account, an appropriate design solution will be selected for each basement situation. As previously stated, it is difficult to generalise when discussing the construction of basements. However, for convenience we will consider three broad categories of approach to basement construction. These are:

- Open excavations
- Excavation with temporary support
- Excavation with permanent retaining walls formed prior to the main excavation.

Within these categories some forms are more appropriate for the formation of shallow basements whilst others are best utilised in the construction of deep basement forms.

Shallow basements

For most small-scale basements the required excavation depth is likely to be relatively shallow. In such instances the most economical form of construction is to effect an open excavation with sloping sides if the constraints of space on the site permit. Sufficient space must be available around the building to allow for the creation of a stable slope to the sides of the excavation. The angle of this slope, and hence the required area of ground, will depend upon the nature of the soil and its moisture content. Cohesionless soils will be more prone to collapse than cohesive soils and will display a shallower **angle of repose**. There is also considerable risk of collapse of the excavated sides where there is high moisture content or where there is flowing water. In such circumstances the use of this type of construction would be inappropriate. Where space permits and the soil type is suitable the excavation of basements using this method is economical and relatively speedy for shallow basements.

Figure 5.6 illustrates the typical sequence of operations for the formation of a shallow basement using open excavations. It will be seen that the sequence is essentially the same as for the construction of a building enclosure above ground with the exception of the initial excavation, the waterproofing and the backfilling following completion of the structure. Although this technique may be feasible in some instances, most situations are restricted by the amount of space surrounding the basement area; hence alternative approaches must be adopted which rely on the use of temporary or permanent ground support, as in the case of deep basement construction.

Deep basements

Where deep basements are to be constructed beneath commercial buildings a method of construction is often needed that minimises the extent of disruption to surrounding areas. The approach adopted for the construction of shallow basements on relatively open sites, as described earlier, is generally inappropriate for a number of reasons. Firstly, the extent of excavation required around the building to facilitate construction at depth would be impractical. Secondly, there is the risk of affecting the degree of ground support

Angle of repose is the term used to describe the angle at which soil of a given type will remain stable if formed into a pile.

Figure 5.6
Formation of a shallow
basement.

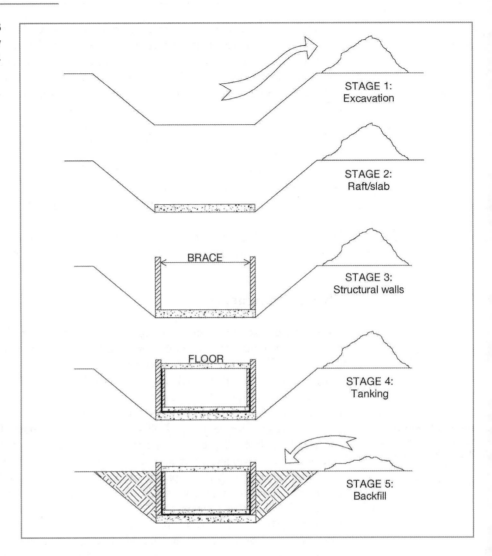

provided to adjacent buildings and the consequent risk of subsidence and damage to their structures. Thirdly, there is the cost associated with such large-scale works. There is also the potential for disruption to the works caused by the presence of high water tables and the constant need to remove ground water.

For these and other reasons the approach taken to construction of deep basements tends to be different from that for shallow forms. This is primarily a consequence of the need to construct walls of considerable depth within the ground. Two principal techniques are commonly used: excavations with temporary support and excavations supported by permanent retaining walls embedded in the ground.

Temporary support to excavations

This approach is generally considered to be appropriate where sites provide limited access around the excavation, preventing the type of open excavation previously described. The

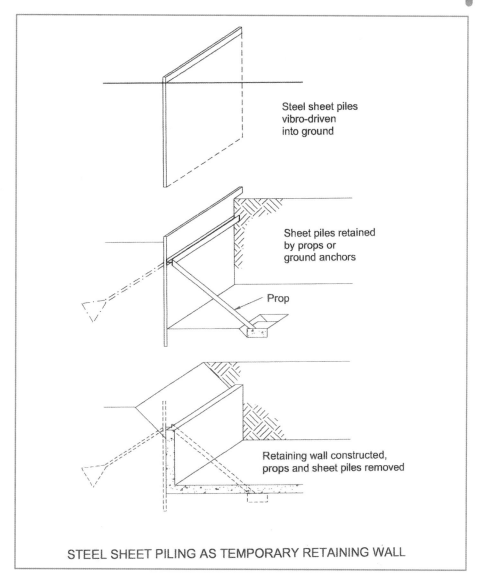

Figure 5.7
Excavation using
temporary support.

Steel sheet piles
vibro-driven
into ground

Sheet piles retained
by props or
ground anchors

Prop

Retaining wall constructed,
props and sheet piles removed

STEEL SHEET PILING AS TEMPORARY RETAINING WALL

principle is that temporary restraining elements are placed within the ground around the perimeter of the intended area to be excavated (Figure 5.7). Once this is done the earth is excavated to create a void which will become the basement area. One of the most common solutions to the creation of the temporary supporting walls is to use steel sheet piles driven or vibrated into the ground. These are then propped with temporary supports to prevent them from collapsing as the earth is excavated from the intended basement area. Permanent basement walls are then constructed and the temporary support is removed, with any resultant voids filled as required. The removal of the temporary supports often results in minor subsidence in the immediate area, and this makes such an approach unsuitable where adjacent structures are present. Thus in inner city sites this approach is uncommon.

Figure 5.8
Sectional precast
concrete.

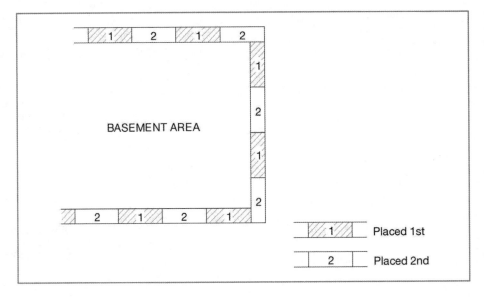

Sectional precast concrete.

Sectional precast concrete

The use of precast concrete panels has become relatively unusual in modern basement construction as the adoption of piled and diaphragm walls has become more commonplace. However, the principles of sectional construction are considered here for completeness. The technique is most appropriate where basement depths are relatively shallow and relies on the sequential excavation of alternate sections around the perimeter of the basement enclosure (Figure 5.8). Temporary support is provided to allow alternate sections of excavation to be undertaken. The temporary support is often achieved using steel sheeting supported by diagonal bracing or ground anchors. Reinforced concrete upstand sections are then placed, with the sectional wall panels connected and secured to them. Once these are in place the temporary support is removed and the excavation backfilled prior to moving on to the next, alternate, section where the process is repeated and the resulting adjacent wall panels are structurally connected to form a continuous perimeter wall to the basement enclosure. Once again the temporary support from the second set of excavations is removed and the excavation is backfilled. At this point the basement void can be excavated from within the enclosing wall, and the structural basement floor cast to form the final element in what becomes the 'box structure' of the basement.

Permanent embedded retaining walls to deep basements

The construction of deep basement walls can be costly if traditional excavation and construction techniques are used. Limitations on access to most commercial sites also restrict the possibility of such an approach. Hence a series of techniques have been developed which allow construction of the walls within the ground prior to the full excavation of the basement void. These techniques provide the potential to create a perimeter retaining wall around the basement area from within which the soil can be excavated to form the basement void. An essential element of this approach is that the retaining wall forms part of the permanent basement structure. The most common

Figure 5.9
Techniques for providing permanent embedded retaining walls to deep basements.

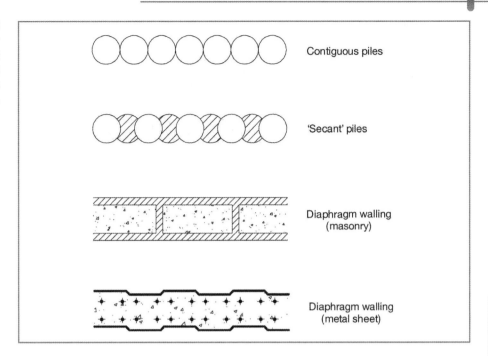

Contiguous piles

'Secant' piles

Diaphragm walling (masonry)

Diaphragm walling (metal sheet)

options available for the formation of deep basement walls using this approach are (Figure 5.9):

- Contiguous piles
- Secant piles
- Diaphragm walls.

Contiguous piles

This approach relies on the placement of many piles immediately adjacent to each other all the way around the perimeter of the basement area (Figure 5.10). The piles are most likely to be bored replacement piles and will extend to a depth that exceeds the intended lowest point of the completed basement. Once placed, the piles form the perimeter enclosure of the basement and act to prevent the surrounding soil from falling into the enclosure as excavation proceeds. Soil is excavated from within the contained area to create the void which will become the basement. As this occurs the lateral pressure from the surrounding ground acts against the piles. Hence bracing must be provided to restrain the piles as the excavation gets deeper. Bracing may be in the form of a steel framework or a series of steel props. Once constructed, the walls will generally be provided with an inner skin of blockwork or masonry to provide for a flat inner face to the basement area.

Secant piles

Secant piling is in some ways similar to contiguous piling. The primary difference results from the configuration and mode of placement of the piles. Unlike contiguous piles, which are sited adjacent to each other, secant piles actually overlap in terms of positioning. The formation of 'male' and 'female' piles allows the formation of a continuous

Figure 5.10
Contiguous piles for
deep basements.

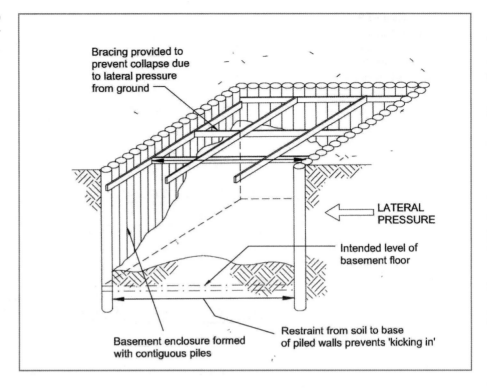

Figure 5.10
Contiguous piles for
deep basements.

Bracing provided to
prevent collapse due
to lateral pressure
from ground

LATERAL
PRESSURE

Intended level of
basement floor

Restraint from soil to base
of piled walls prevents 'kicking in'

Basement enclosure formed
with contiguous piles

wall of piles without the possibility of gaps between adjacent units. The mode of excavation for the creation of the basement void is the same as for contiguous piles and there is the same requirement for bracing of the piled walls as excavation proceeds. Once completed, they will be provided with an inner lining wall, as in the case of contiguous piles. The main benefits of secant piling over contiguous piling are associated with strength and moisture exclusion. Clearly, because the piles are intersected there is less risk of ground water passing through the secant pile structure. In addition, the piles act collectively rather than as individual units as contiguous piles do. This results in a greater strength for a given pile diameter and allows the use of smaller piles in secant walls than might be necessary when adopting contiguous pile construction. For these reasons it is far more common to see the use of secant piles in modern basement construction.

Diaphragm walls

The term *diaphragm wall* derives from the function of the wall in separating two compartments: the inner basement void and the outer surrounding ground. Several techniques are available, but the most common rely on the creation of the diaphragm wall within the ground using formwork or some other support medium to allow excavation without the risk of collapse. One common method adopts the use of **Bentonite** to provide temporary support to the ground as excavation of a trench proceeds (Figure 5.11). As excavation of the trench progresses Bentonite is pumped into the void, providing support to the sides of the excavation until the required depth is reached. At this point concrete is pumped to the bottom of the excavation using a 'tremie' pipe. The concrete displaces the Bentonite which is then collected for subsequent reuse. Alter-

Bentonite, or builders' mud as it is sometimes known, is a thixotropic material that allows the excavation of voids through the Bentonite whilst supporting the excavation.

Figure 5.11
Diaphragm walls
formed using
Bentonite.

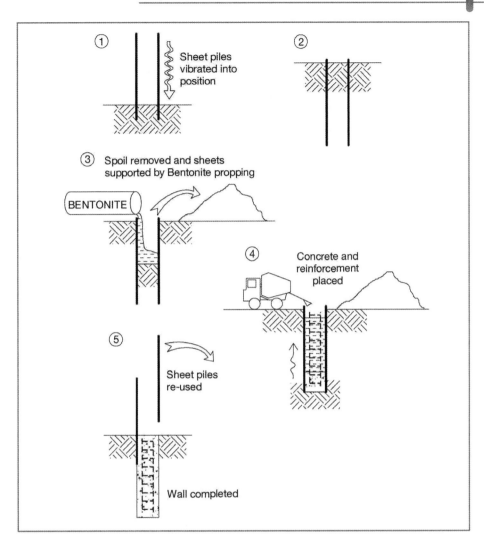

natively, the excavation may be supported using steel sheet piling as the temporary support mechanism.

The term *diaphragm wall* is also used to refer to masonry walls that are constructed to create a wide central void area with inner and outer leaves connected by fins of masonry (Figure 5.12). The resulting void sections are often filled with concrete to create a wide, robust structure. These are often utilised for retaining walls where there is a difference in ground level or for large-scale walls to buildings such as sports halls etc.

Construction process for basements

When applying the principles of permanent embedded retaining walls, two common construction methods are 'top-down' and 'bottom-up' construction. Each of these methods has merits and disadvantages.

Figure 5.12
Masonry diaphragm
walls as retaining walls.

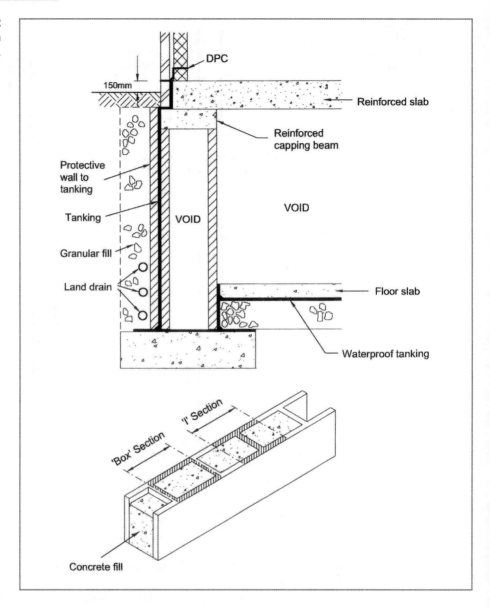

Bottom-up construction

The principle of bottom-up construction relies on a simple sequence of operations as indicated in Figure 5.13.

The first stage in this process is to form perimeter walls to enclose the area to be excavated. These will become the structural walls of the basement on completion of the process. In order to form the enclosure there is a range of techniques that allow the walls to be formed within the body of the ground without the need for excavation around them. These were identified earlier and include contiguous or secant piling, diaphragm walling and steel sheet piling.

Figure 5.13
Bottom-up
construction.

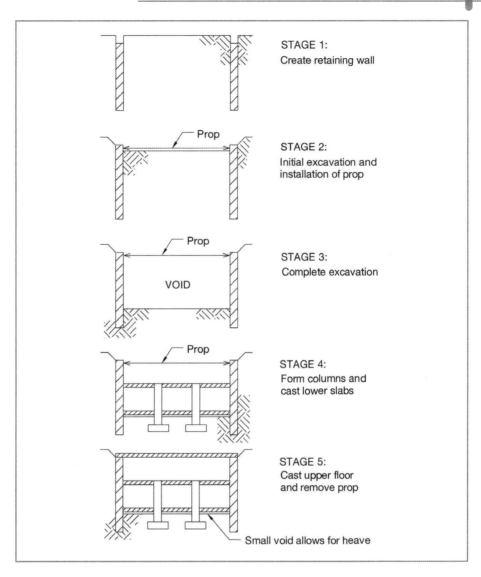

STAGE 1:
Create retaining wall

Prop

STAGE 2:
Initial excavation and
installation of prop

Prop

STAGE 3:
Complete excavation

VOID

Prop

STAGE 4:
Form columns and
cast lower slabs

STAGE 5:
Cast upper floor
and remove prop

Small void allows for heave

Excavation will then take place to a point just below the required level of the completed basement floor. It is important that the retaining walls proceed beyond this to prevent the base of the walls from 'kicking in' as the restraining ground is removed. Naturally, as earth is removed from the basement enclosure, the effects of ground pressure pushing in must be controlled. Hence a system of temporary props and braces must be installed to restrain the walls. Having installed these props, excavation can take place to the desired depth.

At this point the foundations that will support the columns of the building are formed using pad or piled foundations and the lower sections of the columns are formed. The next stage is to form the base slab or basement floor. Although this could be cast as a ground-supported slab, it is more common for a suspended floor to be formed over the top of a void former. This eliminates the effects of ground pressure acting beneath the slab to induce heave. The upper floors are then formed as work proceeds in an upward direction towards ground level. The columns of the building frame are extended as work proceeds.

Although bottom-up construction is still common, it is being replaced in many projects by top-down construction, which has advantages in terms of speed, cost and stability of the ground around the works due to the absence of the need for temporary supporting elements.

Top-down construction

Unlike other construction methods, top-down construction attempts to make use of the permanent elements of the construction works to provide rigidity and bracing against loadings from the ground. This removes the need for temporary supporting structures during excavation and presents a speedy and economical solution to the formation of deep basements. Additionally, it is possible in some cases to commence the formation

Figure 5.14
Sequence of operations for top-down construction.

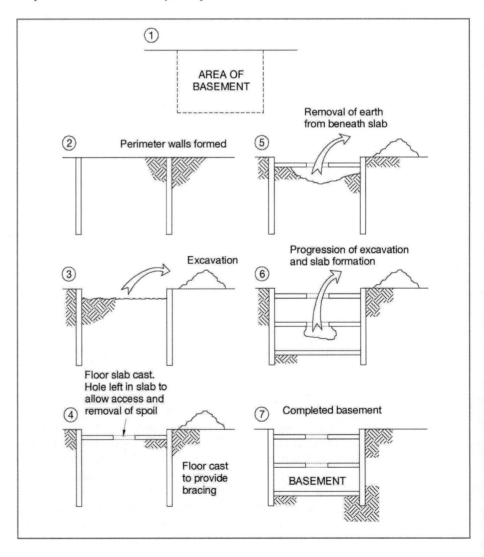

of the superstructure contemporaneously with the basement construction.

The sequence of operations involved in this method of basement construction is illustrated in Figure 5.14.

The first stage of the process is the formation of the permanent retaining walls to the perimeter of the basement area. This may take the form of contiguous or secant piling, diaphragm walling or (more commonly in modern construction) the use of permanent steel sheet piling.

The next stage is the creation of the foundations that will support the columns of the building structure at internal locations. These columns will provide intermediate support for the floors of the basement and the building superstructure as work proceeds. In most cases the foundations to these columns will take the form of CFA piles (see Figure 4.13) with columns attached using pile caps or, increasingly in the case of steel framed buildings, using the **plunge method** (Figure 5.15).

Once the perimeter walls and the internal columns have been formed, the uppermost floor slab is created. This is effected by levelling and compacting the ground to provide

The **plunge method** of column formation allows the positioning of columns that will support the intermediate floors of the basement without the need to undertake excessive excavation. The piled foundation is formed, and while the concrete is still wet a steel column is plunged into it and secured to remain vertical.

Figure 5.15
The plunge method of column/foundation formation.

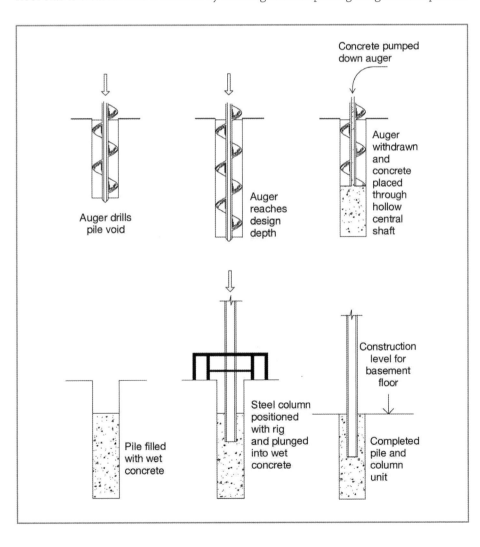

PART 2

formwork for the reinforced slab, which is then cast *in situ* over a temporary concrete and blinding layer topped with Visqueen. Excavation then takes place to remove earth from beneath the slab through appropriate holes in the slab. These may be the positions of stairwells, service shafts etc. Once the Visqueen and temporary base have been removed, the underside of the slab should be smooth and level. Mechanical plant is used to excavate from beneath the slab, often feeding a conveyor to transport the spoil above ground. Having excavated to an appropriate depth, the next floor slab is cast using the same approach, and the process is then repeated until the base slab is reached.

The basement floor or base slab may be cast as a ground-supported floor, as it is in continuous contact with the sub-base. However, more often it will be cast as a suspended floor over a void former in order to remove the risk of heave acting beneath the floor from upward ground pressure.

Basement walls and foundations

There are several approaches to the creation of basement structures, each allowing for different types of foundation to be used. Naturally, the basement 'types' described earlier will tend to favour individual foundation solutions with many detailed variations from site to site. However, the general principle of attempting to keep foundation design as simple as possible will always be adopted. The incorporation of complex details, movement joints and so on reduces buildability and introduces possible areas of difficulty for the process of waterproofing. More complicated foundation options tend to favour internal waterproofing solutions, whilst simpler forms are more accommodating to a range of possible waterproofing options. Although there is an almost infinite variety of individual solutions to basement design, we have restricted our discussion here to the following generic forms (Figure 5.16):

- *Traditional strip foundations and ground beams*: With this option a traditional strip foundation is used in the case of shallow basements where excavation is facilitated and access is good. The external wall of the basement is then formed with masonry construction. In ground conditions where the loading on the soil would be excessive this may take the form of a ring beam or ground beam supported at regular intervals by piles. Once again this relies on the ability to access the area for bricklaying etc. In both cases the basement floor is separate from the wall and foundation elements and there may be complexity in ensuring continuous waterproofing details.
- *Raft foundations*: In this option the structural raft acts as both the foundation to the basement walls and as the structural basement floor. This is a simpler option, as the floor and foundation structure is a single unit; thus there is less potential for movement and less need for complicated waterproofing. This option is often used for deep basements, and the external walls will generally be constructed from masonry with a series of structural columns located so as to support the superstructure above. If these penetrate the basement floor the continuity of waterproofing must be considered.
- *Concrete 'box' construction*: A development of the raft construction, this relies on the creation of a concrete box or shell which is a monolithic structure. The basement slab acts as a raft foundation whilst the perimeter walls act as retaining walls to resist ground pressure and the entry of water from the ground.

- *Piled walls*: The creation of basement walls using piling techniques such as contiguous or secant piles allows for the utilisation of the walls themselves as foundation elements for the superstructure. Naturally, the internal columns must also be supported by independent piles or pad foundations.

Figure 5.16
Basement walls and foundations.

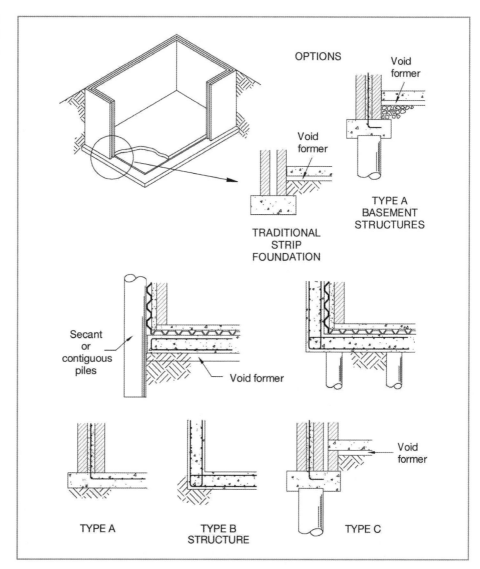

CASE STUDY: WALLS BELOW GROUND

Here we see examples of contiguous and secant piling used to form basement walls. Note the concrete capping/ring beam which connects the piles and provides a sound base upon which to form the superstructure walls.

In this case we also see precast reinforced concrete sections used in conjunction with the secant piles.

The photos above show a masonry diaphragm wall being used as retaining wall between differing ground levels. Note the connecting fins between the inner and outer skin of the wall. The resulting voids are filled with concrete to ensure sufficient strength.

COMPARATIVE STUDY: BASEMENTS

Option	Advantages	Disadvantages	When to use
Tanked basements	Familiar technology Access to internally tanked walls allows remedial treatments	Externally tanked structures difficult to repair Internally tanked structures subject to hydrostatic pressure forcing membranes off walls	Perhaps the most common option of small-scale basements with traditional walls construction
Waterproof structures	Single-element waterproofing and structure Extent of waterproofing detailing is reduced	Need to incorporate water bars etc. in structural elements	Very common for larger-scale basement structures and deep basement forms
Drained cavities	Simplicity of construction	Need to cater for removal of water Possible problems with soil fines and silting of pumps etc.	Used where basement conditions do not need to be totally waterproof
Masonry walls	Simple familiar technology Allows for ready installation of tanking membranes to flat wall surfaces	Costly for deeper basement forms Required to be reinforced or very thick to cope with lateral loads	Shallow basement forms with ready access for bricklaying
Contiguous/ secant piles	Simple operation to form basement walls High strength Ability to produce walls at depth without excavation	Requires heavy plant Possible disturbance issues	Perhaps the most common form of basement wall for deep basements, particularly on congested sites

PART 2

REFLECTIVE SUMMARY

- Basements can be constructed using 'top-down' or 'bottom-up' construction.
- Several options are available for waterproofing, including tanking, waterproof concrete and drained cavities.

REVIEW TASK

- List the advantages and disadvantages of each of the options available for water-proofing basements.
- Visit the companion website at www.palgrave.com/engineering/riley2 to view sample outline answers to the review task.

AIMS

After studying this chapter you should be able to:

■ Demonstrate a knowledge of how methods of building for industrial and commercial forms have evolved
■ Understand the term *system building* and be able to demonstrate an understanding of open and closed building systems
■ Outline the sequence of construction works required for given scenarios
■ Understand to what extent expenditure on building work occurs
■ Appreciate the costs required to manage buildings after completion and be able to demonstrate a knowledge of how life cycle costs are achieved.

This chapter contains the following sections:

INFO POINT

■ BS 8204-2: Screeds, bases and *in situ* floorings. Concrete wearing surfaces – Code of practice [formerly BS 8204-2: 2002] (2003)
■ BRE Report 332: Building elements: floors and flooring (1997)
■ BRE Report 460: Building elements: floors and flooring – performance, diagnosis, maintenance, repair and the avoidance of defects [formerly BRE Report 332: Floors and flooring, 1997] (2003)
■ *Concrete for industrial floors*, British Cement Association (2000)
■ *Concrete for industrial floors: Guidance on specification, mix design*. Good Concrete Guide 1, CS (2007)
■ *Concrete industrial ground floors – a guide to their design and construction*, 3rd edn, Technical Report 34, CS (2003)
■ Gaber, G. (2006) *Design and construction of concrete floors*, 2nd edn, Butterworth–Heinemann [978–0–750–66656–5]
■ *Solid foundations – building guarantees technical manual*. Section 3: Substructure. ZM (2006)

6.1 | Functions of ground floors and selection criteria

Introduction

- After studying this section you should have developed an understanding of the functional requirements of ground floors to industrial and commercial buildings.
- You should appreciate the implications of building scale and proportions upon the design and construction processes.
- In addition, you should appreciate the degree to which the ground conditions and features of the chosen site affect the selection of ground floor options.
- You should also be aware of the need to take into account the potential for movement within large building elements and the ways in which this is dealt with both within the element and at points of contact with other building elements such as frames and foundations.

Overview

The nature of industrial and commercial buildings is such that the design and construction requirements for ground floors, as with other elements, differ considerably from those of domestic buildings. One of the reasons for this is the considerable difference in the size of the building and its elements. Also, the nature and extent of loadings applied to the floors are quite different and more demanding than they are for the floors of houses and small-scale buildings. Notwithstanding this, the criteria which are used for the basis of selection and the generic performance requirements for floors are essentially the same. The manner in which these criteria are satisfied requires a different approach when dealing with large-scale industrial and commercial forms.

Ground floors to these buildings can take a number of forms, depending, among other things, on the nature of the site, the quality of construction and the required speed of erection of the building. However, all of the available design solutions essentially fulfil the same functional requirements. The detailed functional requirements will be examined in detail later within this section and the main options for the construction of floors will also be examined. Ground floors will generally fall into one of two basic types: *suspended floors* and *solid* or *ground-supported floors* (Figure 6.1). The distinction between the two forms when considered in simplified terms is as follows:

- Solid floors are formed such that the underside of the floor is in continuous contact with, and is supported by, the ground. These are generally adopted for large single-storey buildings such as industrial sheds and warehouses.
- Suspended floors are formed such that the structural elements of the floor span between supports, not relying on the ground for support of the floor structure. In industrial and commercial buildings the need to adopt such a form may arise as a result of the need to create a void or basement beneath the floor. This is quite common in multi-storey framed commercial buildings.

Figure 6.1
Ground floor
classification.

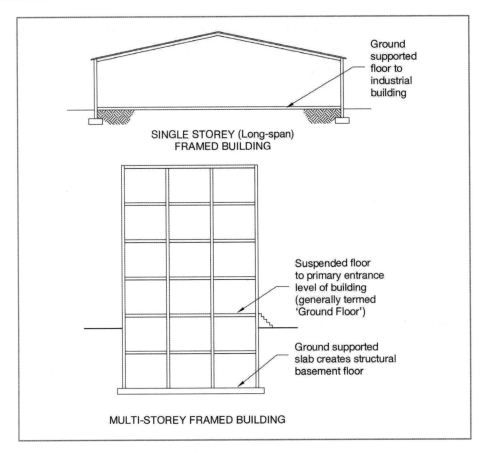

Functional requirements of ground floors

The primary function of the ground floor is to provide a safe and stable platform for the activities that are carried out within the building. It must be remembered that the loads exerted on the floors of large buildings can be considerable, and they differ quite markedly from those that must be considered for smaller buildings such as dwellings. However, in addition to the need to withstand loads there are a number of other equally important functions that must be fulfilled by the floor if it is to satisfy user needs and the requirements of the Building Regulations. The nature of the construction form of ground floors is dictated by the relative importance of these aspects, and it is appropriate that the functional requirements are tightly defined and adopted as criteria for selection from a range of design options. There is also a range of aspects related to the specific nature of the chosen construction site and the general construction form of the building which set these criteria into a context. These will be examined in detail later. Let us first consider the generic performance or functional requirements for ground floors of all types. These functional requirements can be summarised as follows:

- Structural stability
- Thermal insulation

- Exclusion of ground water
- Durability
- Provision of appropriate surface finish.

Structural stability

The ground floor of any industrial or commercial building must be designed and constructed in such a way that it is capable of supporting the dead loads and live loads to which it is likely to be subjected (Figure 6.2). These loads may be considerable, and in the case of industrial buildings particularly there is likely to be a high concentration of loads from machinery etc. In addition, issues such as vibration from machines and dynamic or rolling loads from wheeled vehicles such as fork-lift trucks place demands on the floor construction that must be considered in selecting an appropriate solution. Hence the construction form must be such that the floor is robust enough to resist or transfer these loads without undue deformation or the risk of structural failure. In the case of ground-supported floors this relies on the floor structure, often reinforced with steel, which is in continuous contact with the ground and distributes or dissipates the loads effectively. Suspended floor forms are uncommon in **industrial buildings** but are used extensively in multi-storey **commercial buildings**. In such situations the main entrance floor may be termed the ground floor but is, in structural terms, more often an intermediate floor above lower floor areas or a basement. In these instances the mechanism is one of load transfer to the supporting elements of the structure. Normally these take the form of basement walls at low level with columns of the framed structure providing support at intermediate points for beams which, in turn, support the floor.

Thermal insulation

As with all elements of the external fabric of buildings there is a requirement for ground floors to provide a degree of resistance to the passage of heat. The degree to which this results in the need for the installation of an insulative material varies depending on the floor structure and formation. In the case of large floor areas it is often possible to

Although the generic criteria for selection and the performance requirements are the same for floors to **industrial and commercial buildings**, the specific functional needs differ considerably. Industrial buildings are generally loaded far more heavily and are subject to harsher use.

PART 2

Figure 6.2
Load transfer through framed building.

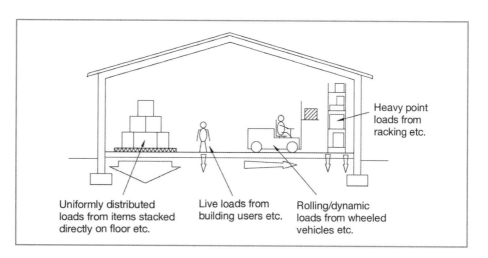

Heavy point loads from racking etc.

Uniformly distributed loads from items stacked directly on floor etc.

Live loads from building users etc.

Rolling/dynamic loads from wheeled vehicles etc.

Figure 6.3

Heat loss through
floors.

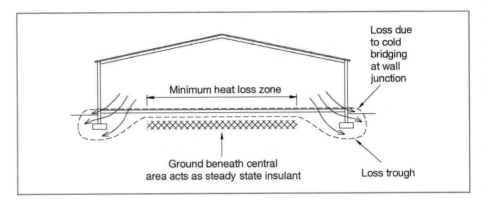

Figure 6.3

Heat loss through
floors.

achieve the required level of resistance to the passage of heat without the addition of insulation material. However, as the required levels of insulation demanded by the Building Regulations have increased, so the need to provide specific insulative materials as part of the floor composition have increased accordingly. Figure 6.3 illustrates the nature of heat loss through large floors.

In the case of solid floor types the formation of the floor in continuous contact with the ground or sub-base results in the creation of a protected central zone to the floor area. This effectively provides a degree of insulation to the inner area of the floor which inhibits heat loss. At the perimeter of the floor there will be a greater level of heat loss as a result of the potential for heat loss through perimeter walls into the exterior. The sub-base beneath the floor will initially absorb heat from the building. After a period of time, however, this will move to a near steady-state condition and the level of heat lost from the building interior will then be minimal.

For this reason the level of heat loss at the edges of the floor will be far greater than at the central area. Hence in some cases it may be economical to insulate the floor to differing levels at different locations. In the past this has taken the form of insulation provision only at the perimeter of the floor, with the central area left uninsulated. The potential cost saving associated with economic use of materials may be significant in large construction projects. Hence the complexity of detailing associated with utilising differing approaches to insulation within a single floor unit is generally justified. In the construction of framed buildings with large floor areas there is an additional issue associated with the potential for movement within the floor and the need to accommodate it. This issue is dealt with in detail later within the chapter.

In the case of suspended floor constructions the creation of a void or basement beneath the floor results in a more uniform level of heat transfer across the floor (Figure 6.4). It is probable that the suspended floors within a framed building will be at an intermediate level, and thus the area beneath it will be within the interior of the building. Hence heat transfer is unlikely to be a major factor unless the floor separates areas that are heated from those that are not. This may be the case where the basement area is used for parking, for example. In such instances the void will be subject to the passage of air, and thus the creation of a steady-state central area as seen in solid floors will not be possible. Hence, it may be essential to provide a uniform level of insulation across the whole of the floor.

Figure 6.4
Heat loss through
suspended floors.

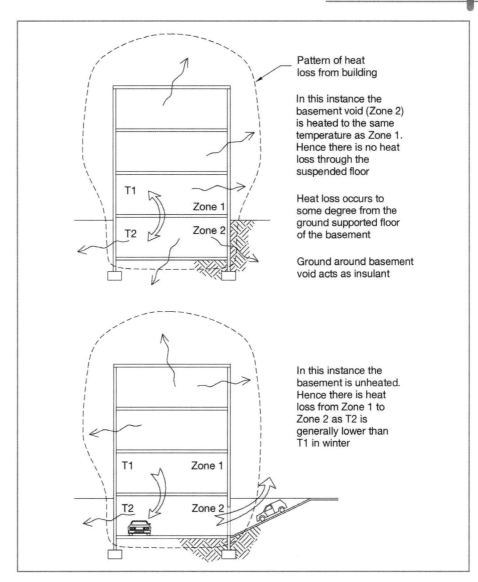

Pattern of heat
loss from building

In this instance the
basement void (Zone 2)
is heated to the same
temperature as Zone 1.
Hence there is no heat
loss through the
suspended floor

Heat loss occurs to
some degree from the
ground supported floor
of the basement

Ground around basement
void acts as insulant

In this instance the
basement is unheated.
Hence there is heat
loss from Zone 1 to
Zone 2 as T2 is
generally lower than
T1 in winter

PART 2

Exclusion of ground water

The issue of preventing the passage of moisture to the interior of a building is dealt with in detail elsewhere within this text (see section 3.3), but it may be useful to remember that this need is the subject of Part C of the Building Regulations. The construction form of the floor, whether ground-supported or suspended, must include details designed to arrest the passage of moisture if there is likely to be such a risk. In the case of industrial and commercial buildings it is quite common for basements to be created beneath the main building. In such instances the structural floor of the basement will be subjected to quite high levels of water pressure, far in excess of those experienced in the construction of floors to dwellings and small buildings.

Figure 6.5
Modes of moisture
entry through floors.

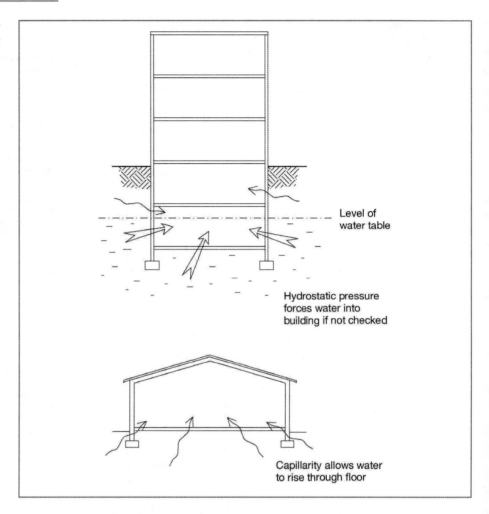

The normal mechanism of moisture ingress through ground floors is that of capillarity (Figure 6.5). Moisture is drawn into the building element as a result of the capillary action within porous materials such as concrete. This must be resisted by the incorporation of impervious materials or capillary breaks within the floor. In situations where the floor is some distance below ground level, as in a basement, there is also the problem of moisture being driven into the structure by positive pressure.

Durability

The nature of the environment within which ground floors must function may be aggressive particularly in the case of industrial buildings where corrosive materials may be used and the nature of the activity within the building demands a robust construction form. Hence the materials that are used must be sufficiently durable to provide a satisfactory lifespan. The selection of appropriate materials is essential for the longevity of the floor, as is the selection of an appropriate design option to cope with the nature of the activity that will be carried out in the building.

Provision of appropriate surface finish

Ground floors must provide a level and smooth surface even though it is quite probable that they will not receive decorative finishes in many cases. In the case of industrial buildings floors are generally left with a concrete or screeded finish which may be sealed or primed to prevent contamination and surface damage from spillages etc. It is unlikely that they will receive any form of decorative surface finish or carpeting for example. In the case of commercial buildings it is increasingly common for raised access floors to be provided; hence the surface of the structural floor is unlikely to be exposed to direct contact with the users of the building. If this is the case it is still essential that the finish is level and true, since minor undulations and irregularities can result in problems with installation of the raised floor. Raised access floors are considered in detail in Chapter 13.

Unlike floors to dwellings, concrete floors to commercial and industrial buildings are often provided with a power floated surface. Alternatively they may be provided with a screed, which will generally be cement and sand based and may be reinforced with steel where heavy loads are envisaged. It is normally the case that the cheapest option that allows the floor to perform its desired function will be adopted. Hence the minimum possible standard of surface finish to the structural floor will be chosen. It is not always necessary or appropriate to finish the floor surface to high levels of tolerance.

Factors affecting the selection of ground floors

A number of factors affect the selection of ground floor alternatives. These relate to the functional performance of the individual options and also take into account the following:

- Construction form of the building
- Nature of the site
- Nature and extent of loadings
- Required surface finish
- Cost.

When attempting to evaluate individual design solutions these factors will be considered. It must always be remembered that where two alternatives are able to satisfy the functional requirements equally well, the cheaper option is almost certain to be selected. In recent years there has been a considerable degree of research and development into the design and construction of concrete floors to industrial and commercial buildings. As a result, techniques have been developed for the formation of large floor sections that owe much to the established principles of road engineering and construction. As a result of the large scale of these floors and the relatively high loadings applied to them it is now almost certainly the case that ground floor options will be restricted to solid or ground-supported types. The construction of suspended floors is generally restricted in these buildings to intermediate or upper floors. The design and construction of these suspended types is dealt with in Chapter 11.

REFLECTIVE SUMMARY

When considering ground floors to industrial and commercial buildings, remember:

■ The nature of floors to industrial buildings is quite different from that of commercial buildings.
■ The scale of operations involved in laying large floors is such that the process is similar in principle to road laying.
■ Floors are required to satisfy a range of generic performance requirements, including strength, durability and thermal insulation.

REVIEW TASKS

■ In what ways are the ground floors to office buildings different from those to industrial buildings?
■ Set out the reasons why these differences are present.
■ Visit the companion website at www.palgrave.com/engineering/riley2 to view sample outline answers to the review tasks.

6.2 | Ground-supported floors

Introduction

■ After studying this section you should have developed an appreciation of the more common options for the formation of ground-supported floors to industrial and commercial buildings.
■ You should appreciate the issues associated with interaction between floors and structural frames and walls of such buildings.
■ In addition, you should be familiar with the components of the various floor formations and you should understand their functions and the mechanisms by which they satisfy them.

Overview

As outlined previously, the essential characteristic of ground-supported floors is that they are in continuous contact with the ground beneath and they transfer their loads through this contact area. The loads applied to floors in non-domestic buildings may be considerable and they may also be highly concentrated where there is provision of storage racking, heavy machinery and so on. The floors are also subject to a range of dynamic loads, such as rolling loads from trolleys and fork-lift trucks. Naturally, these are more common in industrial buildings than in commercial buildings; however, the broad principles of floor design are similar. Ground-supported floors to framed buildings may take a number of forms, although all are similar in principle, differing only in the specific details of design and construction. All of the floor options must fulfil the same performance requirements, and as such the level of flexibility in generic form is limited.

Considerations in solid floor construction

Before selecting a final design solution for a given situation a range of preliminary issues must be considered. Unlike the design of floors to dwellings, where loading and use characteristics vary little from one situation to another, floors to non-domestic buildings vary greatly in the demands placed upon them. For this reason floors are often categorised into broad groupings (Table 6.1) in order to assess the probable design requirements and to allow assessment of available options.

The loads applied to floors of industrial and commercial buildings are likely to fall into one of three broad classifications. These are as follows:

- *Rolling loads* from wheeled vehicles such as trolleys and fork-lift trucks. Heavy dynamic loads have a great effect on floor loading and abrasion of the surface; hence they often necessitate the use of thicker slabs and higher strength grades of concrete.
- *Point loads* from legs of racking or shelving etc. Commonly experienced in industrial buildings, warehouses and so on, but also significant in offices for filing storage racking etc., these loads can result in localised failure of the slab, which can be avoided by increasing its thickness.
- *Uniformly distributed loads* (UDLs) from items stacked directly on the floor. In practice these UDLs are likely to be restricted to zones or areas of the floor, since access is required around them.

It can be seen from Table 6.1 that the level of surface abrasion anticipated is a significant factor in selecting concrete of an appropriate strength. As the strength of the floor increases, so the resistance to surface abrasion improves. The use of appropriate applied surface finishes allows a lower strength of concrete to be used where abrasion is perceived to be an issue. However, the nature of the applied loadings and the internal environment of the building are not the only factors that must be taken into account in selecting an appropriate design option for ground floor construction. A range of other issues must be considered, and a selected summary of such issues is indicated in Table 6.2, although this is by no means exhaustive.

Because there are so many variables associated with the design of ground floors which may result in differences between sites and within a single site, assumptions must be made in order to simplify the process of design and selection.

The structural performance of the floor depends on the ability of the concrete slab to cope with the bending stresses that result from loading. This is a function of the tensile strength of the floor and relies on the specification of an appropriate strength grade of concrete. It is generally the case that floors are now designed as being effectively unreinforced, although steel mesh is often incorporated to control the risk of cracking. The inclusion of reinforcement also assists in reducing the required number of joints within the floor construction. These joints are required to manage the effects of thermal and moisture-induced movement within the construction. Historically this often resulted in the adoption of a 'chequerboard' pattern of floor construction with bays of controlled dimensions making up the totality of the floor area. This approach has now largely lost favour and has been replaced by 'long strip' techniques in which the size of bays is conse-

Table 6.1 Categories of floor conditions for industrial and commercial buildings.

Duty category	Possible finish	Typical concrete strength (N/mm²)
Light traffic (foot and trolley), typically shops and offices	Carpet, tile or sheet material	30
General industrial situation: vehicle traffic (pneumatic tyres), moderate chemical exposure	Structural slab finished as final wearing surface	40
General industrial situation with heavy abrasive conditions, e.g. vehicles with solid wheels, high levels of chemical exposure	Range of applied finishes to suit	35
Heavy industrial use	Structural slab finished as final wearing surface	50
Heavy industrial situation with heavy abrasive conditions, e.g. vehicles with solid/metal wheels, high levels of chemical exposure	Range of applied finishes to suit	35

quent upon the reinforcement present and the practicalities of construction operations.

Formation of ground-supported concrete floors

As previously noted, the design of ground-supported floors has developed in recent years such that there are few alternative forms. Although there are options based on the variation of individual components of the floor, all of the options adopt the same approach to construction with the use of a layered form incorporating specific individual elements. The layers are (Figure 6.6):

■ *Slab*: The main load-supporting element is the floor slab, which will normally take the form of a layer of mass concrete, which may be reinforced as required, cast *in situ* to the desired level. In floors which are to take high loadings, or where the loadbearing capacity of the ground is low, the slab may be reinforced with mild steel bars or mesh. In addition it may be necessary to undertake ground improvement works beneath the slab. In normal circumstances a slab thickness of 150 mm or more would be used.
■ *Wearing surface*: The surface of the floor must usually be suitable to accept a surface finish or to be trafficked directly. Hence the upper surface of a cast slab may be suitably finished or it may receive an appropriate applied topping or surface finishing material.
■ *Slip membrane*: As with domestic floors, the exclusion of moisture from the building interior is of paramount importance. Hence a damp-proof membrane is installed. However,

Table 6.2 Aspects to be considered in selection of floor design.

Issue	Specific factors
Floor surface characteristics	Possibility of impact damage Required resistance to abrasion Intended surface finishes/sealants Consistency/regularity of surface
Loading	Presence of static and dynamic loadings Presence and patterns of point loads/load concentrations Uniformly distributed loads General levels of loading
Intended lifespan	Fixed life building? Maintenance profile Time to first major refurbishment/refit
Site conditions	Safe bearing capacity of site Areas of made-up ground Need for ground consolidation/stabilisation beneath slab? Topography Drainage conditions

since the grade of concrete used for floors to industrial and commercial buildings is often inherently waterproof the main functions of the membrane are to act as a slip plane and to prevent the loss of cement fines to the sub-base. The large scale of the floors means that dimensional variation due to thermal and moisture changes can be significant; the slip plane reduces friction beneath the slab and allows this to be managed.

- *Sub-base*: The provision of a firm level surface on which to cast the slab is aided by the provision of a layer of selected hardcore beneath the floor slab. The sub-base consists of a layer of appropriate material of sufficient thickness to provide a sound base for the floor. In addition, the voids between pieces of hardcore act to break the capillary path of moisture rising from the earth.

Figure 6.6
Components of ground-supported floors.

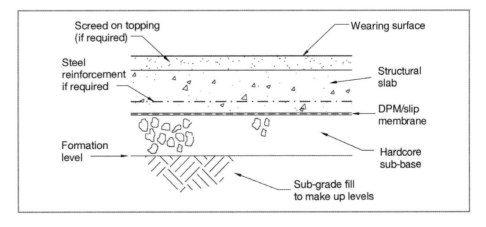

■ *Sub-grade*: This is the ground beneath the slab that is excavated to formation level prior to forming the various layers of the floor construction. In some instances, such as on sloping sites or where the ground is poor, it may be necessary to bring suitable imported material to site. This is termed made-up ground or making-up levels.

Because of the large scale of floors to industrial and commercial buildings it is probable that insulation will not be required beneath the main area of the floor. Indeed, the use of calculations to assess the thermal performance of the building as a totality has tended to result in a situation in which floors may not be insulated at all. However, since the majority of heat loss occurs at the perimeter of the floor, often by cold bridging at the point of contact between the floor and the external wall, insulation may be provided vertically at the junction of the two elements. This also assists in controlling the effects of thermal and moisture-induced movement in large floors.

Casting large floors

It is not feasible to cast floors of the size required for large industrial and commercial buildings in one section. The floor will generally be divided into smaller sections in order to allow for the process of construction with a reasonable degree of practicality. In addition the subdivision of the floor allows for the management of tensile stresses resulting from contraction of the slab due to moisture and thermal changes during curing of the concrete. The provision of joints required for the construction of the slab in sections or bays allows the presence of cracks to be controlled, rather than allowing the random pattern of cracking that would otherwise occur.

The subdivision of the floor area is facilitated by the inclusion into the design of a series of construction and movement accommodation joints. As previously discussed, the construction of ground-supported floors is now most likely to be effected using the long strip method (Figure 6.7). This method has increased in popularity due to its

Figure 6.7
Long strip and chequerboard options.

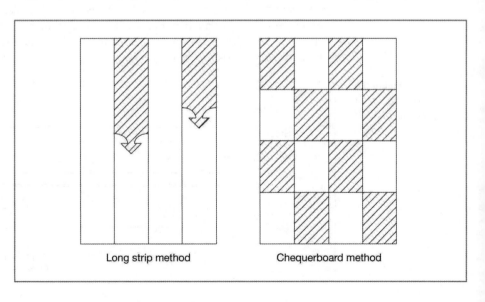

Long strip method   Chequerboard method

capacity to allow for speedy construction and ease of placement of concrete, resulting in improved economics over the chequerboard method, with its complexity of layout and difficulties in placement of concrete.

The chequerboard method

This method relies on the subdivision of the floor into a series of sections of restricted width and length. Alternate sections of the floor are then cast in a chequerboard arrangement. Once these have cured sufficiently the remainder of the sections are cast. One of the difficulties associated with this is the need to create many structural joints running along and across the floor. In addition, there is the potential problem associated with placing concrete in the second phase of bays whilst working over those cast in the first part of the process. Since this method is now little used it is not explored in detail here. However, the joints that are adopted to facilitate the construction of such floors share essentially the same principles as those used for the long strip method.

The long strip method

This method is based on the division of the floor into a series of long strips, typically around 4.5 metres in width, running the full length of the building or up to a selected movement joint or an 'end of day' construction joint. The strips are cast in two phases, with alternate strips cast initially and the infill joints cast several days later (Figure 6.8).

The concrete is laid between formwork which defines the strips and is placed in a continuous process starting at one end and working towards the other or to a selected joint, as previously described. The surface is levelled manually using a simple compacting beam and following the top edge of the formwork as a guide. Narrow edge strips (typically 600–1000 mm wide) are formed adjacent to perimeter walls to allow ready access for the compacting beam across full width strips. Once the first set of strips have been

PART 2

Figure 6.8
Long strip floor layout.

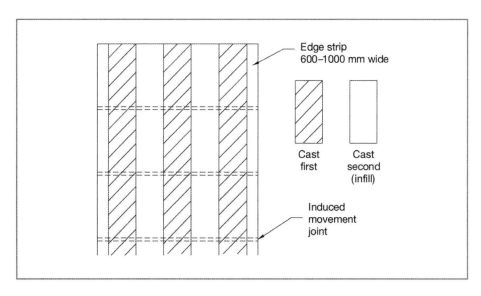

created and the concrete has cured sufficiently to prevent damage from adjacent working, the remaining or infill strips are placed.

The long strips are likely to suffer from shrinkage cracking as the concrete cures. In order to prevent random cracking the strips are normally divided into bays to control the location of cracks. This is effected by sawing grooves in the surface of the floor as it cures or by placing proprietary crack inducers that create a focus for cracking in the material. As the concrete of the slab contracts, cracks form at the weakened locations in a controlled pattern. The need for the inclusion of these joints is related to the length of the strips and the presence or otherwise of reinforcement.

Joints in large floors

We have already seen that the processes of expansion and contraction that occur naturally within any large building component must be accommodated within large floors by the incorporation of a range of joints. The design of such joints is important, since they must be able to accommodate the desired range of movement, while at the same time transferring loads between adjacent bays and restricting vertical differential movement between them. If this is not achieved the slab will be subject to variation in the extent of vertical movement between bays, which will result in the formation of steps in the slab. This is generally resisted by the inclusion of a series of tie bars set horizontally at mid-depth and spanning across the joint.

The joints fall into one of four broad categories:

'**De-bonding**' is important in the provision of reinforcement that is intended to prevent sections of the floor from moving vertically in relation to each other while allowing the natural process of expansion and contraction to take place without inducing stresses in the floor.

- *Longitudinal joints* (Figure 6.9): These are the main construction joints in the floor which separate the slab into the long strip pattern described earlier. The joints generally incorporate tie bars as discussed, which may sometimes be '**de-bonded**' to allow for a degree of contraction in the slab across the joint. If the tie bars are not de-bonded, contraction is restricted. In either case, vertical loads are transferred between adjacent bays through the bars.
- *Induced joints*: These are used to control bay length within the long strips by providing a mechanism for controlled cracking of the slab as it cures. Created by means of a saw cut in the surface or the use of a crack former, these joints rely on friction between the sides of the joint for restriction of vertical movement. Generally the interlocking of the exposed aggregate at the crack faces will be sufficient.
- *Movement joints* (Figure 6.10): These are included within the slab to accommodate the naturally occurring movement of the material when subjected to changes of temperature or ambient moisture level. Contraction joints are most common as the concrete has a natural tendency to shrink following initial placement. However, in some instances expansion joints are also required.
- *Isolation joints* (Figure 6.10): Although these are technically a sub-group of movement joints, they warrant individual consideration here. Their purpose is to allow for unhindered movement of the slab around fixings etc. that may penetrate the slab, without causing damage at points of contact. In framed buildings they are most commonly found around the bases of columns or stanchions.

Figure 6.9
Longitudinal joints in
large floors.

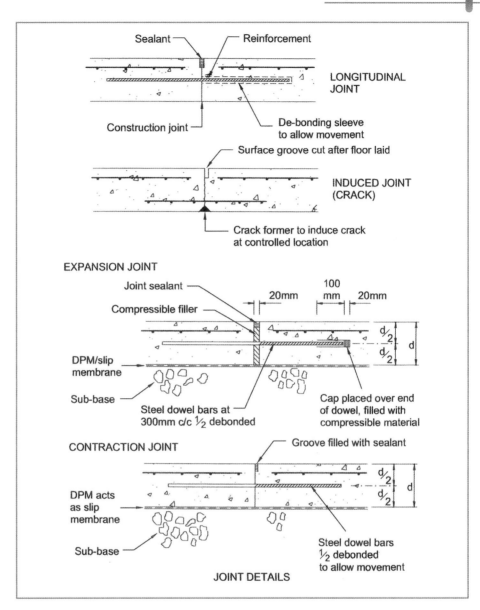

Reinforced concrete floors

In large floors, as described in this section, the use of reinforcement is really intended
to control the risk of cracking rather than to act as tensile reinforcement in the traditional
fashion of reinforced concrete elements. Steel mesh or bars will be introduced to counter
the tensile forces set up as a consequence of contraction in the slab, rather than to resist
the tensile forces created due to bending of the slab under load. Such forces should be
dealt with by adequate slab thickness, appropriate strength of concrete and sufficient
support from the sub-base beneath the floor. The utilisation of steel reinforcement allows

Figure 6.10
Isolation and movement
joints in large floors.

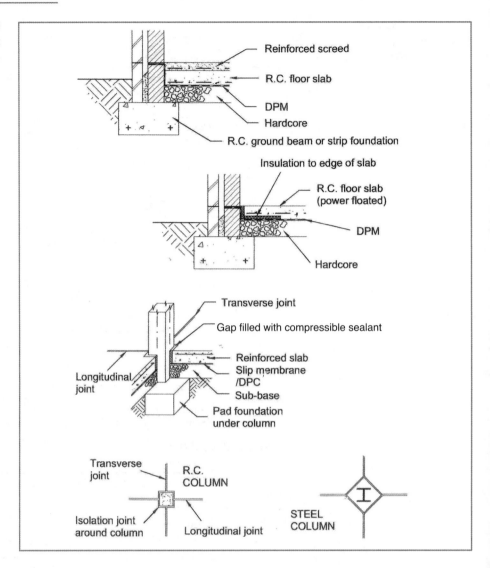

greater distances between contraction joints and tends to limit any chance of random cracking in the slab.

REFLECTIVE SUMMARY

■ Solid ground floors to industrial and commercial buildings are designed in accordance with duty categories ranging from light traffic to heavy industrial situations.

■ The selection of an appropriate floor design takes into account surface finish, loading, lifespan and site conditions.

■ The most common method for creating large ground-supported floors is the long bay method, derived from road-laying techniques.

■ The scale of these floors demands incorporation of a range of movement and isolation joints.

> **REVIEW TASK**
> ■ With reference to a building you know, generate a simple sketch indicating the probable layout of bays, structural columns and movement/isolation joints in accordance with the guidance noted in this section.
> ■ Visit the companion website at www.palgrave.com/engineering/riley2 to view sample outline answers to the review task.

6.3 | Suspended floors

Introduction

- After studying this section you should be familiar with the various forms of suspended floor available for use in the formation of ground floors to commercial and industrial buildings.
- You should understand the circumstances in which suspended floor constructions may be adopted and the reasons underlying the decision to use them.

Overview

As previously noted, it is generally the case that ground-supported floors will be adopted for the construction of industrial and commercial buildings. In some circumstances, however, the utilisation of ground-supported floors may be unsuitable or impractical as a result of functional factors. In most situations a combination of high loadings and large distances between possible support position (i.e. long spans) limits the potential options for suspended floors. Very often in framed buildings we use the term 'ground floor' to mean the primary or main entrance floor if there is a basement area or large void beneath. In fact this may not be a ground floor in the true sense, but rather is likely to be an intermediate floor in terms of its construction. Since the alternative construction forms of intermediate or upper floors are dealt with in some detail later in the book, it is not appropriate to examine them in detail here. However, there are a number of issues that are specific to the adoption of suspended floor forms close to ground level, which are considered briefly in the following subsection.

In circumstances where it is necessary to provide a large void beneath the 'ground' floor, the use of a suspended floor system is necessary. In commercial and industrial buildings this will generally be in the form of a precast or *in-situ* formation capable of carrying great loads and without being subject to loss of performance over time. Such systems are now most commonly based on the use of precast reinforced concrete units. Although the use of simple reinforced concrete slabs cast *in situ* is also still an option, this tends to be less popular due to slower construction and the need to provide suitable formwork and temporary support for the slab.

Beam and block floors

The systems used for the formation of beam and block type floors in these buildings are essentially the same as those used in dwellings, relying on the provision of a series of profiled reinforced beams, set at relatively close spacings (typically 600 mm), spanning between supporting walls (Figure 6.11). The beams are often an inverted T shape to allow support to be given to the infill blocks that are dropped in between the beams. A sand/cement screed is then applied over the floor to give a smooth surface finish to receive carpet and other finishes. The only real difference between industrial and commercial systems is in the scale or size of the supporting beams and the strength of the infill blocks, which are required to cope with much greater, and often more concentrated, loads than in a domestic situation.

These systems have the advantage that they comprise a series of small units, which can be readily manipulated on-site without the aid of large-scale mechanised plant. However, it must be noted that beams intended to cope with large spans will require assisted lifting due to size and weight, although when constructing large buildings the presence of lifting equipment and mechanised plant is ubiquitous in any case.

As shown in Figure 6.12, it is acceptable to support this type of floor on the inner skin of a cavity external wall or directly off a ring beam or ground beam. Where supported by walls, whether external or internal, a DPC will always be located below the floor units as illustrated.

Infill blocks are dropped between the reinforced concrete beams and it is easy to finish off the blocks at the end of the beams by cutting blocks to length. At the other floor edge

Figure 6.11
Suspended, precast concrete (beam and block) flooring systems.

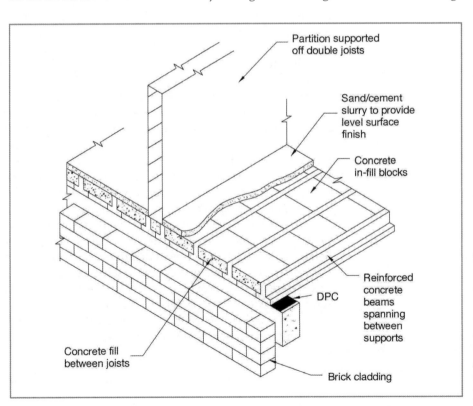

Partition supported off double joists

Sand/cement slurry to provide level surface finish

Concrete in-fill blocks

Reinforced concrete beams spanning between supports

DPC

Concrete fill between joists

Brick cladding

Figure 6.12
Edge detailing – beam
and block floors.

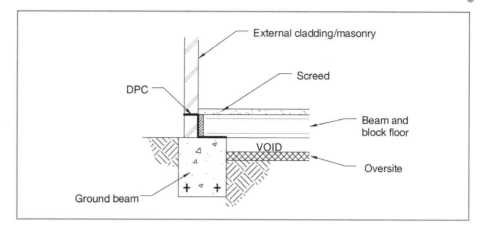

there needs to be careful consideration of the finishing detail to ensure that the floor is continuously supported up to the perimeter.

'Plank' floors

The details of plank or wide slab floors are dealt with in Chapter 11. The details of these systems are essentially the same when used as a ground floor element. However, as in the case of beam and block floors, care must be taken when dealing with edge and support details. The resistance to passage of moisture is achieved by incorporation of a DPC, whilst cold bridging can be avoided by careful insulation provision at the abutments with external walls etc.

PART 2

REFLECTIVE SUMMARY
- The use of suspended floor solutions for industrial and commercial buildings is quite rare.
- The common forms of suspended floor in use are essentially the same as those for upper or intermediate floors.
- Beam and block and plank floor options exist, but are uncommon in use.

REVIEW TASK
- Set out the reasons why you think that suspended ground floors are uncommon in industrial and commercial buildings.
- Visit the companion website at www.palgrave.com/engineering/riley2 to view sample outline answers to the review task.

COMPARATIVE STUDY: GROUND FLOORS

Option	Advantages	Disadvantages	When to use
Solid concrete Chequerboard	Cheap Familiar technology Easy detailing to insulate and resist moisture Good loadbearing performance	Several operations involved and complexity introduced in phasing of bay laying	Less common method of forming large ground floor Used on relatively level sites with large areas
Solid concrete Long bay	Cheap Familiar technology Easy detailing to insulate and resist moisture Good loadbearing performance Rapid progress during construction	Need to incorporate a range of construction joints during laying Timing of material delivery crucial to continuity on-site	Most common form of large-scale ground floor construction Used on relatively level sites with large floor areas
Beam and block/plank floors	Cheap Fast for small areas Copes with sloping sites and deep voids	More expensive than solid floors Relatively low loadbearing capacity	Although very common in house building an unusual choice for industrial and commercial buildings

Building superstructure

7

High-rise buildings

AIMS

After studying this chapter you should be able to:

- Appreciate the functional performance characteristics of frames for multi-storey buildings
- Relate functional requirements to the design alternatives for frames suitable for the construction of high-rise buildings
- Appreciate the details required to ensure functional stability
- Be able to appreciate the reasons why particular frame types are chosen in given situations

This chapter contains the following sections:

INFO POINT

- Building Regulations Approved Document A: Structural strength (2004 including 2010 amendments)
- BS 4: Structural steel sections. Specification for hot-rolled sections (2005)
- BS 449: Specification for the use of structural steel in building (1969)
- BS 1305: Specification for batch type concrete mixers (1974)
- BS 4466: Specification for scheduling, dimensioning, bending and cutting steel reinforcement for concrete (1989)
- BS 5628: Code of practice for use of masonry (2005)
- BS 5642: Precast concrete sills and copings (1983)
- BS 5950: Structural use of steelwork in building (2000)
- BS 8110: Structural use of concrete (1997)
- BS DD ENV 1993: Design of steel structures
- BS EN 934: Admixtures for concrete, mortar and grout (2001)
- BRE Digest 496: Timber frame buildings. A guide to the construction process (2005)
- *BRE guidance on timber frame design for multi-storey buildings* (2004)
- BRE Report 454: Multi-storey timber framed buildings: a design guide (2003)
- DD ENV 1992: Precast concrete – design elements and structures
- *Design of multi storey braced frames*, Publication 334, Steel Construction Institute (2004)
- *Manual for the design of steelwork building structures*, 3rd edn, Institution of Structural Engineers [0–901–29726–7] (2008)
- Martin, L. and Purkiss, J. A. (2005) *Concrete Design to EN 1992*, 2nd edn, Elsevier Butterworth–Heinemann [0–760–65059–1]

PART 3

7.1 | Frames: functions and selection criteria

Introduction

- After studying this section you should understand why the use of frames for multi-storey buildings is now prevalent.
- You should be able to appreciate the functional requirements of frames for multi-storey buildings and be able to understand their implications for the construction form.

Overview

Modern **multi-storey commercial buildings** need to be designed for the needs of the proposed occupier, but also need to consider the requirements of future occupiers.

Increased demands for efficient use of space and cost-effective building practices have resulted in the adoption of framed structures for most **multi-storey commercial buildings**. The use of a structural frame allows the maximum provision of usable space with the minimum space taken up by the structure. Advances in construction technology have allowed buildings of great height with great spans to be erected with the adoption of structural frames.

Functions of framed buildings

Framed buildings differ from traditional loadbearing structures in that the loads of the building are transmitted through a structural frame. The frame supports the external weatherproof walling and roof, and also internal walls and floor plus loadings from fixtures and fittings. Framed buildings are particularly suitable for medium to high-rise structures, and the main type of frame used for this form of structure is the skeleton frame. There are several benefits to their use, which include:

- The small area of the frame maximises the usable floor area of the building.
- The lack of massive structure, which would normally restrict the use of floor areas, allows greater flexibility in the use of the building by providing large areas of open floor.
- It is normally accepted that the use of framed structures results in savings in time and cost over loadbearing forms.

The construction of buildings of great height necessitates the use of a structural frame because a loadbearing structure would need to be enormous at the lower levels to support the upper levels. A number of different approaches to the design of high-rise buildings are common, but all share the same basic advantages over loadbearing forms of construction:

- The use of a frame formed from a strong durable material allows the loadings from the building fabric to be safely transmitted to the ground at selected points.
- The reduction of overall building mass allows the construction of bigger buildings on a given site.

- The use of a frame reduces the size of the building structure and therefore makes the most effective and flexible use of space. Changing occupation through the building life requires a great level of flexibility in the use of space if the building is to be economically viable over its predicted lifespan.
- Using a durable frame onto which components of the building fabric are fixed allows for the replacement of components, such as external cladding, during building refurbishment with little effect on the frame itself. The design life of a frame will far exceed that of the cladding, as generally the frame is not exposed to the elements or damage from vandalism. The changing demands of building users, together with advances in technology and variation in design fashions, impose requirements for building design flexibility, and hence this aspect is of great importance.

Skeleton frames

These frames are a series of rectangular frames placed at right angles to one another, so that the loads are transmitted from member to member until they are transferred through the foundations to the subsoil. These frames can be constructed using different materials. The type of material used depends on many factors, some of which are site conditions, economics, availability of labour and materials, time factor, health and safety, environmental preference, capital costs, maintenance costs and personal preference. Figure 7.1 shows a typical skeleton frame arrangement.

Figure 7.1
Traditional skeleton frame.

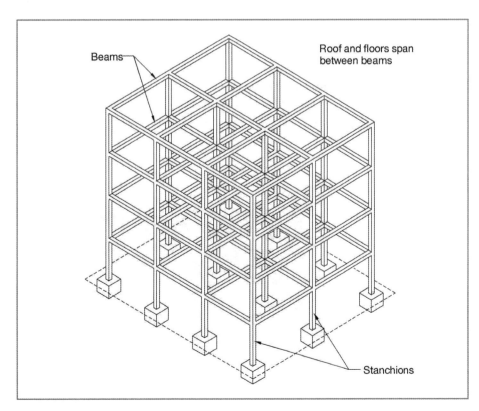

Beams

Roof and floors span between beams

Stanchions

PART 3

Figure 7.2
Alternative skeleton
frame.

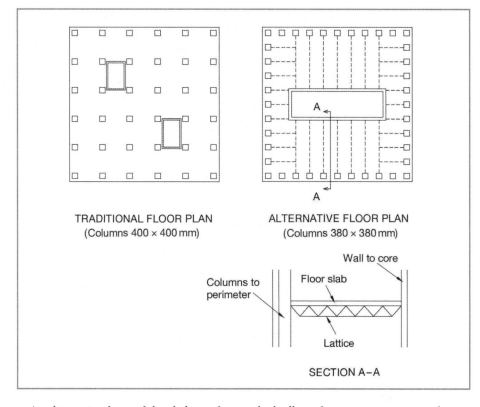

TRADITIONAL FLOOR PLAN
(Columns 400 × 400 mm)

ALTERNATIVE FLOOR PLAN
(Columns 380 × 380 mm)

SECTION A–A

An alternative form of the skeleton frame which allows for a greater amount of unrestricted floor space is shown in Figure 7.2. In this system, some of the internal columns are moved to the perimeter of the building to form an external tube. A central core is constructed using concrete walls to form another tube. The two tubes are then linked using underfloor lattice beams. The two tubes are rigid and the connecting beams allow

Figure 7.3
Skeleton frame
construction.

for movement. The cross-sectional area of the columns in this type of frame is the same as the cross-sectional area of the columns in the traditional skeleton frame, but the columns do not encroach into the usable floor area.

The main materials used for frame construction are reinforced *in-situ* concrete, reinforced precast concrete, structural steel and, increasingly for specific types of structure, timber. Figure 7.3 shows how much greater uninterrupted space can be achieved using this layout.

REFLECTIVE SUMMARY

- The use of a structural frame allows the maximum provision of usable space with the minimum space taken up by the structure.
- Framed buildings differ from traditional loadbearing structures in that the loads of the building are transmitted through a structural frame.
- The use of a frame formed from a strong durable material allows the loadings from the building fabric to be safely transmitted to the ground at selected points.
- The reduction of overall building mass allows the construction of bigger buildings on a given site.
- The use of a frame reduces the size of the building structure and therefore makes the most effective and flexible use of space.
- Using a durable frame onto which components of the building fabric are fixed allows for the replacement of components.
- Skeleton frames are a series of rectangular frames placed at right angles to one another, so that the loads are transmitted from member to member until they are transferred through the foundations to the subsoil.
- The most common type of frame used is the skeleton frame, which can take different forms.

REVIEW TASK

- What are the main reasons for choosing a frame structure as opposed to a load-bearing form of construction for multi-storey buildings?
- Visit the companion website at www.palgrave.com/engineering/riley2 to view sample outline answers to the review task.

7.2 | Frames: forms and materials

Introduction

- After studying this section you should have developed a knowledge of the different materials and systems that are used to construct multi-storey frames and the methods of construction.
- In addition, you should have gained an increased awareness of the aspects of the particular type of system that will need to be considered in given situations, and understand the health, safety and environmental implications of each system.

PART 3

■ Given specific scenarios you should be able to set out a series of criteria for the selection of appropriate multi-storey frame selection.

Overview

The materials used for the construction of high-rise buildings are *in-situ* reinforced concrete, precast reinforced concrete, structural steel and timber. The choice of material will depend on many factors, including the cost and availability of materials at the proposed time of construction, buildability factors, availability of skilled labour, cash flow situation of the client, design flexibility requirement during the construction phase, number of floors, and proposed use of the building, plus maintenance requirements. In addition to these, many clients are insisting that buildings are constructed using sustainable construction and design principles, and are requiring designers and contractors to ensure that all alternatives have been considered in order to produce as 'green' a building as possible. Finally, and most importantly, health and safety need to be foremost in the mind of designers when choosing a particular frame system. If a system is specified that is known to be dangerous and an accident occurs, the designer could be deemed liable if an alternative, safer system could have produced the same end product.

In-situ reinforced concrete frames

In situ, in construction terms, means built at an element's final proposed position, as opposed to being manufactured elsewhere and then being brought to site and just fixed in position.

In-situ reinforced concrete frames are frames in which all the members are constructed on-site using steel reinforcement bars and wet concrete. Formwork is constructed to the required sizes and in the correct positions and then the steel is fixed and the concrete placed *in situ*.

Formwork

Formwork (sometimes referred to as 'shuttering') is a structure, usually temporary, which is designed to contain fluid concrete. It must be designed to allow the concrete to form into the required shape and dimensions, and support the concrete, the people working on the structure, the plant required for placing concrete, and weather loadings until the concrete cures sufficiently to become self-supporting. The term *formwork* includes the materials actually in contact with the concrete and all necessary supporting structure.

The functional requirements of formwork are:

■ *Strength and containment*: The loads and pressures that the formwork must withstand take the form of:

  – Dead loads that consist of the self-weight of the formwork, concrete and any reinforcement
  – Imposed loads that include impact and surge loads of the concrete being placed, the operatives and concreting equipment, storage of materials such as reinforcement and timber, and forces from the permanent structure
  – Wind/weather loadings

– Accidental loadings
– Hydrostatic pressure that is caused by the fluid concrete acting against the sides of vertical or steeply sloping formwork.

■ *Resistance to leakage*: If grout leaks out of the formwork, the concrete in that area will be of reduced quality.

■ *Accuracy*: The degree of accuracy required will depend on the item being cast, i.e. foundation formwork may not need to be as accurate as formwork for, say, a concrete frame structure. The site engineer must check all levels and dimensions before any reinforcement is fixed and then again after the reinforcement is in place.

■ *Ease of handling*: Panels and units must be designed so that their maximum size does not exceed that which can be easily handled safely. All formwork must be so designed and constructed to include facilities for adjustment, levelling, easing and striking without damage to the formwork or the concrete. Currently there has been an increase in the use of proprietary formwork systems that have safety elements already incorporated, such as handrails and toe boards.

■ *Access for concrete*: The extent of the access required will depend on the particular contract, the amount of concrete being poured or placed at any one time, and the ease of carrying out the concreting operations. Access for concrete in *in-situ* concrete frames is usually achieved using either a concrete skip that is lifted by a crane or by a concrete pump (Figure 7.4). The use of a pump is quicker than the use of skips, but when pumping up to any great height the pump can get blocked. A way around this is to use a light aggregate concrete, which is usually a lot wetter than normal concrete. If this is to be used, then attention to resistance to leakage would need to be greater.

■ *Reuse*: For *in-situ* concrete to be an economical form of construction, the formwork needs to be reused as many times as possible. The layouts of multi-storey buildings

Figure 7.4
A concrete pump being used to facilitate the casting of concrete floors to a steel-framed building.

tend to be the same for every floor, and therefore the formwork required will be the same for every floor. After casting concrete slabs, the formwork and props must be left in position for between 4 and 7 days, depending on the temperature. To avoid delays in the contract, if a second set of formwork is purchased the work to the next floor can be started as soon as the concrete is hard enough to walk on without causing damage. Therefore if two sets of formwork are purchased the sequence for the use of each set is as follows:

- Set A: 1st floor
- Set B: 2nd floor
- Set A: 3rd floor
- Set B: 4th floor
- Set A: 5th floor
- Set B: 6th floor etc.

If either set is damaged then additional costs are incurred to replace it. The formwork facing material must be chosen to ensure that the desired concrete finish is consistently achieved.

Thirty-five per cent of the total cost of any finished *in-situ* concrete element is generated by the construction of the formwork.

Formwork is expensive. On average, about **35 per cent** of the total cost of any finished concrete unit can be attributed to its formwork, of which just over 40 per cent is for materials while nearly 60 per cent is for labour costs. Reducing the complexity of the formwork will reduce erection times and therefore labour costs.

Materials used for formwork

There are a number of materials which may be used for formwork construction. These vary from the most common, timber and steel, to plastic and rubber, and in some circumstances even concrete if the formwork is permanent. The selection of the materials to be used in any particular situation will be based on the following factors:

- Strength of the materials
- Economic use of the materials
- Ease of handling, working and erection
- Ability to form the desired shape
- Facilities for adjustment, levelling, easing and striking
- Quality of finish required.

Timber

Softwood is the most commonly used material for formwork. The reasons for this are:

- *Availability*: softwood is readily available from a wide number of outlets
- *Economy*: softwood is relatively inexpensive
- *Structural properties*: softwood has good compressive and tensile strength and also low weight
- *Ease of work and handling*: softwood is easily worked using basic tools and equipment
- *Complicated formwork shapes*: can be built up as required

- *Insulation properties*: in cold weather softwood helps to retain the heat in curing concrete
- *Sustainability*: softwood that comes from renewable sources is an environmentally friendly option.

Metal

Steel and aluminium can both be used in proprietary and special purpose-made formwork. The systems are constructed so that their component parts can be simply bolted or clipped to each other and any supporting formwork. Where there is a high degree of repetitive work their use can be an economical alternative to traditional timber and plywood formwork. However, from an environmental perspective these two materials are not good, as the embodied energy required to produce them is very high.

Plastic

The use of both sheet and formed plastics for formwork can be an economical consideration where high-quality finishes coupled with repetitive use are required and where complicated shapes are needed.

The advantages of using plastic are low weight, dimensional stability, high potential reuse value, the ability to produce complex shapes and the ability to produce a high-quality defect-free finish.

The disadvantages are the very high manufacturing costs, susceptibility to impact damage and surface scratching, the possibility of deflection under load (which can occur unless the formwork is adequately stiffened with ribs), and, from a sustainability perspective, the damage to the environment caused by the production of plastics.

Glass-reinforced plastic

Glass-reinforced plastics are used in specific situations, such as forming troughs and waffle moulds in slab construction (see Chapter 11) and for making profiled form liners, mainly for wall construction, which can provide a variety of intricate moulded patterned finishes.

Expanded polystyrene

Foamed plastic is used as a disposable formwork material and also for forming voids or holes for service entries. It has the advantage of being easily removable by poking out, burning out or being dissolved by petrol, although the latter two methods are not recommended from both health and safety and environmental perspectives!

Reinforced cardboard

The system shown in Figures 7.5 and 7.6 is a very simple yet effective method that can be used for any shape of column, but is best utilised for forming circular columns where the building of timber formwork would be complex and therefore costly. The card tubes are fixed in place and the concrete poured; the card can then be removed by tearing. The finish is excellent and requires no further treatment. The columns can simply be painted.

PART 3

Figure 7.5
These photographs show round section columns being cast *in situ* with the aid of single-use cardboard formwork. The cardboard tube is lined with smooth PVC to create a finish of high quality. Figure 7.6 indicates the components of this system.

Figure 7.6
In-situ columns cast using single-use tubular formwork.

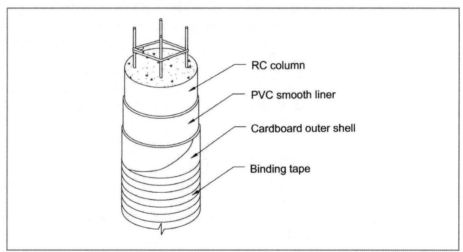

RC column

PVC smooth liner

Cardboard outer shell

Binding tape

Mould oils and **emulsions**

Emulsions are either drops of oil in water or drops of water in oil. They should not be used in conjunction with steel as they cause rusting.

Two defects which can occur on the surface of finished concrete are:

- Uneven colour, which is caused by the water in the concrete being absorbed irregularly into the formwork. Using different materials or a mixture of old and new materials can make the problem more likely to occur.
- Blow holes that are caused by air being trapped between the concrete and the formwork.

Mould oils can be applied to the inside surface of the formwork to prevent uneven colouring problems. The use of a hermetic layer inside the formwork can help to avoid this problem, but it can increase the chance of blow holes occurring.

These oils are designed to alleviate the problems by encouraging the air to slide up the formwork face. Neat oil will encourage blow holes but will discourage uneven colouring. A mould oil which incorporates an emulsifying agent will discourage blow holes and reduce uneven colouring. Care must be taken when applying mould oil, as too much oil in the concrete will cause retardation in the setting, and if the oil coats the steel reinforcement the concrete/steel bond will be of reduced quality.

Types of formwork

Column formwork

A column form or box consists of a vertical mould that has to resist considerable horizontal pressures in the early stages of casting. The column box must be fixed over a 75 mm kicker (Figure 7.7) that will have been cast as **monolithic** with the base or floor. Reinforcement bars connected to the previous slab or foundation will be left exposed so that the column reinforcement can be fixed onto it to give a more secure fixing. The exposed steel bars are sometimes called 'starter bars'. The kicker not only acts as a method of accurately positioning the column but also stops the loss of grout from the bottom of the shutter. The side panels of the column box are usually strengthened using horizontal cleats. The form can be taken to full column height and cut-outs can be left to take an adjacent beam form. However it is more common to cast the column up to the bottom of beam level, leaving reinforcing bars projecting, and then cast the top of the column with the beam. The only advantage in casting full height columns is that the beam forms act as lateral restraint; however, this can be overtaken by the complexity of formwork involved.

Metal clamps that are adjustable within the limits of the blades hold the column forms together (Figure 7.8). The whole column structure is then checked for horizontal accuracy and adjustable props are used to give lateral restraint.

Monolithic basically means 'all in one'. There will be no joints between components.

PART 3

Figure 7.7
Column formwork.

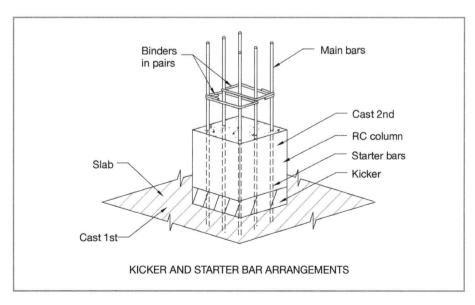

KICKER AND STARTER BAR ARRANGEMENTS

Figure 7.8
Column formwork
support.

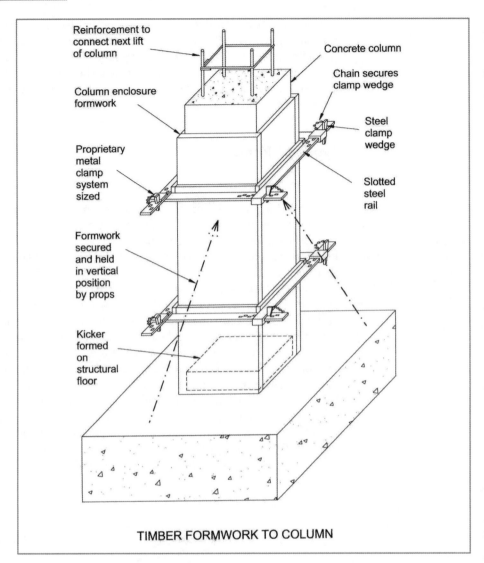

Reinforcement to connect next lift of column

Concrete column

Chain secures clamp wedge

Column enclosure formwork

Steel clamp wedge

Proprietary metal clamp system sized

Slotted steel rail

Formwork secured and held in vertical position by props

Kicker formed on structural floor

TIMBER FORMWORK TO COLUMN

If the kicker is not formed while the previous floor or foundation is being poured, the concrete will have to be 'scabbled' using a mechanical breaker and chisel in order for the column concrete to bond with the slab concrete. If the starter bars are missing, holes will have to be drilled in the correct position and then 'glued in' using epoxy resin. Both of these practices will lead to a poorer quality of connection and care should be taken to ensure that this situation does not arise.

Proprietary formwork systems can be used if the shape of the columns required is difficult to achieve using timber, such as circular columns (Figure 7.9).

Beam and slab formwork

A beam form consists of a three-sided box which is supported by cross-members that are propped to the underside of the soffit board (Figure 7.10). In the case of framed

Figure 7.9
Proprietary formwork used for circular columns.

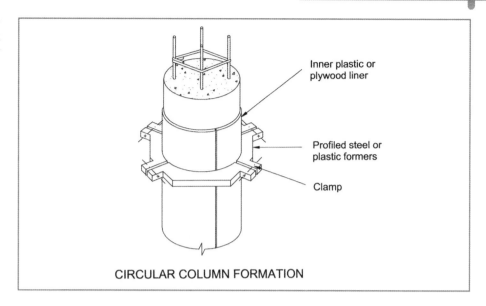

Inner plastic or plywood liner

Profiled steel or plastic formers

Clamp

CIRCULAR COLUMN FORMATION

buildings the support to the beam box is also obtained from the columns. The soffit board should be thicker than the side boards since this is the member that will carry the dead load until the beam has developed enough strength to be self-supporting. Soffit boards should be fixed inside the side boards, so that the side boards can be removed first. Removal of these boards will give a greater surface area exposed to air and increase curing times and will also release shuttering earlier for reuse. Generally beams and slabs are cast monolithically, so beam and slab formwork are erected together. The main advantage of this is that there is only one concreting operation involved. When the two are cast separately there is a possibility of shear failure between the two members. The concrete protects the steel against fire and corrosion.

Floor or slab formwork consists of panels that can be easily manhandled. Shuttering plywood comes in sheets of 1220 mm × 2440 mm, which can then be cut to size on-site. The panels can be framed or joisted and supported by the beam forms, with any inter-

PART 3

Figure 7.10
Beam and slab form-work arrangements.

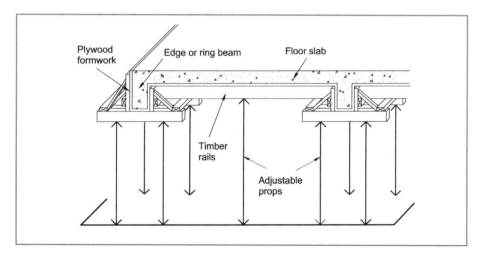

Plywood formwork

Edge or ring beam

Floor slab

Timber rails

Adjustable props

Figure 7.11
Sliding anchors.

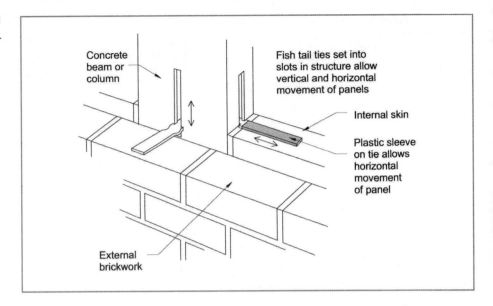

Concrete beam or column

Fish tail ties set into slots in structure allow vertical and horizontal movement of panels

Internal skin

Plastic sleeve on tie allows horizontal movement of panel

External brickwork

mediate supports where required. Adjustments for levelling are usually made by using adjustable props which can be wound up or down.

If the building is to be clad in brickwork, equal angle sections need to be attached to the beams at each level of the building for the brickwork to sit on. The holes for the angle bolts can be drilled when the concrete is cured, but problems can occur if the drill hits the steel reinforcement. There is little tolerance in the steel angles, and diamond drilling may be required, which is expensive and damages the continuity of the steel reinforcement. A better alternative is to fix the shoes for the angle fixing bolts to the beam formwork and cast them into the concrete.

An example of a suitable system is shown in Figure 7.11.

Fixings for other types of cladding can also be set into the *in-situ* concrete.

Wall formwork

Wall formwork is very similar to column formwork. A timber frame is made which sits on a kicker that has been preformed (Figure 7.12). The two sides of the wall formwork are tied together using steel coil wall ties and then fixed firmly into position using steel props. The middle section of the coil tie is left in the wall after the concrete has been poured.

Figure 7.13 shows how formwork has been constructed to take concrete to form walls. The photograph on the left shows the formwork before casting of the wall. The photo on the right shows it after casting and with the formwork partially removed.

Formwork on-site

When the formwork has been constructed and assembled the interior of the forms should be clear and free from dirt, grease and rubbish before the application of the mould oil. All joints and holes should be checked to ensure they are grout tight. Care must be taken when placing the concrete to avoid displacement of the reinforcement. The depth

PART 3

Figure 7.12
Wall formwork.

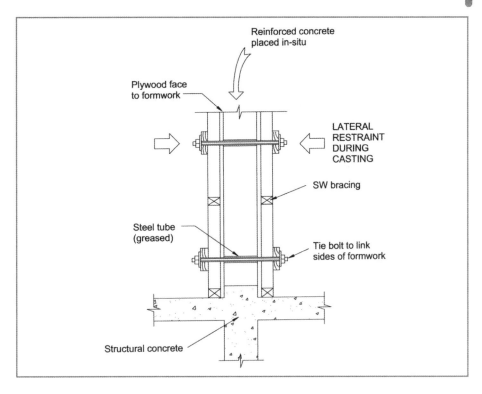

Reinforced concrete placed in-situ

Plywood face to formwork

LATERAL RESTRAINT DURING CASTING

SW bracing

Steel tube (greased)

Tie bolt to link sides of formwork

Structural concrete

of concrete that can be placed in one pour will depend on the mix and section size. If vibrators are used to compact the concrete this vibration should be continuous throughout the placement of the concrete, but care should be taken not to over-vibrate, so that segregation does not occur.

The appropriate time for striking of formwork may be determined by the results obtained from test cubes, or Table 7.1 may be used as a guide.

In very cold weather the minimum periods should be doubled; when using Rapid Hardening Portland Cement the times can be halved.

Formwork must be removed slowly, as quick removal will have the same effect as a shock load being placed on the member. Heavy loads should not be placed on the new concrete until the engineer has specified that no damage will occur.

Figure 7.13
Creating concrete walls
using formwork.

Table 7.1 Concrete curing times.

Location	Surface air temperature (°C)	
	16	7
Vertical formwork	9 hours	12 hours
Slab soffits (props left under)	4 days	7 days
Removal of props	11 days	14 days
Beam soffits (props left under)	8 days	14 days
Removal of props	15 days	21 days

The formwork should be cleaned immediately after striking and should be stacked neatly in piles until reused.

To ensure that the concrete cures without cracking or distortion occurring, it may sometimes be necessary to cover the concrete with tarpaulins to avoid rapid evaporation of the water in the mix. This is usually only necessary in times of very hot weather.

Speeding up *in-situ* concrete construction

Tunnel form construction

Tunnel form construction is a method of *in-situ* concrete frame construction that can be used in buildings that require a modular layout. Figure 7.14 shows tunnel form formwork being lifted into place by a crane.

Figure 7.14
Tunnel formwork.

Figure 7.15
A completed set of
'tunnels'.

PART 3

As the figure demonstrates, it has been delivered to site already constructed to form one-half of a box. Once this is fixed in place the formwork for the other side of the box is lifted into place and fixed. A series of these boxes are fixed in place with gaps between them where the wall is to be formed, and concrete is poured in to form the walls and slab.

Figure 7.15 shows a completed set. Although these may look like a series of boxes, what is actually formed is a series of tunnels, hence the name 'tunnel form'. These form the walls and floor to rooms in the building.

Jump form construction

Jump form *in-situ* construction is a derivative of slip form construction, but whereas with slip form construction the formwork moves continuously as the concrete is being poured, in jump form the formwork section is installed, the concrete poured and then when the concrete is cured the whole section of formwork 'jumps' up a level with the help of jacks and cranes.

This form of *in-situ* concrete construction is ideal for forming elements of the building that require a 'core', such as lift and service ducts. Figure 7.16 shows two building cores being constructed. The jump form is fixed in place and ready for concreting for the core in the foreground, and the formwork is in the process of 'jumping' for the other core.

Both tunnel form and jump form construction enable the process of constructing *in-situ* concrete multi-storey buildings to be undertaken more quickly than with traditional types of formwork. However, the downside of using these forms is that they have basically evolved from the principles of system building and this means that design flexibility is reduced.

Figure 7.16
Jump form construction.

Formwork safety

General

One of the main dangers with using **formwork** is making alterations or removing props on an *ad hoc* basis.

In design, safety is of prime importance. **Formwork** arrangements must provide for safe access, and those who have to use equipment must be fully instructed in the safe procedures to be adopted. Work that is carried out in an unplanned and haphazard way during construction spells danger. Even work that has been carefully planned can be made hazardous by *ad hoc* alterations. These days, many proprietary formwork systems have safety systems already installed, such as guard rails and toe boards.

Even the best planned and safest systems can fail because of some breakdown in communication between the designer and the user. Accidents can be classified as those caused by bad design (where an incorrect assessment of the loads to be imposed is undertaken) and those caused by unsound construction and misuse of materials (which can occur even after a lot of effort has been expended on careful preparation of a method statement). They happen where, inadvertently or through lack of training, substandard methods have been adopted or where materials have been assembled incorrectly, or where major components have been removed so that other parts of the system have become overstressed.

Accidents that are caused by poor assembly are those in which adjustable props have been badly located or erected out of plumb and/or the prop pin has been misplaced and replaced unsuitably (with, say, a piece of reinforcing steel that has been found lying around the site or one or more six inch nails!). The pin in the prop is vitally important and takes the majority of the loading. The pin is manufactured from steel that is strong enough to take the loads. If the original pin is missing, the prop should be discarded until replacement pins can be procured.

Accidents caused by factors normally considered beyond the control of the formwork designer or supervisor include props being removed to ease access to other areas of work; formwork connected with nails being left on the site and people standing on the nails and suffering foot injuries; reinforcing steel being used instead of the pin in the prop, which can extend from the prop by up to 400 mm and can hit someone in the eye; and overloading of the formwork by the loading out of materials before the concrete is completely cured.

The CDM Regulations have now placed greater responsibility on the designer with regard to health and safety on-site. As it is known that the erection and use of formwork is a high-risk activity, this may affect the choice of *in-situ* reinforced concrete as a framing solution in the future.

In-situ concrete

Concrete is a mixture of cement, coarse aggregates, fine aggregates and water. Concrete can either be mixed on-site (a batching plant would be required for large quantities) or can be purchased ready mixed in loads of up to 6 m^3. Concrete is designed to acquire its working strength at 28 days after placing and is excellent at withstanding compressive stresses.

Admixtures can be added to concrete to accelerate the curing rate, but care must be taken when using these, as if incorrectly used they can have serious adverse effects on the hardened concrete.

In situ means that the concrete is poured in the location where it is to remain. A general rule of thumb is that the quality of site-produced components is not as good as that of factory-produced components. The quality of the concrete poured on-site depends on various factors and good practices being undertaken at site level:

- Concrete must not be watered down to improve workability
- Concrete should be placed in a reasonable time – not too slow, as curing will occur at the lower levels and lead to stratification
- Concrete should not be poured from a great height, as separation will occur
- In cold or hot weather the curing concrete will need protection
- Vibration of the concrete must be thorough
- Cubes must be taken and the results carefully monitored.

All of the above quality assurance mechanisms are carried out on-site; therefore a knowledgeable management team is essential.

Reinforcement

Steel reinforcement is used in *in-situ* concrete frames to increase the tensile strength of members. Concrete is not good at withstanding tensile stresses, and when structural members are loaded one side is generally in compression and one side in tension. Steel bars are placed in the side of the member in which the tensile stresses will occur. Figure 7.17 shows an example of the simply supported beam, where the compression/tension relationship is constant along the beam length, and an example of the contiguous beam, where the compression/tension relationship changes along the length of the beam. In the example of the contiguous beam, steel bars are required in the top section of the beam, but will only be stressed at the position where the top of the beam is in tension.

The steel specified for use in structural frames must fulfil certain requirements if an economic structural member is to be constructed. The basic requirements are:

Figure 7.17
Compression and
tension in reinforced
concrete beams.

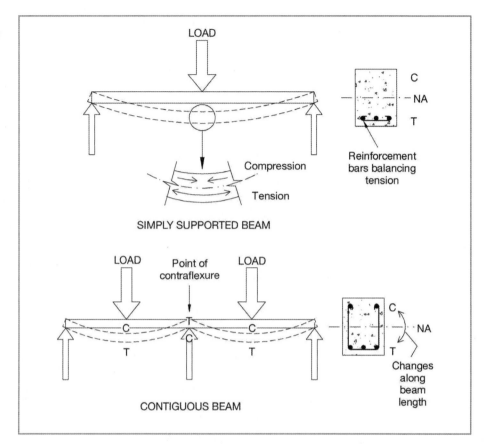

- The tensile strength must be suitable for the required situation
- The tensile strength must be achievable without undue deflection occurring
- It must be possible to bend the steel to any required shape
- The surface of the steel must be able to bond itself to the concrete
- The steel must be available at reasonable cost. Steel, like any other metal or metal alloy, is very prone to changes in cost.

Steel bars can be procured to satisfy the above requirements and are supplied in two basic types, namely mild steel and high-yield steel. Bars are supplied in a number of diameters and are usually supplied already shaped. The bending of steel bars on-site is a costly process, and is usually only undertaken if the steel has been supplied incorrectly, or more usually if the correct section has been lost and a new one needs to be made. Bending steel on-site also creates quality issues, as the accuracy that can be achieved on-site is not as good as the accuracy that can be achieved in the factory environment. Shaped or bent reinforcement can be tied on-site to form mats for slabs, or cages for columns and beams.

The most common profiles for steel bars are shown in Figure 7.18.

Steel mesh (Figure 7.19) is used in the top third of slab or beam sections to prevent cracking. It is often referred to as 'anti-cracking mesh' and is welded together in the factory and then cut to size on-site.

Figure 7.18
Steel bar profiles.

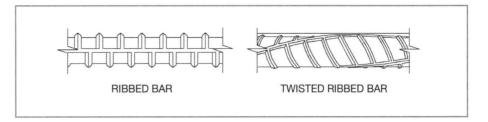

RIBBED BAR   TWISTED RIBBED BAR

Figure 7.19
Steel mesh (after BS
8110).

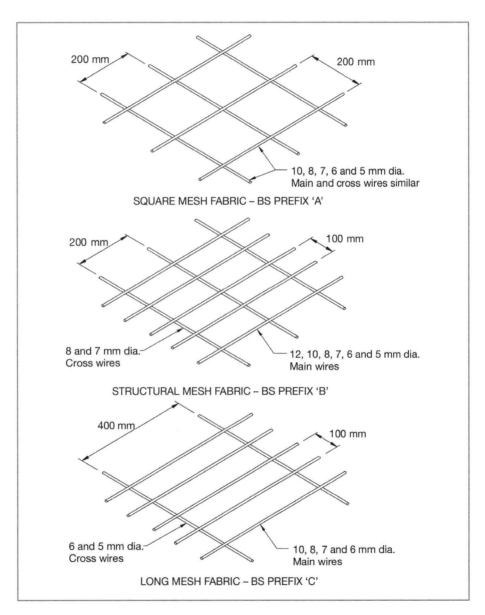

200 mm   200 mm

10, 8, 7, 6 and 5 mm dia.
Main and cross wires similar

SQUARE MESH FABRIC – BS PREFIX 'A'

200 mm   100 mm

8 and 7 mm dia.
Cross wires

12, 10, 8, 7, 6 and 5 mm dia.
Main wires

STRUCTURAL MESH FABRIC – BS PREFIX 'B'

400 mm   100 mm

6 and 5 mm dia.
Cross wires

10, 8, 7 and 6 mm dia.
Main wires

LONG MESH FABRIC – BS PREFIX 'C'

PART 3

The most common dimension between the main and cross wires is 200 mm, and this can be supplied with bars of 10, 8, 7, 6 and 5 mm diameter.

The steelworker uses drawings and bending schedules that are issued by the structural engineer in order to ensure that the steel is placed correctly. Figure 7.20 gives an example of this.

After the formwork has been erected, the surface must be thoroughly cleaned and the mould oil applied. The steel reinforcement is then fixed into position. Spacers are used to keep the reinforcement at a suitable position from the timber formwork. These spacers can be either steel or concrete. The **quality** of the formwork is then checked for line, level and stability and the wet concrete poured. After a suitable curing period the formwork is then stripped and further members prepared for casting.

The **quality** of finished *in-situ* concrete elements will depend on the care taken in preparing the formwork before the concrete is poured.

Figure 7.20
Bending schedule example.

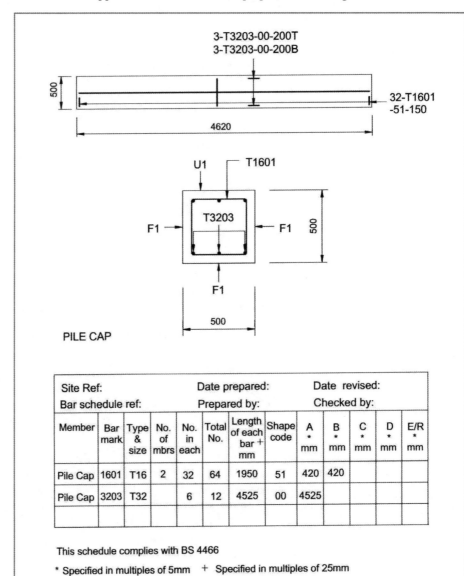

Site Ref:					Date prepared:			Date revised:				
Bar schedule ref:					Prepared by:			Checked by:				
Member	Bar mark	Type & size	No. of mbrs	No. in each	Total No.	Length of each bar + mm	Shape code	A * mm	B * mm	C * mm	D * mm	E/R * mm
Pile Cap	1601	T16	2	32	64	1950	51	420	420			
Pile Cap	3203	T32		6	12	4525	00	4525				

This schedule complies with BS 4466

* Specified in multiples of 5mm   + Specified in multiples of 25mm

Precast reinforced concrete frames

The overall concept of a precast concrete frame is the same as any other framing material. Multi-storey frames can be produced on the skeleton principle for a single building, but the most cost-effective use of precast concrete frames is as part of a 'system' building where a large number of identical components are to be used. If this is the case, the manufacturer only needs to design one set of component formwork for each structural element, which is then used and reused.

Precast units are produced in a factory to the exact sizes required. They are then delivered to site and erected, using different methods for connection.

Precast concrete is viewed unfavourably by many in the UK due to the problems with multi-storey residential buildings that occurred in the 1950s and 1960s, many of which have had to be demolished due to them being structurally unsound. The collapse of a section of the Ronan Point tower block in 1968 prompted an investigation into these forms of construction which found that the main quality issue was the corrosion of the steel joints between structural elements. This corrosion was largely due to the poor quality of the *in-situ* protection to the preformed joints, which led to the corrosion of the connecting steelwork, the spalling of the protective concrete and the further corrosion of the joints, which rendered them useless. Stricter quality control, better supervision and a more experienced workforce have alleviated the majority of these problems, but precast concrete is still not a common choice for a framing material.

Possibly this is due to some architects' perspectives that using precast elements for any part of the structure leads to a lack of flexibility of design. Precast work is used extensively in mainland Europe, even for domestic construction, and the need for better quality buildings constructed in a short duration may produce a rise in the popularity of precast concrete for framing over the coming years.

Figure 7.21 shows a precast concrete frame being erected. Notice the angles fixed to the sides of the columns. These will support further cross-members of the frame. The reinforcing bars extending from the tops of the columns will be housed into the bottom of the next columns up.

The columns will need extra support while the 'wet' concrete joint encasing the steel column connection cures.

Methods of connection for precast concrete frames

Foundations

Precast columns are connected to their foundations by one of two methods, depending on the size of the loads. For light and medium loads (Figure 7.22) the foot of the column can be placed in a pocket left in the foundation. The column can then be plumbed and positioned by fixing a collar around its perimeter and temporarily supporting the column from this collar by propping. The gap around the collar can then be grouted.

The alternative (Figure 7.23) is to cast or weld a base plate to the foot of the column and use holding down bolts to secure the column to its foundation.

Columns

The main principle involved in making column connections is to ensure continuity, which can be achieved by different methods. In simple connections a direct bearing and grouted dowel joint can be used, the dowel being positioned in either the upper or lower column.

Figure 7.21
Erecting a precast
concrete frame.

Figure 7.21
Erecting a precast
concrete frame.

Where continuity of reinforcement is required, the reinforcement from both columns is left exposed and either lapped or welded together before completing the joint with an *in-situ* concrete joint. A more complex method is to use a stud and plate connection (Figure 7.24), where a set of threaded bars are connected through a steel plate welded to a set of bars projecting from the lower column; again the connection is completed using *in-situ* concrete. The exposed steelwork then has to be encased in *in-situ* concrete to protect the steel from corrosion and fire damage. It is in this element of column connections that quality issues arise. If this concrete is not placed correctly then water will be allowed in, which will corrode the steel.

Beams

The important factor when considering beam connections is to try to achieve continuity within the joint. Two basic methods used are (Figure 7.25):

- A projecting concrete haunch is cast on the column with a locating dowel or stud bolt to fix the beam
- A projecting metal corbel is fixed to the column and the beam is bolted to the corbel.

Column and beam reinforcements, generally in the form of hooks, are left exposed. The two members are hooked together and covered with *in-situ* concrete to complete the joint.

With most beam-to-column connections, lateral restraint is provided by leaving projecting reinforcement from the beam sides to bond with the floor slab or precast concrete floor units.

Figure 7.22
Column for foundation
connection method 1.

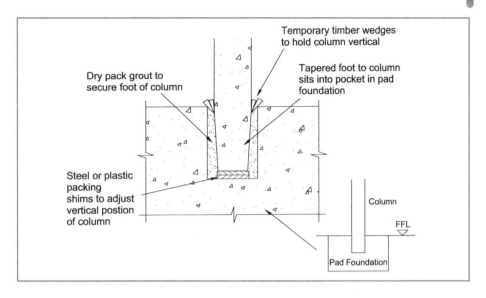

Figure 7.23
Column for foundation
connection method 2.

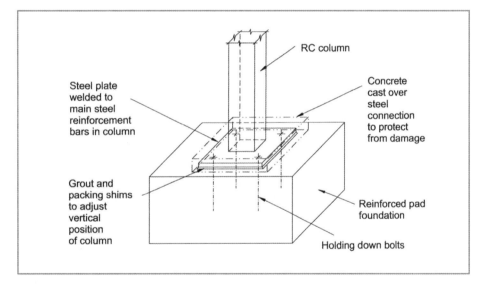

Structural steel frames

Structural steel frames are used frequently these days as a major form of construction. The design, fabrication, supply and erection of a steel frame are usually carried out by a specialist subcontractor. The main contractor's responsibility is to fix frame holding down bolts in the correct position and at the correct level before the steel subcontractor arrives on-site.

There are a number of steel **sections**, which are given in British Standard 4, Part 1. These are the most commonly used sections. They can be categorised as Universal beams, Joists, Universal columns, Channels, Angles and T-bars (Figure 7.26). The sizes are standardised, which makes them economical to produce, but will generally mean that the section is over-designed. As the assessment of section sizes required is determined by calculation of the loads to be placed on the structural member, the designer will consult standard steel tables to assess which is the most suitable size. In extremely rare cases an

The manufacture of structural steel is made economical by the use of standard **section** sizes by manufacturers. This means that standard moulds are used repeatedly, as opposed to one-off moulds being required.

Figure 7.24
Precast concrete
column-to-column
connections.

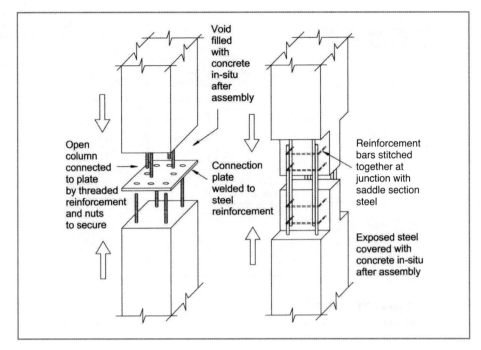

exact match will be found, so the designer will go to the next section size up. Factors of safety are also incorporated, a method of calculation which is known as 'limit state design'. Steel is an excellent material for withstanding tensile stresses, but not so good at withstanding compressive stresses. Therefore the section size tends to be over-designed to withstand those compressive stresses.

Structural steel connections

Connections in structural steelwork are classified as either 'shop connections' or 'site connections', and can be made by using bolts, rivets or by welding.

There are various different types of bolt that can be used, the most common being black bolts. These can be either hot or cold forged, with the thread being machined onto the shank. This type of bolt does not have a high shear strength, and this should be taken into consideration during design. The term 'black' does not refer to the colour of the bolt, but is used to indicate the comparatively wide tolerances to which these bolts are generally made.

Bright bolts have machined shanks and are therefore dimensionally more accurate.

High-strength friction bolts are manufactured from high-tensile steel and are therefore much stronger.

Rivets have been superseded by bolt or weld connections, but this type of connection may be required on refurbishment projects if existing steel is to be refurbished.

Weld connections are generally considered to be a shop connection, since the cost, together with the need for inspection (which can be difficult on-site), generally make this method uneconomic for site connections. If mistakes are made, however, welded joints may be required. These are generally unplanned and expensive. For example, if steel is delivered undersized then extra plates may be needed to thicken the section, and these will have to be welded. Welds can be either butt or fillet, each used in appropriate

Figure 7.25
Precast concrete beam-
to-column connections.

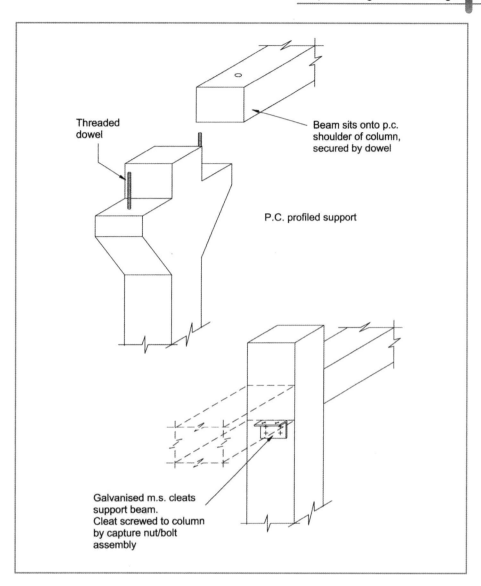

Threaded
dowel

Beam sits onto p.c.
shoulder of column,
secured by dowel

P.C. profiled support

Galvanised m.s. cleats
support beam.
Cleat screwed to column
by capture nut/bolt
assembly

PART 3

positions. Welding on-site can create health and safety hazards. If people stare at the
welding flame for too long they can be afflicted by arc eye, which is a very unpleasant
condition that makes sufferers feel like they have sand grains under the eyelid. This
symptom can last for several days, and thus it is essential that if welding is being undertaken the area of work must be shielded from site operatives. Also welders must use
adequate protection.

Frame erection

Before the foundation or pile cap is concreted, the holding down bolts for the frame must
be fixed to enable them to be attached solidly to the base slab. The tolerance for these

Figure 7.26
Standard steel sections.

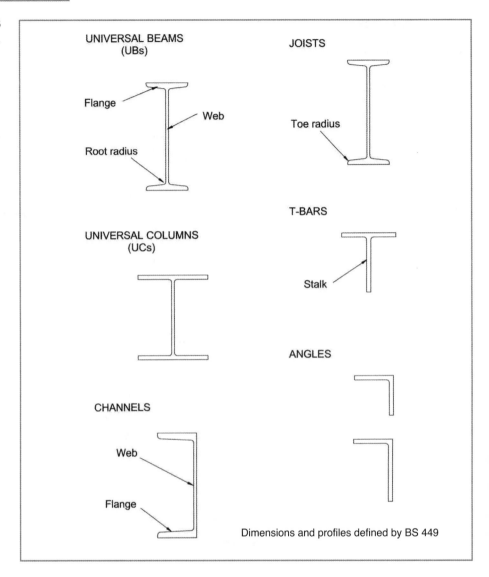

bolts is ±3 mm, and therefore the accuracy of the setting out is of paramount importance. Either the steel manufacturer supplies, or the site joiners make, plywood 'jigs' the same size as the steel base plate. The jigs will have the gridline positions marked. The bolts are then fitted to the jig with the correct dimension from the gridlines in both directions. These jigs are then fixed in position, either to steel reinforcement or formwork, and must be checked by the site engineer for both line and level before concreting commences. Bolt boxes are used to increase the tolerance of the bolts. During curing of the concrete the site engineer or other operatives will need to rotate the bolts to ensure that they do not fix too solidly. After the concrete is cured the timber jigs are removed and the column bases can be bolted to the bolts.

If the bolts are incorrectly positioned there are two alternatives:

■ Blow out the bolts using high-pressure water jets and replace them, which is very costly
■ Drill holes into the slab in the correct position and fix them with **epoxy resin**, which is not as costly (Figure 7.27). Some structural engineers will not allow this, as the joint between the bolt and the concrete is not as strong.

Epoxy resins are thermosetting synthetic resins containing epoxy groups. They work by the mixing of two different parts that produce a reaction that creates a strong bond.

If the bolts are set in at the incorrect level the alternatives are:

■ If they are too high and the thread of the bolt is too high for the nut, the whole level of the frame may be increased if it is agreed with the engineer. Alternatively, the concrete may be broken out to reduce the level of the pad foundation and the bolts drilled into the concrete. This is shown in Figure 7.28.
■ If they are too low, bolt extenders may be used, but these can be costly. Figure 7.29 gives details of bolt extenders.

Figure 7.27
Drilled bolt details.

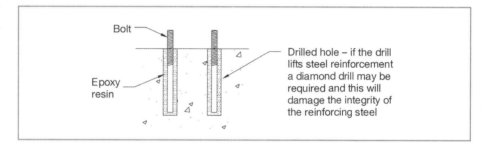

Figure 7.28
Reducing the level of a pad foundation by breaking out concrete.

PART 3

Figure 7.29
Bolt extenders.

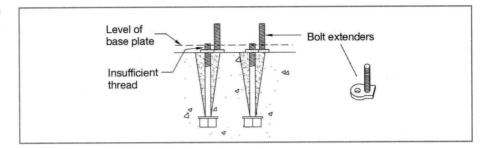

Figure 7.30
Bolt boxes.

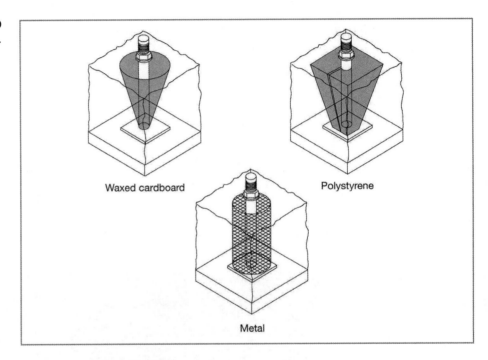

The frame erection commences when all the bases have been cast and the holding down bolts checked for accuracy. Access into and on the site must be clear to enable the subcontractor to work efficiently.

The usual site procedure is to erect two storeys of the frame and then check the frame for level and alignment. When the two storeys of the frame are deemed to be in the correct position, the holding down bolts are then grouted in. The grout is usually a mix of neat cement and water to a specified mix or a purposely manufactured material such as 5* grout.

The remainder of the steel frame is then erected and checked at each level for accuracy.

To protect the steel frame from corrosion at ground level the columns should be painted with a bitumen-based primer before being cased in concrete, with at least 75 mm cover.

Figure 7.30 shows the alternative types of bolt boxes available.

Figure 7.31 shows an example of how the bolts are fixed into position before concreting.

Figure 7.32 illustrates how the column is fixed firmly once the correct position is established using shims and grouting.

Figure 7.31
Steel base plate fixed
to concrete base.

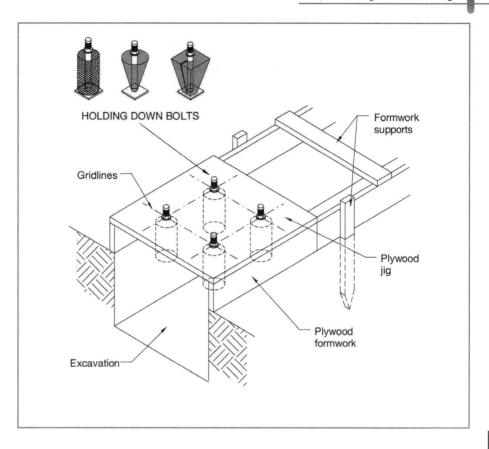

Figure 7.32
The base plate grouted
firmly in position.

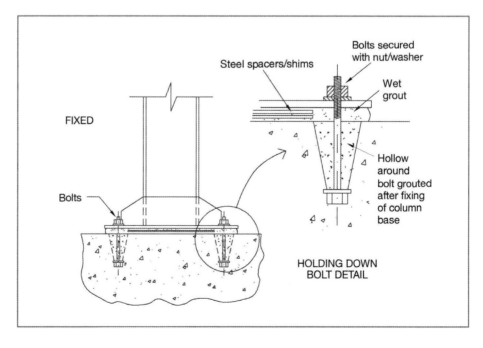

Figure 7.33
Typical holding down
bolts fixed to plate.

Figure 7.34
Column-to-base plate
connection details.

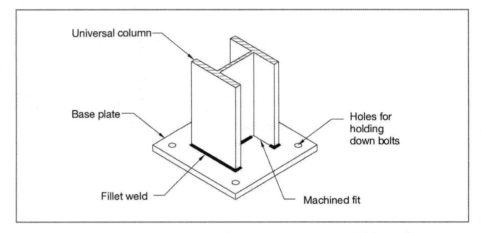

Figure 7.33 shows the bolts fixed to the timber plate ready for placing.

Figures 7.34–7.36 show typical steel frame connection details for ground floor columns to base plates, columns to columns, and beams to columns.

Another benefit of using steel frames is that they can be manufactured with provision to fix the scaffolding directly to the frame. This reduces the need for a full scaffold and the frame has holes in the beams to take bolts which attach guard rails to the frame directly. This results in a cost-saving for the scaffolding element. This could be done for precast and *in-situ* concrete frames but would be much more difficult, especially with *in-situ* concrete where the accuracy that would be needed in casting in bolts would be difficult to achieve on-site. Figure 7.37 shows the guard rails attached directly to the frame, and the holes in the steel frame to fix the bolts through.

Structural timber frames

Timber frames are used extensively for domestic construction, and are increasing in popularity for the construction of multi-storey buildings, specifically those that have a modular design such as student halls of residence or multi-storey apartments.

Figure 7.35
Column-to-column
connection details.

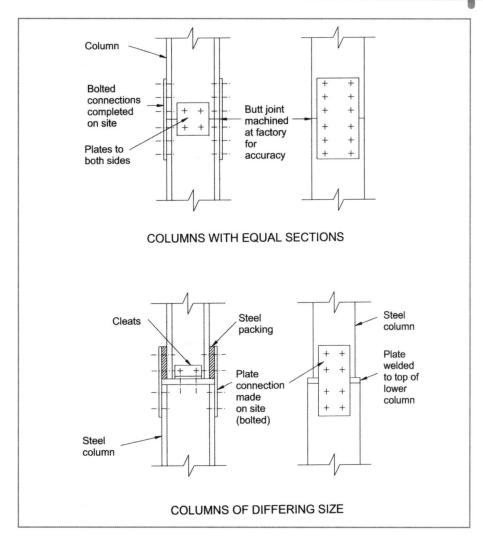

Column

Bolted
connections
completed
on site

Plates to
both sides

Butt joint
machined
at factory
for
accuracy

COLUMNS WITH EQUAL SECTIONS

Cleats

Steel
packing

Plate
connection
made
on site
(bolted)

Steel
column

Steel
column

Plate
welded
to top of
lower
column

COLUMNS OF DIFFERING SIZE

PART 3

The **BRE** recommends
that timber-framed
buildings can be used
for multi-storey
structures up to six
storeys in height.

The **BRE** tested a six-storey construction in which a building was constructed, which it then tried to blow up and burn down! However, there is no recommendation as yet to build higher than six storeys due to a recent issue that arose regarding fire safety, specifically during the construction process.

The method of construction is usually the same as for a single-storey building, using a panel or post form of construction. The most common method used in the UK is *panelised* construction (Figure 7.38). Panels are prefabricated with light timber sections stiffened with a wood-based board material. The erection operation on-site is very fast and the accuracy and reliability of the components are outstanding. Modern timber-framed buildings have the reputation of being very well engineered.

Another method of construction that is increasing in popularity is *volumetric* construction (Figure 7.39). Volumetric units are not often used for dwellings in the UK, but are widely used for temporary buildings and more recently for hotels and other buildings with repetitive space, such as apartment blocks, where each room can be constructed as a single

Figure 7.36
Beam-to-column
connection details.

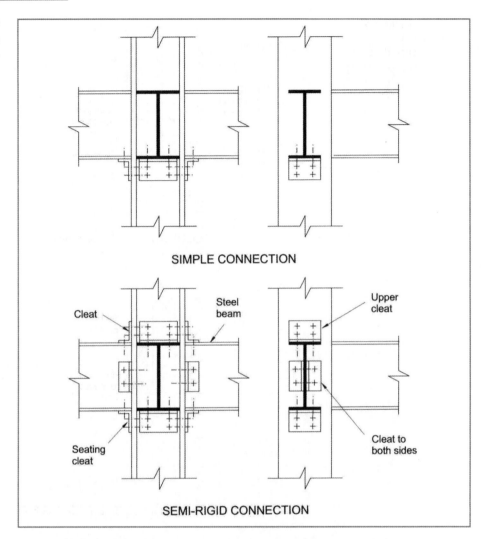

SIMPLE CONNECTION

Cleat

Steel
beam

Upper
cleat

Seating
cleat

Cleat to
both sides

SEMI-RIGID CONNECTION

Figure 7.37
System for attaching
guard rails directly to
the structural frame.

Figure 7.38
Panelised construction.

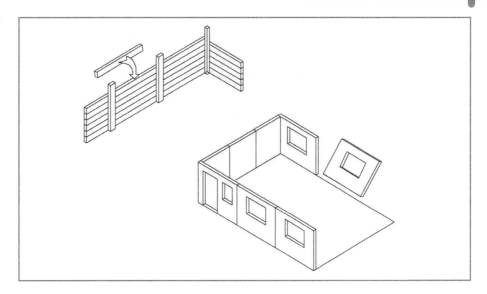

unit. The advantages are that factory-produced units can be highly finished, with both internal and external finishes in place before being transported to site. This type of modular construction is not new, but the use of timber as the main framing material is very recent.

Another increasingly popular trend is the construction of multi-storey buildings where the ground floor and maybe some of the higher floors are constructed using steel or reinforced concrete floors and then the upper floor frame is constructed using timber. This type of approach works well for buildings such as hotels and student accommodation. For example, the building shown in Figure 7.40 is student accommodation where the ground floor frame is being constructed using a steel frame which allows for larger spans. The ground floor will be used for retail units and a large reception area so larger spans are required. The upper floors are being constructed using timber frame because these are student accommodation flats which are uniformly

Figure 7.39
Volumetric
construction.

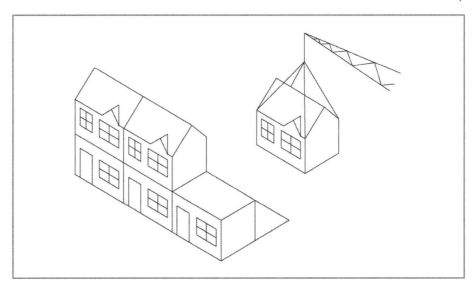

PART 3

Figure 7.40
Timber frame connected to steel frame.

Figure 7.41
General arrangement of timber-framed building.

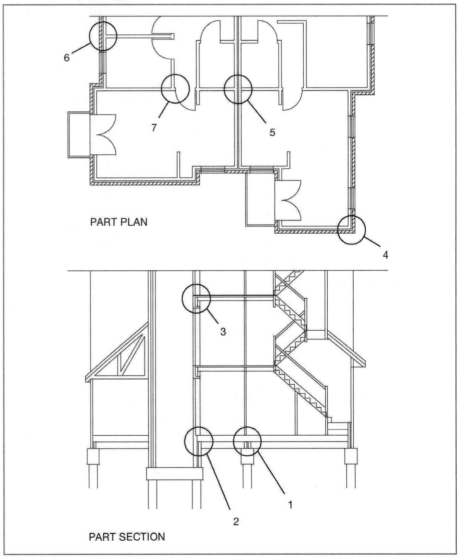

PART PLAN

PART SECTION

Figure 7.42
Internal loadbearing panels.

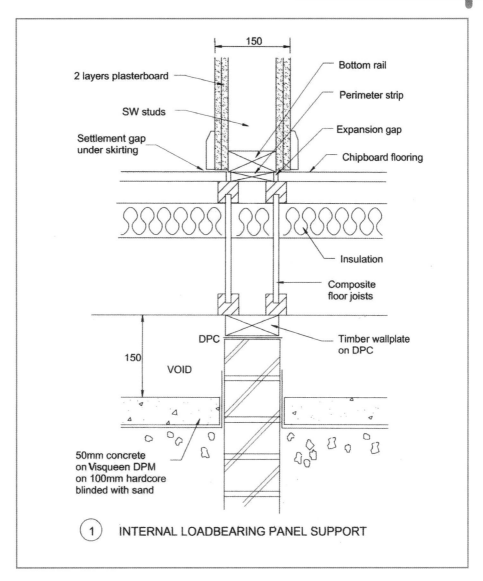

2 layers plasterboard

SW studs

Settlement gap
under skirting

Bottom rail

Perimeter strip

Expansion gap

Chipboard flooring

150

Insulation

Composite
floor joists

DPC

150

VOID

Timber wallplate
on DPC

50mm concrete
on Visqueen DPM
on 100mm hardcore
blinded with sand

(1) INTERNAL LOADBEARING PANEL SUPPORT

volumetric units. This approach also works well for hotel construction where the lower floors are usually used for the reception, restaurants, conference facilities etc. and the upper floors are generally all uniform-size hotel rooms. Figure 7.40 shows the timber frame volumetric construction taken from the future lift shaft which is constructed with a steel frame.

The method of construction for a multi-storey timber frame is as follows:

■ Concrete slab: the level of the slab must be incredibly accurate, as any 'packing up' will reduce stability
■ Fix sole plate to slab: this may be timber or a metal channel with timber infill
■ Nail/bolt in wall panels

- Drop in prefabricated floor if applicable
- Fix sole plate to 1st floor internally and externally, and so on up to six storeys.

The main differences between multi- and single-storey frames are:

- Wall panels may be bigger: 3, 4 or 5 m long
- Bigger section sizes will be needed at the lower levels to take the increased loads
- 6 mm settlement is expected for *each* storey; therefore some flexibility of joints is required
- Posts will need to be used in addition to beams to create vertical stability
- Bolt connections as opposed to nailing may be used to increase stability

Figure 7.43
Upper floor edge detail.

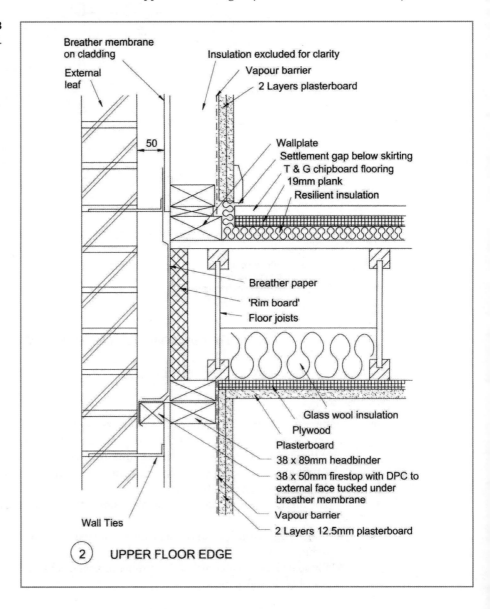

2 UPPER FLOOR EDGE

Figure 7.44
Suspended ground
floor detail.

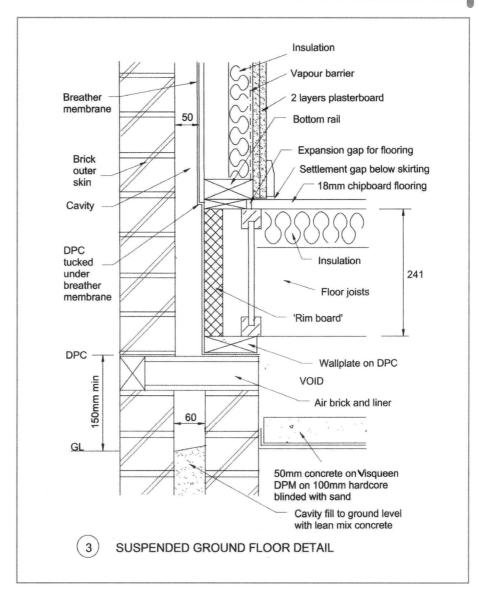

Breather membrane

50

Brick outer skin

Cavity

DPC tucked under breather membrane

DPC

150mm min

60

GL

Insulation

Vapour barrier

2 layers plasterboard

Bottom rail

Expansion gap for flooring

Settlement gap below skirting

18mm chipboard flooring

241

Insulation

Floor joists

'Rim board'

Wallplate on DPC

VOID

Air brick and liner

50mm concrete on Visqueen
DPM on 100mm hardcore
blinded with sand

Cavity fill to ground level
with lean mix concrete

(3) SUSPENDED GROUND FLOOR DETAIL

PART 3

■ Above four storeys a ring beam will be required and this may be made of glulam or steel.

This ensures that if, for instance, a gas boiler explodes, it will only blow out a small section of the building, and total collapse will not occur. Figures 7.41–7.48 show typical connection details for multi-storey timber frames.

One of the main differences between constructing a multi-storey timber frame building and reinforced concrete/steel-framed buildings is that the scaffold is erected before the frame. When constructing reinforced concrete/steel frames, the frame is erected first and then the scaffold is erected and tied back into the frame to ensure it is

Figure 7.45
External corner
junction.

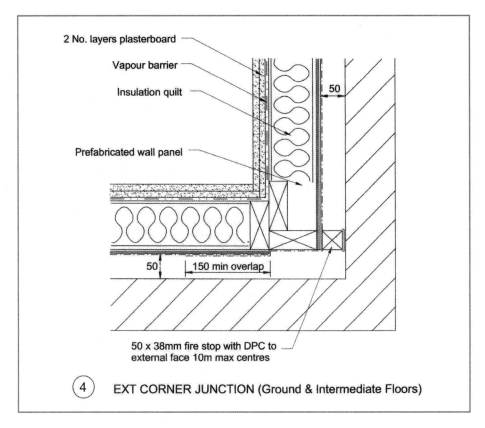

2 No. layers plasterboard

Vapour barrier

Insulation quilt

Prefabricated wall panel

50

50 | 150 min overlap

50 x 38mm fire stop with DPC to
external face 10m max centres

(4) EXT CORNER JUNCTION (Ground & Intermediate Floors)

stable. Timber frames are not strong enough to withstand the weight of the scaffold and any subsequent loads placed on it until the frame is complete. Therefore the scaffold is erected first as free standing (i.e. it does not need to be tied to the frame for stability) and then the frame is constructed inside the scaffold. Figure 7.49 shows the scaffold being erected before the frame erection commences.

As timber is deemed to be a much more sustainable material than concrete and steel, this type of frame construction is increasing in popularity and is likely to be considered as a viable option more often in the future.

Figure 7.46
Party wall detail.

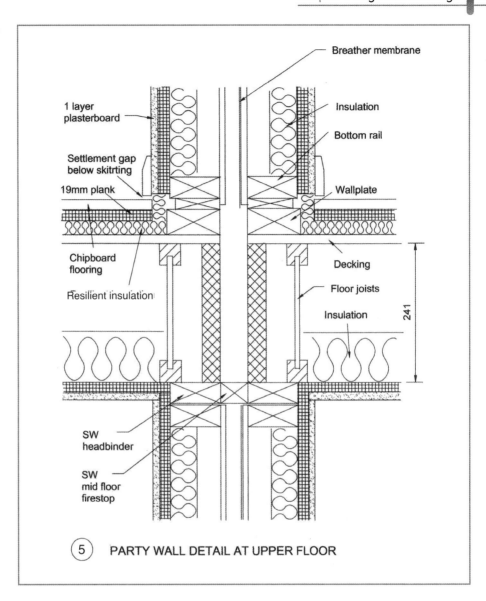

Breather membrane

1 layer
plasterboard

Insulation

Bottom rail

Settlement gap
below skitrting

19mm plank

Wallplate

Chipboard
flooring

Decking

Resilient insulation

Floor joists

Insulation

241

SW
headbinder

SW
mid floor
firestop

(5) PARTY WALL DETAIL AT UPPER FLOOR

PART 3

Figure 7.47
Wall connections.

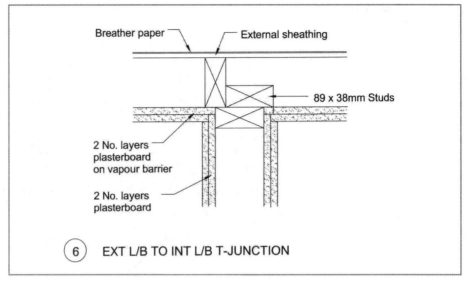

Breather paper

External sheathing

89 x 38mm Studs

2 No. layers
plasterboard
on vapour barrier

2 No. layers
plasterboard

6 EXT L/B TO INT L/B T-JUNCTION

Figure 7.48
Corner details.

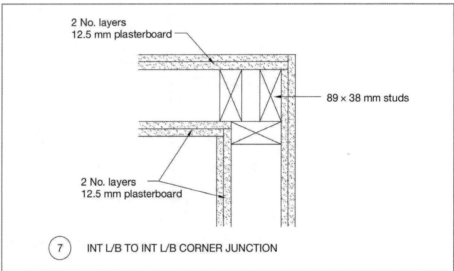

2 No. layers
12.5 mm plasterboard

89 × 38 mm studs

2 No. layers
12.5 mm plasterboard

7 INT L/B TO INT L/B CORNER JUNCTION

Figure 7.49
Scaffold being erected
before commencement
of the timber frame
construction.

Precast concrete panelised construction

A method of construction for multi-storey commercial buildings that is increasing in popularity in the UK is precast concrete panelised construction. The principles of panelised construction have been discussed earlier in relation to timber frame construction and the benefits of precast concrete in this type of construction are very similar, but much taller buildings can be constructed than when using timber.

As with all precast construction, the quality of the manufacture of the panels is as high as when they are factory produced. However, quality of the finished building may be compromised if the on-site work is not undertaken correctly. The principles and benefits of panelised precast concrete construction are the same as those for precast concrete frames, but an additional benefit is the reduction of the need for columns which increases the nett lettable area of a commercial building. Larger uninterrupted floor layouts can also be achieved which enhances the flexibility of the space for multiple concurrent users. Figure 7.50 shows the external walls and the circular internal core walls being constructed using precast concrete panels. The panels have been pre-insulated: the insulation has been fixed in place in the factory before concreting. Undertaking this process in factory conditions should ensure that the insulation is undamaged.

Figure 7.50
External walls and circular internal core walls constructed using precast concrete panels.

Precast columns have been used to support the internal floors, but the number required is significantly less than in a traditional framed building. The floor construction is being undertaken using the waffle floor *in-situ* concrete method, therefore illustrating the growing use of mixed method, multi-storey building construction. This approach enhances the benefits of different methods and reduces the shortfalls of others. Speed of construction and improvements in quality may also be seen.

7.3 | Comparative study

Introduction

- This section offers a comparison of *in-situ* concrete, precast concrete, structural steel and timber frames for multi-storey buildings.

Site costs

Building owners want to obtain a financial return on their capital investments as soon as possible, and therefore speed of construction is of the utmost importance. The use of steel or precast concrete frames will enable the maximum amount of fabrication off-site, during which time the general contractor can be constructing the foundations in preparation for erection of the frame, thereby increasing speed of construction. Also, structural steel members tend to be of a smaller size than their *in-situ* concrete counterparts and of equal strength, thus yielding a higher level of usable floor area upon completion and requiring smaller foundations. However, one issue that may determine the choice of frame may be the client's cash flow situation. Using a slower form of construction will reduce the cash requirements from the client in the early months of construction. If the client does not have all the money required for the building before construction work starts and is relying on income generated from business to pay for the works month by month, he or she may prefer to use *in-situ* concrete as opposed to steel or precast concrete. The cash flow diagram of Figure 7.51 shows the difference in cash flow at the beginning of the contract for *in-situ* concrete and for structural steelwork.

Figure 7.51
Comparison of cash flow requirements for *in-situ* concrete and steel frame buildings.

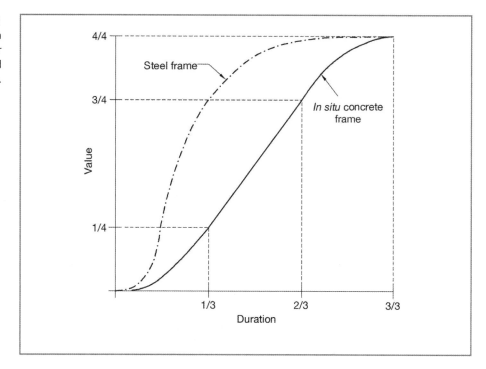

Figure 7.51
Comparison of cash flow requirements for *in-situ* concrete and steel frame buildings.

Construction costs

The main factors are design considerations, availability of labour and site conditions. Concrete allows for more flexibility in design, but for more complicated shapes and sizes formwork costs can rise dramatically. For erection of a steel frame semi-skilled labour is required with a competent foreman, but for precast or *in-situ* concrete frames the labour needs to be highly skilled. Availability of labour in the area of construction may then be a determining factor in the choice of frame to be used.

The availability of materials depends on current market trends and can therefore only be assessed at the time of design, but the cost of concrete does not tend to fluctuate anywhere near as much as that of steel.

Maintenance costs

The maintenance costs, either short or long term, are negligible for steel, *in-situ* concrete frames or timber frames if the design and workmanship are sound. However, for precast concrete some form of regular inspection of joints may be required to ensure that there has been no corrosion.

Quality

As a general statement, anything that is manufactured off-site will tend to be of a higher quality than anything that is manufactured on-site. This is due to the controlled environment within manufacturing plants as opposed to the very diverse and changeable weather conditions on-site.

The quality of steel production is highly regulated and very few problems with steel frames occur, although they are not unknown. However, if the design is incorrect this may lead to the burning of new bolt holes, which will weaken the steel, and to welding of joints/plates to get the frame to 'work'. Welding on-site can lead to quality issues and all welds need to be tested *in situ* before the structural engineer will pass the work.

The quality of precast concrete produced in factories is again generally excellent. The main quality issue with precast concrete frames is the quality of the *in-situ* concrete element of the joints, which if not constructed correctly will form weak points that will allow moisture in. The steel reinforcement will then corrode and the building will ultimately fail.

Timber frame quality is also excellent, as the majority of components are manufactured off-site. However, the quality of the fixings on-site dictates the overall quality of the building.

Safety

There are many causes of accidents on-site, but falling objects are the biggest cause of fatalities in the construction industry. The benefit of pre-made structural frames (especially steel and timber) is the speed of construction. Because steel and timber are so quick to erect, it is feasible to 'clear' the site when the frame is being erected, thus reducing the risk to other workers from falling objects. However, the erection of these types of frame is inherently dangerous to the operatives carrying out the work and it must be ensured that they use safety equipment and work to detailed method statements.

The main dangers involved with *in-situ* concrete frames are those related to the use of formwork. However, most formwork systems these days have in-built guard rails and platforms to improve safety.

Environmental issues

The production of concrete (extraction) is very damaging to the environment, as is that of steel (fuel requirements). Timber used from sustainable sources is a reasonable option and this would lead us to believe that the use of timber as a structural framing material as much as is possible is the best alternative. However, this is not feasible for buildings over six floors at this time. An Environmental Impact Assessment should be carried out by the designer when making the ultimate decision on type of frame, alongside health and safety provision.

Table 7.2 is a comparison matrix for the four types of frame material. Where figures are used, 1 indicates the best performance for the particular criterion.

Table 7.2 Comparison matrix.

Criteria	*In-situ* concrete	Precast concrete	Steel	Timber
Possible number of floors	Unlimited	Unlimited	Unlimited	6 max.
Low initial cash flow	1	2	3	2
Low weight – reduced foundation size	4	3	2	1
Fire protection	Part of structure	Part of structure	Essential	Required
Corrosion	Unlikely	Possible at joints	Possible where not concreted	N/A
Maintenance	Little	Moderate	Little	Moderate
Handling	Difficult	Difficult	Easy	Easy
Quality control	Difficult to achieve	Difficult at joints	Easy if fabricated correctly	Easy if fabricated correctly
Change of use	Difficult	Difficult	Easy	Easy
Shrinkage/settlement	Moderate	Not on-site	None	6 mm per storey
Availability	No problems	Possible problems	Possible problems	No real problems
Design freedom after work starts	Possible	Very little	None	None
Speed of works	3	2	1	1
Formwork requirements	High	Low	Medium	N/A
Health and safety risk if managed correctly	High	Medium	Low	Low
Environmental impact	High	High	Very high	Low
Propping requirements	A lot	Little	Little	N/A
Insulating properties	Good	Good	Can have thermal bridging effect	Good
Grid sizes	Medium	Medium	High	Medium
Element size	Big	Big	Small	Small

(continued overleaf)

PART 3

Table 7.2 Comparison matrix (*continued*).

Criteria	*In-situ* concrete	Precast concrete	Steel	Timber
Working area	Medium	Medium	Low	Medium
Plant requirement	Crane/concrete pump	Very large crane	Medium size crane/concrete pump	Medium size crane
Skill of labour	Highly skilled	Semi-skilled	Semi-skilled	Semi-skilled
Sound insulation	Good	Good	Possibly some structure-borne sound	Medium
Effects of bad weather during construction	Stop work	Reasonable progress	Reasonable progress	May carry on if rain not too heavy
Programme of works	Slow	Medium	Quick	Quick
Lead-in time	Medium	High	High	Short
Foundation size	Large	Large	Medium	Small

COMPARATIVE STUDY: HIGH-RISE FRAMES

Option	Advantages	Disadvantages	When to use
Concrete *in situ*	Flexibility of design Allows free-form building layout Cash flow advantages due to low initial costs Performance in fire	Highly skilled labour required Slower construction Potential for quality issues on-site	Any framed building, but particularly suited to one-off projects and buildings of non-standard shape
Concrete precast	Quality assurance Fire performance Efficient use of reinforced concrete	Weight of sections to be lifted Really needs mass production for cost effectiveness Lack of favour by designers in UK	Anywhere, but particularly suited to instances where there is much repetition
Steel	Familiar technology Easy to produce Relatively light sections to be lifted High strength-to-weight ratio Ease of formation of non-standard components	Performance in fire if unprotected Production of steel is environmentally damaging	Almost ubiquitous in the UK
Timber	Sustainable material Lightweight Cheap Reduced lead-in time	Limited to six storeys Design scepticism	Repetitive modular buildings such as hotels and apartments are particularly suitable

PART 3

CASE STUDY: STEEL FRAMES

This Case study is focused on the erection of a seven-storey steel frame that is to be used for student accommodation.

The photograph shows the holding down bolts cast into the concrete. The base plates are then connected to the bolts.

Once the steel is checked for accuracy both vertically and horizontally the base plates are grouted. This photograph shows the grouted plate, and also lightning conductors attached to the steel frame. These will be earthed via a connection through a concrete base.

This photograph shows how the size of steel section can be reduced as the building increases in height, using steel packing plates.

The columns and beams are then bolted together. The photographs show column-to-beam, column-to-column and beam-to-bracing connections.

This photograph shows an overview of the partially completed frame. It is being erected using mobile cranes.

Long-span frames

AIMS

After studying this chapter you should be able to:

- Appreciate the functional performance characteristics of frames for multi-storey buildings
- Relate functional requirements to the design alternatives for frames suitable for the construction of high-rise buildings
- Appreciate the details required to ensure functional stability
- Appreciate which type of frame is most suitable for use in given scenarios

This chapter contains the following sections:

8.1   Frames: functions and selection criteria
8.2   Frames and structures: forms and materials

INFO POINT

- Building Regulations Approved Document A: Stuctural strength (2004 including 2010 amendments)
- BS 449: Structural steel (2002)
- BS 1305: Concrete batch mixes (1996)
- BS 5628: Prestressed structural masonry (2005)
- BS 5642: Precast concrete sills and copings (1983)
- BS 5950: Steelwork in building (2001)
- BS 8110: Structural use of concrete (1997)
- BS EN 934: Concrete admixtures (2009)
- DD ENV 1992: Precast concrete – design elements and structures
- DD ENV 1993: Steel structure design
- *BRE guidance on timber frame design for large span buildings* (2004)
- *Corus Guide to the construction of space frames* (2003)

8.1 | Frames: functions and selection criteria

Introduction

- After studying this section you should have gained an understanding of the reasons why large-span buildings are used and what they are most suitable for as a form of construction.
- You should be able to detail the functions of frames and have an understanding of the different types of frames and the impact of loadings on given frame types.

Overview

The **use of structural frames** as the main supporting structure for long-span building allows for greater uninterrupted internal layouts.

The **use of structural frames** for large, single-storey buildings such as warehouses and commercial and industrial units allows for the creation of buildings with large, uninterrupted floor areas, with the possibility of providing great floor-to-ceiling heights. The most common form of single-storey building is that which follows the form of a large shed and is referred to as a 'long-span' building. Such buildings are often erected from a standard selection of parts, being then fitted out and adapted to suit specific purposes. Although several structural forms exist, the steel portal frame is by far the most popular for long-span buildings.

Functions of frames

The purpose of any framed building is to transfer the loads of the structure plus any imposed loads through the members of the frame to a suitable foundation.

Framed structures can be clad externally, with lightweight non-loadbearing walls to provide the necessary protection from the elements and to give the required degree of comfort in terms of sound and thermal insulation. Framed buildings are very suitable for long-span single-storey units.

The main types of frame used for large-span single-storey buildings are:

- *Plane frames*, which are fabricated in a flat plane and are usually called 'trusses' or 'girders', according to their elevation shape. They are designed as a series of triangles, which gives a lightweight structural member, and uses the minimal amount of materials. The main uses are for roof construction and long-span beams of light loading.
- *Space frames*, which are similar to plane frames, but span in two directions as opposed to single spanning. They are used in a variation of the plane frame as a series of linked pyramid frames for forming lightweight roof structures.
- *Structural frames*, which are used for large single-storey buildings that can be used for industrial or commercial purposes. They are used when large uninterrupted floor spaces are required and can also provide large floor-to-ceiling heights. Several forms of structure can be used.

8.2 | Frames and structures: forms and materials

Introduction

■ After studying this section you should have developed a knowledge of the different materials and systems that are used to construct long-span single-storey frames and the methods of construction.
■ In addition, you should have gained an increased awareness of the aspects of the particular type of system that will need to be considered in given situations, and understand the health, safety and environmental implications of each system.
■ You will have gained an understanding of the method of construction of this type of frame and additionally, given specific scenarios, you should be able to set out a series of criteria for the selection of an appropriate multi-storey frame.

PART 3

Overview

The use of **timber** for frames to this type of building will probably increase due to the increasing awareness of the need to build 'green'.

The materials used for the construction of large-span single-storey buildings are usually precast reinforced concrete, structural steel and **timber**. The choice of material will depend on many factors, including the cost and availability of materials at the proposed time of construction, buildability factors, availability of skilled labour, cash flow situation of the client, design flexibility requirement during the construction phase, number of floors and the proposed use of the building, together with maintenance requirements. In addition to these, many clients are insisting that buildings are constructed using sustainable construction and design principles, and are requiring designers and contractors to ensure that all alternatives have been considered in order to produce as 'green' a building as possible. Finally, and most importantly, health and safety need to be foremost in the mind of designers when choosing a particular frame system. If a system is specified that is known to be dangerous and an accident occurs, the designer could be deemed liable if an alternative, safer system could have produced the same end product.

Castellated and lattice beam framed construction

Castellated and lattice beams can be used for flat and pitch roofed framed buildings. The use of lattice beams allows for the greatest spans and also for the weight of beam sections to be reduced.

A lattice beam is a structure built up of three or more members that are normally *pinned* or *hinged* at the joints. Loads that are applied are usually transmitted at the joints and the individual frame members are in pure tension and compression.

Figure 8.1 indicates which of the members of the frame are in tension and which are in compression.

Most frames are simple frames that are built up as a series of triangles. A perfect frame has just enough members to prevent the frame from being unstable; an imperfect frame contains too few members to be stable; and a redundant frame contains more members than are required to make it a perfect frame.

The redundant members may be stressed, and if this is the case it is statically indeterminate: normal methods of resolution will not be applicable. They may not be stressed, and in this case normal methods of resolution apply. Redundant members may be incorporated for aesthetic purposes or for fixing of specific elements of the internal or external cladding, such as signs.

Figure 8.2 shows lattice beams being used to form the frame for a long-span building.

It is important to understand the ways in which structural elements will deflect under certain loading conditions and whether the members are in tension or compression (Figure 8.3). Steel is an excellent material in tension but not so good in compression, whereas concrete is very good in compression but not in tension. Hence reinforced concrete is used to take advantage of the properties of both materials.

However, the use of **reinforced concrete** is not popular for this type of construction. *In-situ* concrete is impractical, as only solid sections can be formed, and precast concrete is very heavy and will require large foundations; it is also difficult to construct and requires large cranes to allow for the lifting of sections.

For flat roof construction using lattice and castellated beams, Figures 8.4 to 8.6 illustrate the different spans that can be achieved.

Figures 8.7 and 8.8 illustrate how lattice beams can be used for pitched roof framed buildings.

Reinforced concrete is not used very often for long-span buildings, even though it combines the benefits of the best properties of steel and concrete. However, there are many examples of older buildings that use this form of construction.

Figure 8.1
Tension and compression zones.

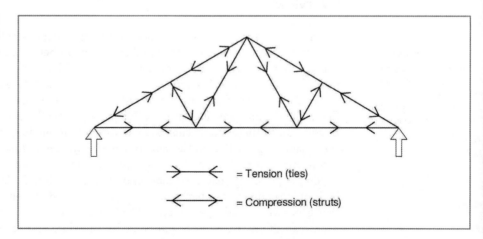

= Tension (ties)

= Compression (struts)

Figure 8.2
Lattice beams in a
long-span building.

Figure 8.3
The benefits of using
reinforced concrete.

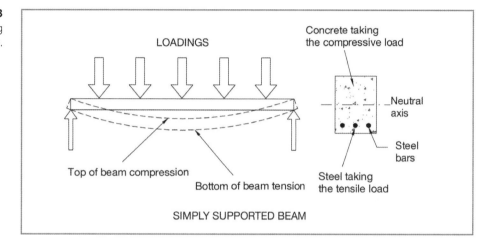

Figure 8.4
Plain beam.

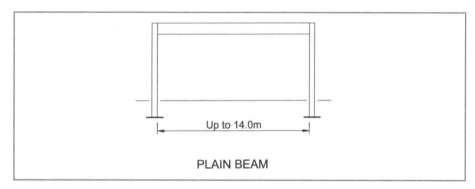

Figure 8.5
Castellated beam.

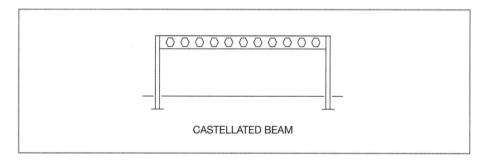

Figure 8.6
Lattice girder.

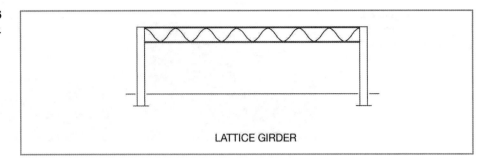

LATTICE GIRDER

Figure 8.7
Symmetrical lattice
trusses on steel
columns.

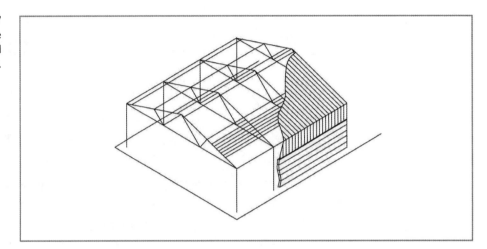

Figure 8.8
Symmetrical lattice
beams on steel
columns.

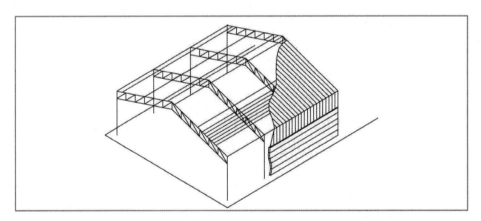

Bracing is essential in
long-span buildings to
ensure stability.

Bracing is used in the plane of the main ties to transfer wind loadings to the end walls, as shown in Figure 8.9, or in the plane of the main ties to transfer the wind load to the side walls, as shown in Figure 8.10.

As has been stated previously, reinforced concrete is not popular for this type of construction. Steel is the most common framing material, and although larger sections may be required in the members that are taking compressive stresses, it is still cost-effective and easy to erect on-site.

Figure 8.9
Bracing in the plane of the main ties to the end walls.

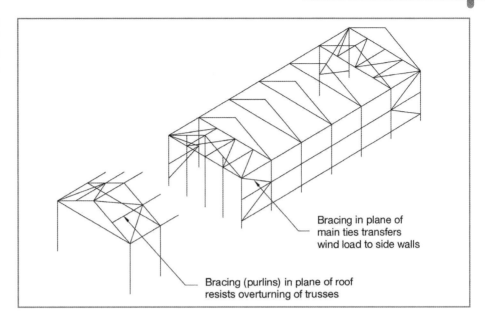

Bracing in plane of
main ties transfers
wind load to side walls

Bracing (purlins) in plane of roof
resists overturning of trusses

Figure 8.10
Bracing in the plane of the main ties to the side walls.

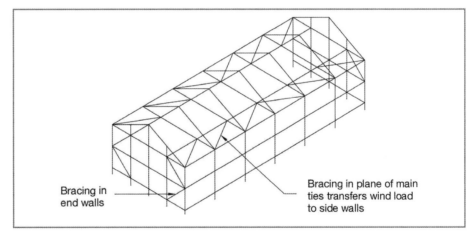

Bracing in
end walls

Bracing in plane of main
ties transfers wind load
to side walls

PART 3

Portal framed buildings

Portal framed buildings (Figure 8.11) can span between 20 and 60 m, and at these spans are practical, easy to construct and cost-effective.

Portal frames are constructed using supporting columns that have rigid connections to raking roof members. The distance between the 'portal units' can be between 4.5 and 7.5 m and these are supplemented by the use of bracing sections, both horizontal and vertical, to form the full structure (Figure 8.12).

The reason for the use of rigid connections is to ensure that the structure acts as a single entity, therefore reducing the bending moments in the roof members and beams. If simply supported beams were used, extremely large structural members would be required in order to prevent excessive deflection at the mid-span. For large spans the

Figure 8.11
Portal frame.

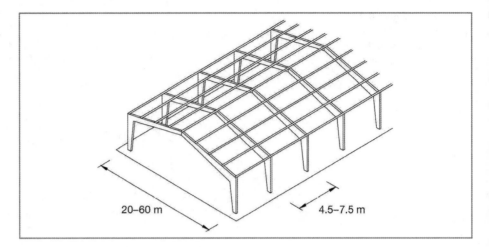

20–60 m   4.5–7.5 m

Figure 8.12
Bracing to portal frames.

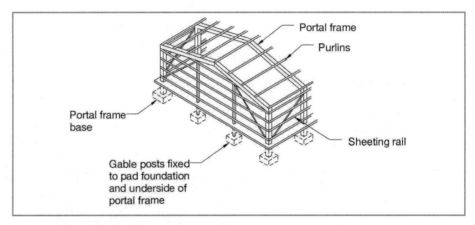

Portal frame

Purlins

Portal frame base

Gable posts fixed to pad foundation and underside of portal frame

Sheeting rail

portal frame is the most effective option, but for smaller spans simply supported frames may be the most cost-effective.

Although the simple portal shown in Figure 8.11 is the most common type of portal frame, there are other types that may be used. Figure 8.13 illustrates these.

Precast reinforced concrete can be used for portal frame construction although it is not common these days; the details are shown in Figure 8.14.

Space frames

When an extremely large span is required, space frames can be used for the construction of large-span buildings. A typical steel space deck module is shown in Figure 8.15.

Space deck roof structures use factory-produced steel pyramid structures bolted together on-site and nodes connected with tie bars to form a double-layer roof construction (Figure 8.16).

The pyramids themselves are of prefabricated welded construction using a square mild steel angle section forming the top, with steel diagonals welded to each corner of the

Figure 8.13
Alternative portal frame
types.

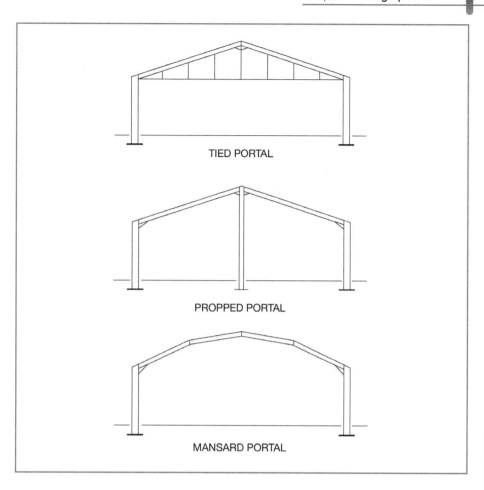

TIED PORTAL

PROPPED PORTAL

MANSARD PORTAL

Figure 8.14
Precast concrete portal
frame.

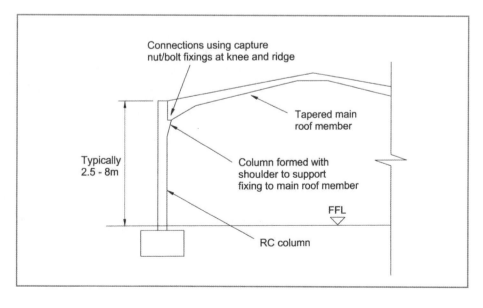

Connections using capture
nut/bolt fixings at knee and ridge

Tapered main
roof member

Column formed with
shoulder to support
fixing to main roof member

Typically
2.5 - 8m

FFL

RC column

PART 3

Figure 8.15
Typical space frame modules.

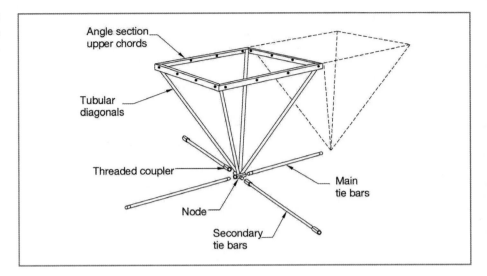

frame and to the steel node, threaded to take adjustable steel tie bars that form the bottom cords. The steel is protected using one coat of zinc phosphate epoxy blast primer to provide protection against corrosion.

Module sizes can vary, but examples of some of the standard sizes are:

- 1200 × 1200 × 750 mm deep
- 1200 × 1200 × 1200 mm deep
- 1500 × 1500 × 1500 mm deep
- 1500 × 1500 × 1200 mm deep
- 2000 × 2000 × 2000 mm deep

Figure 8.16
Space frame.

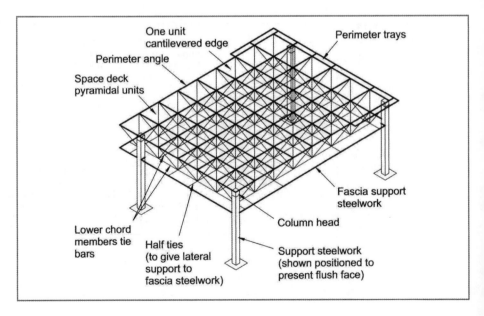

These units are lightweight and use material very economically; hence their use for very large-span buildings. The main advantages of space decks, after their ability to span long distances, are:

- Usually available off the shelf
- Excellent span-to-depth ratio
- Excellent facility for services and ducting
- No purlins are required: roof decking can be fixed directly onto the frame
- Suitable for structures that have irregular plan shapes
- Fully adjustable cambering facilities
- Easy to transport, handle and stack
- Easy to erect
- Come with a standard range of accessories.

Different spans can be achieved using different module sizes. Figure 8.17 indicates span lengths that can be achieved with these different modules.

Figure 8.17
Possible space frame spans.

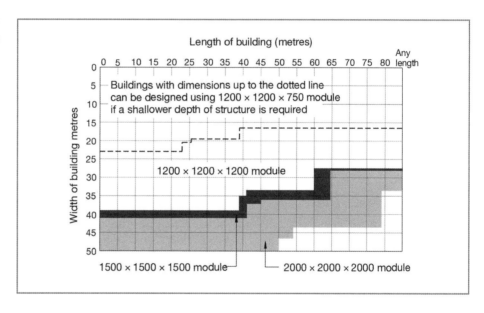

Space frames allow for the longest uninterrupted spans in single-storey buildings.

The **space frame** can be simply supported or can be cantilevered at the edge of the building. Details of the different structures are shown in Figures 8.18 to 8.21.

The maximum and minimum lengths of the cantilever will depend on the module size of the pyramids.

Typical details of steel space frames fixed to steel supporting columns are shown in Figures 8.22 to 8.24.

Space frames can also be connected to precast concrete columns, as shown in Figure 8.25.

Because space frames tend to be large span, they are prone to movement and expansion joints are needed. Typical expansion joint details are shown in Figure 8.26.

Space frames can either be flat or pitched. Details of a 10° pitched roof frame are shown in Figure 8.27.

Figure 8.18
Space frame edge
detail – perimeter
support.

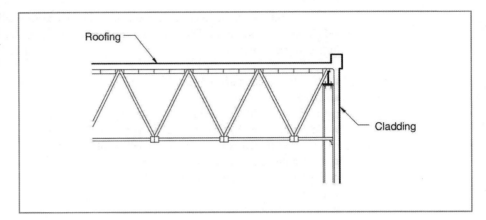

Figure 8.19
Space frame edge
detail – mansard
support.

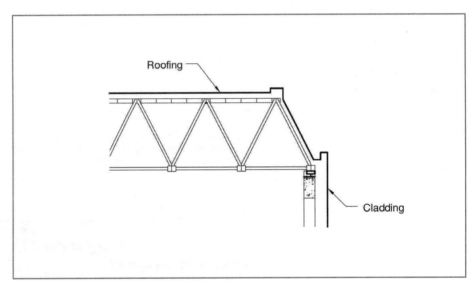

Figure 8.20
Space frame edge
detail – cantilever with
cornice.

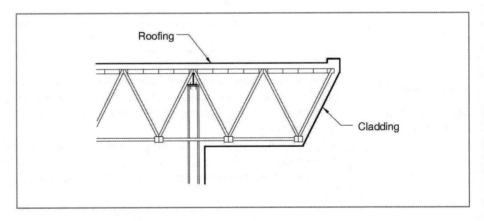

Figure 8.21
Space frame edge
detail – cantilever on
concrete column.

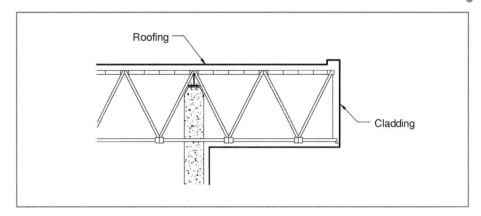

Figure 8.22
External column head
detail.

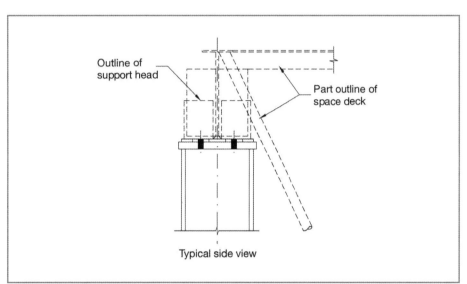

Figure 8.23
Internal column head
detail.

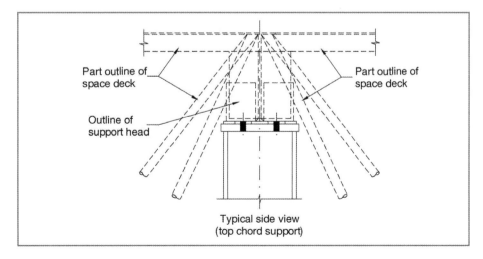

Figure 8.24
Internal column head
support detail to under-
side of frame.

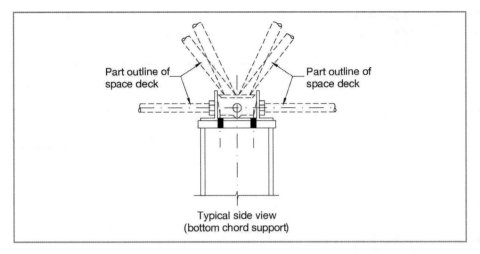

Figure 8.25
Space frame
connection to precast
concrete column.

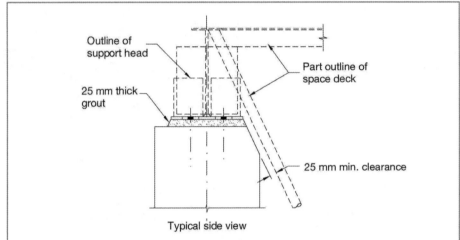

Figure 8.26
Expansion joints in
space frames.

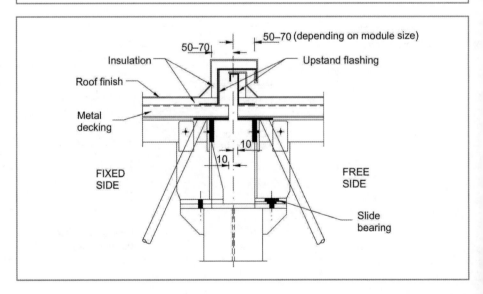

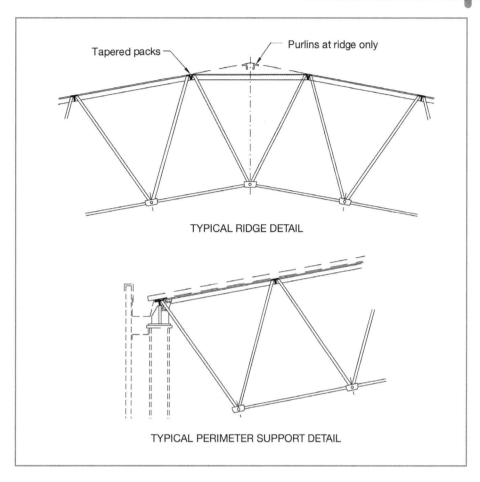

Figure 8.27
Ten degree space
frame detail.

Structural timber frames

Timber frames have been extensively used for domestic construction, and are increasing in popularity. Recently timber frame construction has been used more widely for large-span single-storey buildings.

The columns and portals can be either solid or laminated timber. The advantage of laminated timber is that the quality may be better, as thin sections of timber are used, as opposed to large sections where any knots or defective areas cannot be seen. If using solid timber then a high factor of safety would need to be included in the design. Laminating timber gives greater strength to the section size than solid timber can achieve.

Timber is an excellent material which has good tensile and compressive stress-resisting properties; it is also very sustainable if it comes from renewable sources. The main advantages of timber are the short lead-in time required to procure the frame and the elaborate shapes that the beams can be sculpted into, making them an attractive alternative if the beams are to be left exposed.

Timber portal frames are usually either glue laminated, plywood-faced or frames that use solid members connected with timber gussets.

PART 3

Glued laminated portal frames are usually manufactured in two halves that connect together in the middle. Rigidity is achieved by using timber purlins to connect the portals.

Plywood-faced portals are used for spans of approximately 10 m. The portal contains a skeleton of softwood which is then cased in plywood.

In solid timber and plywood gusset construction the closely spaced frames are clad in plywood, forming a lightweight building shell that is very strong and rigid.

Figure 8.28
Laminated timber portal frame.

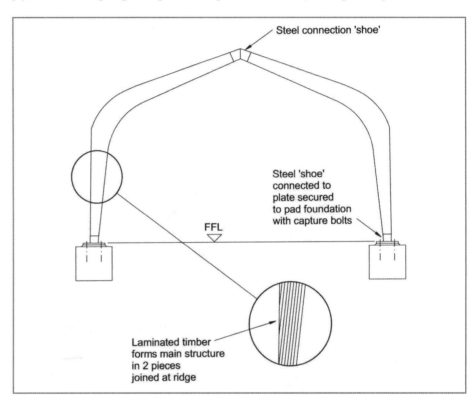

Steel connection 'shoe'

Steel 'shoe' connected to plate secured to pad foundation with capture bolts

FFL

Laminated timber forms main structure in 2 pieces joined at ridge

Figure 8.29
Constructing a laminated timber parabolic curve frame.

Details of the three forms of construction are shown in Figures 8.28 to 8.32.

Figure 8.28 and 8.29 show the construction of a laminated timber parabolic curve frame. The frame is constructed using 30 mm timber strips 'glued' together to form the structural element. The frames were delivered to site in sections and fixed together using steel plates and bolts. The right-hand photograph in Figure 8.29 shows the frame internally after completion, with the glass cladding in place.

Figure 8.30
Ply-faced boxed portal frame.

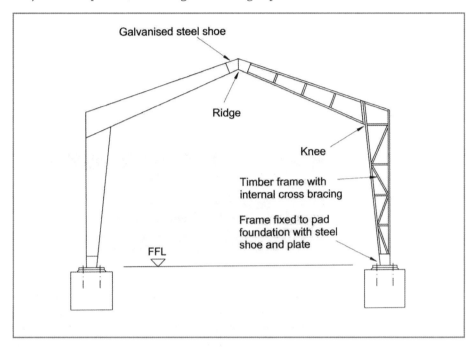

Figure 8.31
Solid timber portal frame.

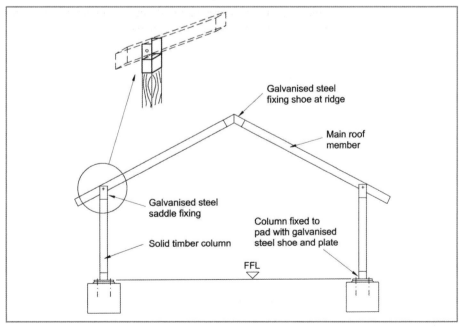

PART 3

Figure 8.32
Low-rise long-span
building with solid
timber frame.

Figure 8.30 shows a ply-faced boxed portal frame and Figures 8.31 and 8.32 show the use of solid timber frame construction for a single-storey building. The photographs in Figure 8.32 show the frame during construction, the connection detail for beams and columns and the finished building. The use of timber frame in this case is more sustainable than other framing materials and requires no further treatment after construction from fire-proofing and aesthetic perspectives.

Method of construction

The method of construction for single-storey large-span timber frames is as follows:

- Concrete slab – the level of the slab must be incredibly accurate, as any 'packing up' will reduce stability
- Fix sole plate to slab – the sole plate may be timber or a metal shoe
- Nail-in wall panels and/or supporting columns
- Drop-in prefabricated floor (if being used)
- Fix sole plate to top of columns
- Fix portals.

The main differences between large-span single-storey and domestic construction are:

- The wall panels may be bigger: 3, 4 or 5 m long
- Bigger section sizes will be needed to take the increased loads of the roof covering

- 6 mm settlement is expected; therefore some flexibility of joints is required
- Additional posts will need to be used to create vertical stability
- Bolt connections, as opposed to nailing, may be used to increase stability.

REFLECTIVE SUMMARY

- Large-span frames can be classified under three main headings:
  - Castellated or lattice beam construction
  - Portal frames
  - Space frames.
- The most common form of frame used for large-span single-storey buildings is the structural steel portal frame.
- The largest uninterrupted spans are possible using space frame construction.
- Bracing is extremely important when constructing a large-span single-storey frame, to prevent collapse.
- *In-situ* concrete is the only system that is not generally used for large-span buildings, but is used for high-rise construction.
- The use of timber framing is becoming more popular for environmental reasons.

REVIEW TASKS

- Detail the main differences between *lattice beam frames*, *portal frames* and *space frames*.
- Why is it advantageous to use a framed structure for a single-storey commercial building and not for a single-storey domestic property, e.g. a bungalow?
- Visit the companion website at www.palgrave.com/engineering/riley2 to view sample outline answers to the review tasks.

PART 3

Frame characteristics

Quality

As a general statement (as we have stated before), anything that is manufactured off-site will tend to be of a higher quality than anything that is manufactured on-site. This is due to the controlled environment within manufacturing plants as opposed to the very diverse and changeable weather conditions on-site.

The quality of steel production is highly regulated and very few problems with steel occur, although they are not unknown. However, if the design is incorrect, this may lead to the burning of new bolt holes, which will weaken the steel, and to welding of joints/plates to get the frame to 'work'. Welding on-site can lead to quality issues, and all welds need to be tested *in situ* before the structural engineer will pass the work.

The quality of precast concrete produced in factories is again generally excellent. The main quality issue with precast concrete frames is the quality of the *in-situ* concrete element of the joints, which if not constructed correctly will form weak points that will allow in moisture; the steel reinforcement will then corrode and the building will ultimately fail.

Timber frame quality is also excellent, as the majority of components are manufactured off-site. However, the quality of the fixings on-site dictates the overall quality of the building.

Safety

The benefit of precast/preformed structural frames elements (especially steel and timber) is the speed of construction. Because steel and timber are so quick to erect, it is feasible to 'clear' the site when the frame is being erected, thus reducing the risk to other workers from falling objects. However, the erection of these types of frame is inherently dangerous to the operatives carrying out the work, and it must be ensured that they use safety equipment and work to detailed Method Statements.

Environmental issues

The production of concrete (extraction) is very damaging to the environment, as is that of steel (fuel requirements). Timber used from sustainable sources is a reasonable option and this would lead us to believe that the use of timber as a structural framing material as much as is possible is the best alternative. However, this is not feasible for very large-span buildings and is not commonplace, although the trend is increasing. An Environ-

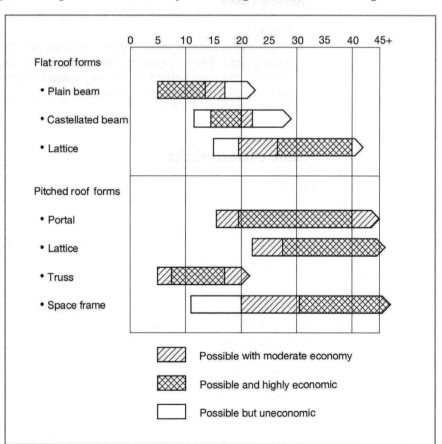

Figure 8.33
Effective and economical spans for long-span frame types.

mental Impact Assessment should be carried out by the designer when making the ultimate decision on type of frame, alongside health and safety provision.

Effective spans

The majority of criteria that can be used for comparing different forms and materials that can be used for long-span multi-storey buildings are the same as were used in Chapter 7. Further to this, Figure 8.33 illustrates for which spans each frame is suitable and where the use of each frame is most economical.

REVIEW TASK

- Determine a set of criteria that the different forms of construction and materials for frames to large-span single-storey buildings can be compared against.
- Visit the companion website at www. palgrave.com/engineering/riley2 to view sample outline answers to the review task.

CASE STUDY: PORTAL FRAMES

Steel is the most commonly used material for the construction of portal frames and the method of construction is similar to that of any other steel frame. For the building shown in Figure 8.34, and the photographs on the right, the method of construction is as follows:

1. Excavate foundations.
2. Fix holding down bolts using timber jigs.
3. Concrete foundations.
4. Fix frame columns that are welded to the gusset plate to the holding down bolts (Figure 8.35).
5. Level the frame to the correct position using steel or plastic packers.
6. Fix the portal beams to the tops of the columns (Figure 8.36).
7. Check that the frame members are at the correct level and that the horizontal positioning of the frame is correct.
8. If some minor adjustments are required the steel erectors should be able to jack the frame in its entirety into the correct position.
9. Grout between the base plate and the concrete foundation.

(continued)

CASE STUDY: PORTAL FRAMES (*continued*)

Figure 8.34
Portal frame building.

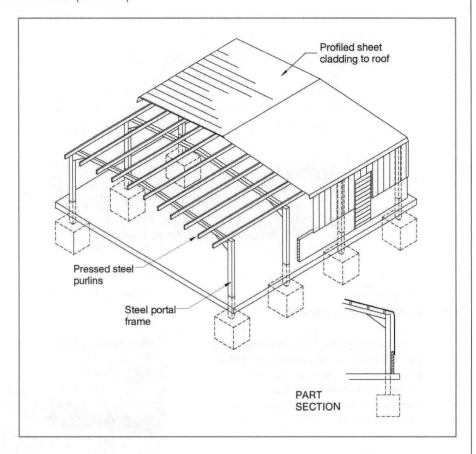

Profiled sheet
cladding to roof

Pressed steel
purlins

Steel portal
frame

PART
SECTION

Figure 8.35
Column and base plate
detail.

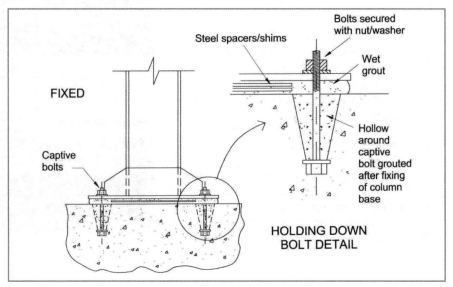

Steel spacers/shims

Bolts secured
with nut/washer

Wet
grout

FIXED

Captive
bolts

Hollow
around
captive
bolt grouted
after fixing
of column
base

HOLDING DOWN
BOLT DETAIL

(*continued*)

CASE STUDY: PORTAL FRAMES (*continued*)

Figure 8.36
Portal beam
connection to
columns.

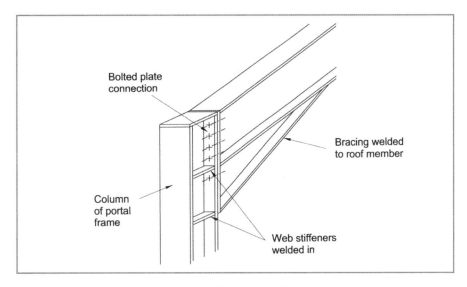

The next activity at this point could be either the construction of the ground floor slab after the internal drainage connections are in place, or to fix the roof cladding panels.

The advantage of fixing the roof panels is that the slab can be constructed in a weathertight environment, which is especially useful if a power float finish to the concrete slab is required. From a health and safety perspective this sequencing of construction is especially useful because the frame and roof cladding are activities with short durations and operatives can be directed away for the potential hazard of falling materials until construction is complete. At this stage of the construction process the only people on the site would normally be groundworkers and the steel erectors, followed by the roof cladders. The groundworkers could potentially be deployed to undertake the external drainage works during construction and hence reduce the risk of accidents to them.

Figure 8.37 illustrates the roof sheeting fixing details for portal frames.

Figure 8.37
Roof sheeting
attached to portal
frame.

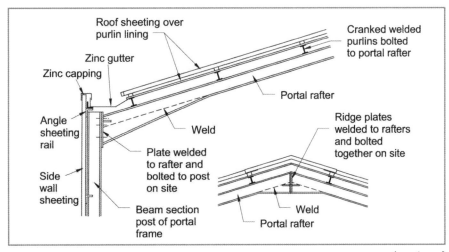

(continued)

PART 3

CASE STUDY: PORTAL FRAMES (*continued*)

Figure 8.38 shows the sequence of construction for a steel portal frame industrial/commercial building with brick cladding at the lower level and metal cladding for the higher level. The roof is metal sheeting. This sequence of construction is suitable when a power float finish to the concrete slab is required.

Figure 8.38
Sequence of construction for steel portal frame.

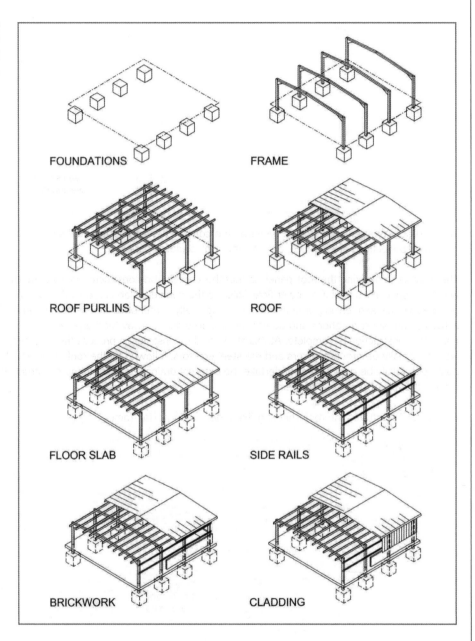

COMPARATIVE STUDY: LONG-SPAN FRAMES

Option	Advantages	Disadvantages	When to use
Concrete precast	Quality assurance Fire performance Efficient use of reinforced concrete Resistance to corrosive substances	Weight of sections to be lifted Really needs mass production for cost-effectiveness Lack of favour by designers in UK	Anywhere, but particularly suited to agricultural buildings and areas where chemical exposure may be an issue
Steel	Familiar technology Easy to produce Relatively light sections to be lifted High strength-to-weight ratio Ease of formation of non-standard components	Performance in fire if unprotected Production of steel is environmentally damaging	Almost ubiquitous in the UK
Timber	Sustainable material Lightweight Cheap Reduced lead-in time	Limited spans, even with 'glulam' forms Design scepticism	Often used where aesthetic requirements dictate, e.g. swimming pools

Fire engineering design

AIMS

After studying this chapter you will be able to:

■ Determine the need for fire protection in high-rise and large-span buildings
■ Identify the main methods used for fire protection
■ Appreciate the different materials and systems that are used to protect structural elements
■ Relate these systems to given building scenarios

This chapter contains the following section:

9.1   Fire engineering design and the fire protection of structural elements

INFO POINT

■ Building Regulations Approved Document B: Fire safety (2010)
■ BS 476: Fire resistance in construction elements (1987)
■ BS 5268, Part 4: Structural use of timber. Fire resistance of timber structures (2002)
■ BS 5588: Fire precautions in the design, construction and use of buildings (2004)
■ BS 8202: Coatings for fire protection of building elements (1987)
■ BS 9999: Code of practice for fire safety in the design, management and use of buildings (2008)
■ BS DD ENV 1992: Fire resistance in concrete structures
■ BS EN 1991-1: Eurocode 1: Actions on structures
■ BS EN 1993-1.2: Eurocode 3: Design of steel structures. General rules – Structural fire design
■ BS EN 1994-1.2: Eurocode 4: Design of composite steel and concrete structures. General rules – Structural fire design
■ BS EN 13501-2: Fire classification of construction products and building elements. Classification using data from fire resistance tests, excluding fire ventilation services (2007)
■ BRE Digest 487, Part 1: Structural fire engineering design: materials behaviour – concrete (2004)
■ BRE Digest 487, Part 2: Structural fire engineering design: materials behaviour – steel (2004)
■ *CIBSE Guide E, Fire engineering*, 2nd edn (2003)
■ *Fire protection for structural steel in buildings* (Yellow Book), 4th edn, Association for Specialist Fire Protection [0–862–00027–0] (2008)
■ Lataille, J. (2003) *Fire Protection Engineering in Building Design*, Butterworth [0–750–67497–0]
■ Muir, P. (2006) *The New Fire Safety Legislation*, RICS [978–1–842–19309–9]

9.1 | Fire engineering design and the fire protection of structural elements

Introduction

- After studying this section you should have developed an understanding of the need for fire protection and the different forms it takes within buildings.
- You should also have developed an appreciation of the different methods and materials used to protect structural elements from the effects of heat produced by fire.
- You should also be able to decide what type of fire protection is the most effective and economical in given building scenarios.

Overview

In terms of fire engineering design, means of escape is probably the most important issue and also the most difficult to design for. This is largely due to the fact that building occupants rarely use the means of escape allocated as a fire escape, instead using their preferred exit, which is usually the one that they enter a building through. Safety of occupiers is of prime importance, but there are other factors that need to be considered when undertaking fire engineering design. The three main aims are:

- Systems and methods of building that allow time for escape from a building
- Reducing the spread of fire
- Preventing structural damage and/or collapse due to fire damage.

Allowing time for escape from a building

Fire detection and alarm systems

Fire alarms can be divided into two main categories:

- *Manual systems*, such as 'break glass' call points
- *Automatic systems*, such as smoke or heat detection.

Automatic fire alarm systems are essential if a building is used for transient occupation.

The main use of **automatic systems** is where the building occupiers are not regular occupiers or the building is very complex and people may be unaware of where the manual systems are located. Detection systems may include conventional smoke detectors, beam detection, and heat and flame detectors. The choice of these systems depends on the room where the detector is allocated. For example, smoke and heat detectors may be wholly unsuitable for kitchen areas. Aspiring smoke detectors are up to 200 times more sensitive than traditional detectors and are therefore used in critical areas such as IT suites. The use of these systems allows the occupants to be alerted early to the potential of a fire and allows for a longer duration for evacuation.

Reducing the spread of fire

Compartmentation

The use of compartments in building design is beneficial in preventing the spread of fire from one area to another. The building is divided into a number of fire-proof boxes in effect, and if a fire starts in one of the boxes it must be contained there. This reduces damage from fire and again ensures that occupants have plenty of time to leave the building. In order for each compartment to contain the fire there must be no 'gaps' in the fire protection. In the majority of multi-storey commercial buildings, raised floors and suspended ceilings are used, along with partitions, to form rooms. Fire-resisting curtains need to be installed within ceiling and underfloor voids to prevent the passage of fire under the floor or above the ceiling to another area of the building. Sometimes these systems are installed correctly, but after years of maintenance to the services enclosed in these voids, the blankets can become damaged and the compartment is then obsolete, allowing for easy spread of fire and smoke. Partitions need to be constructed to give a suitable fire-resistant time to allow for escape from the building, and for the fire service to arrive to extinguish any fire before further damage is caused. Using compartmentation therefore leads to localised damage as opposed to widespread damage.

Figure 9.1 illustrates the benefits of compartmentation in preventing the spread of fire.

Figure 9.1
Compartmentation as a means of preventing the spread of fire.

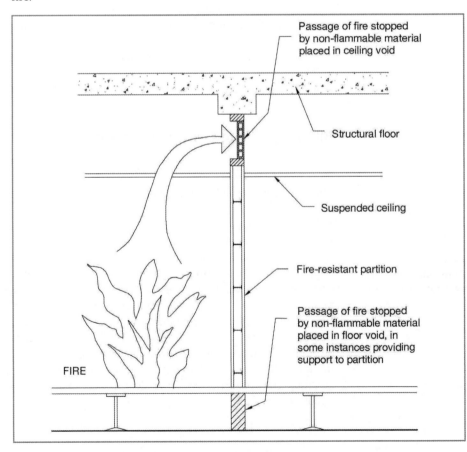

Sprinkler systems

As part of the mechanical services installation, sprinkler systems may be incorporated. These may be triggered manually or automatically, in zones or throughout the building. These systems are very effective in putting out fires quickly, but can potentially cause more damage than the fire itself, especially in critical areas such as computer rooms. Sprinkler systems can reduce the spread of fire and also allow a greater time for evacuation of the building.

Preventing structural damage and/or collapse due to fire damage

The Building Regulations require elements of building structures to have fire resistance. An element will require fire resistance dependent on the size of the element, the use of the building and the function of the element. When building elements are exposed to fire they will lose some of their strength. For example:

- Concrete can fall or spall and steel reinforcement will be left exposed
- Steel loses its strength
- Timber sections reduce in effective size due to charring.

In the UK, the most common method of determining a building's necessary level of fire resistance is from Approved Document B.

There are four common ways of providing elements to steelwork:

- Encasing in concrete
- Spray-applied systems
- Boarded systems
- Intumescent coatings.

In-situ and precast concrete frames

One of the benefits of using precast reinforced concrete is that the fire protection to the frame is intrinsic to the structure itself. The concrete protects the steel reinforcement from heat and thus prevents buckling of the structure. Therefore with *in-situ* concrete and precast concrete frames the steelwork is protected by the concrete which forms the overall construction. However, the concrete must have sufficient cover to prevent spalling, which would allow the steel reinforcement to be exposed, resulting in heating and deformation and/or collapse.

Structural steel frames

For structural steel frames Part B of the Building Regulations, together with schedule 8, gives the minimum fire resistance periods and methods of protection of steel structures. The traditional method is to encase the frame in concrete, which requires formwork and adds to the load of the structure. Many 'dry' techniques are available, but these are not suitable for exposed conditions as damage can occur which is both unsightly and removes

There are many trade names of products that provide **fire protection** to steel frames, but they are classed under three main headings: boards, sprays and intumescent coatings.

the fire protection efficiency. It is possible to protect the outer skin members with concrete and the inner building columns and beams with a 'dry' method, such as:

- *Spray-applied systems*: This type of **fire protection** is based on a spray-applied cement-based material that provides direct insulation to the element it is applied to, thus reducing the heating rate of the steel member so that its limiting temperature is not exceeded for a required period. This system is relatively cheap and can be applied to complex shapes easily. However, it is difficult to ensure that the correct thickness is applied over the whole surface of the structural element, and overspraying can lead to a great amount of cleaning being required. The coating can also be easily damaged and/or knocked off during use or maintenance of the building.
- *Boarded systems*: The principle of using boarded systems is basically to box in the section that is to be protected. Boarding works in a similar way to spray systems, by protecting the element from the heat of a fire for longer. The main advantages of boarded systems over sprayed systems are that a better finish is achieved and uniformity of thickness is easier to control. However, poor workmanship and damage can reduce the effectiveness of the boarding. The process is also labour-intensive and therefore costlier than sprayed protection.
- *Intumescent coatings*: These work by 'swelling' under the application of heat and producing a char layer which insulates the steel, thus preventing it reaching its failure temperature within a specified time. They are spray-applied and can therefore cover complex shapes easily, and there is the option of off-site covering, which allows for better quality assurance in the process. The disadvantages of this system are that sometimes an unsightly orange peel effect is created on the surface of the steel, and exposed steel may also be unsightly.

There is a growing lobby that believes that the use of fire protection is less effective than was thought previously, and the most effective way to provide fire protection to steel sections is simply to oversize them, so that they can withstand the loads of the building at raised temperatures safely. This method reduces the problems of poor workmanship on-site and potential damage during occupancy and maintenance operations.

Figure 9.2 shows the appearance of steel sections with the different fire protection coatings applied.

Figure 9.3 illustrates some of the complex shapes that may need to be coated in a given building. It is for this reason that quality issues can arise, especially with boarded systems. Gaps can be left in the protection that allow heat damage to the section. It is also difficult to construct formwork on-site in these section shapes. The best alternatives for these sections would therefore be sprayed protection or intumescent coatings.

A typical example of fire protection used in steel-framed buildings would be that all the columns are cased in concrete and one of the other methods used to encase the beams. This is usually because the beams will be concealed by the use of a suspended ceiling and therefore the poor aesthetic quality of the fire protection is concealed.

Timber frames

When **timber** burns, the external charring provides some fire protection to the internal core and can prevent collapse.

Fire resistance is an important consideration of using timber. Although wood is used as a fuel, large sections are difficult to ignite and the charcoal produced in the surface provides protection to the wood underneath. The rate at which **timber** chars is

Figure 9.2
Steel sections with
applied fire protection.

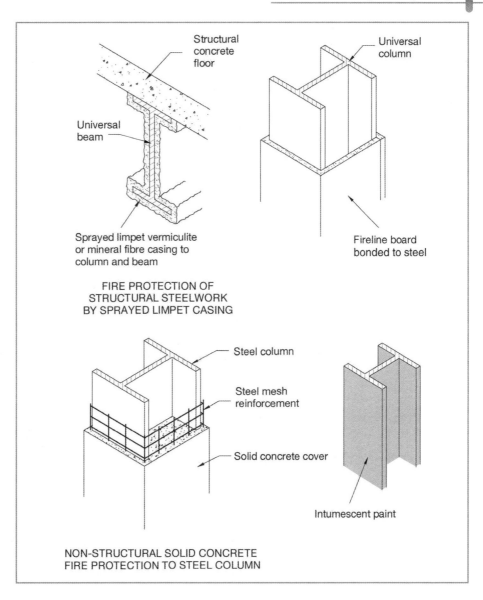

Structural
concrete
floor

Universal
column

Universal
beam

Sprayed limpet vermiculite
or mineral fibre casing to
column and beam

Fireline board
bonded to steel

FIRE PROTECTION OF
STRUCTURAL STEELWORK
BY SPRAYED LIMPET CASING

Steel column

Steel mesh
reinforcement

Solid concrete cover

Intumescent paint

NON-STRUCTURAL SOLID CONCRETE
FIRE PROTECTION TO STEEL COLUMN

Figure 9.3
Fire protection to
complex-shaped steel
sections
(after Corus).

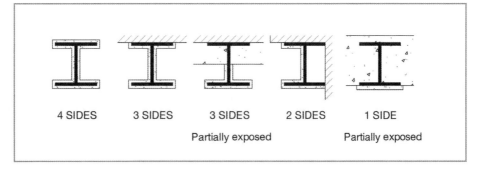

4 SIDES   3 SIDES   3 SIDES   2 SIDES   1 SIDE

                              Partially exposed   Partially exposed

NHBC is the National House Building Council. **TRADA** is the Timber Research and Development Association.

predictable and is little affected by the density and moisture content of the timber section. The uncharred sections retain their integrity and mechanical properties – unlike steel, timber does not soften or expand in fire. However, to add further fire protection, timber-frame buildings will usually have to ensure that staircases, party walls and internal partitions comply with Building Regulations and have the same fire rating as any other framed buildings. This can be achieved by using plasterboard cladding, cavity barriers etc. The **NHBC** and **TRADA** set the minimum regulations for the fire protection of timber frames.

REFLECTIVE SUMMARY

- Fire protection is extremely important in multi-storey and large-span commercial and industrial buildings because:
  - It allows time for occupiers to escape from the building
  - It can prevent the spread of fire and smoke, thus reducing damage
  - It can protect the structure and prevent collapse.
- Compartmentation is a common method of preventing the spread of fire.
- Sprinkler systems can only really be used if the prevention of damage to an area of the building is not critical.
- Fire protection for *in-situ* and precast concrete frames is intrinsic to the structure itself.
- There are four main methods for protecting structural steelwork from fire:
  - Concrete encasement
  - Sprayed protection
  - Boarded protection
  - Intumescent coatings.
- Large sections of timber do not burn easily as charring protects the inner core. However, in order to comply with British Standards, extra protection is required in critical areas.

REVIEW TASKS

- What are the main areas that need to be considered in fire engineering design?
- What systems and materials can be used to protect structural steelwork from the effects of heat produced by fire?
- Visit the companion website at www.palgrave.com/engineering/riley2 to view sample outline answers to the review tasks.

COMPARATIVE STUDY: FIRE PROTECTION SYSTEMS

	Spray-applied systems (Mandolite HS3/Monokote MK-6/ HY)	Boarded systems (Conlit 150/ Vermiculux/ Vicuciad)	Intumescent coatings (Steelguard 2550 FMA/ Steelmaster 120)
Relative cost	Low to medium	Low to high	Medium to high
Wet or dry trade	Wet	Mainly dry	Wet
Cleanliness of application	Messy, with protection required to adjacent surfaces	Relatively clean	Protection required to adjacent surfaces if done on-site; clean if applied off-site
Internal/external use	Internal and external Some coatings unsuitable for use within plenum ceilings or in clean room environments	Internal use. Additional protection required for external use	Internal with some external systems
Robustness	Relatively robust, but brittle nature can make them vulnerable to some mechanical damage	Some rigid boards are relatively brittle and can be vulnerable to mechanical damage Mineral fibre boards may require additional covering	Thinner applications prone to transport damage if coated off-site
Finish	Texture finish	Variable: boards are mainly smooth with joints visible. Mineral fibre boards are textured with fixings often visible	Smooth-paint effect that can have a coloured decorative finish applied
Thickness range (approx.)	10–75 mm	Multiple layers used Boards 6–100 mm	Thin film: 0.3–3.5 mm Thick film: 2.0–32 mm
Minimum fire resistance	240 minutes	240 minutes	120 minutes

PART 3

10 External walls and claddings for multi-storey and large-span commercial buildings

AIMS

After studying this section you should be able to:

- Appreciate the functional performance characteristics required of external walls for high-rise and large-span buildings
- Relate functional requirements to design alternatives for external walls
- Understand the reasons why materials are chosen for cladding, and which materials are best suited for particular cladding forms
- Understand the relationship between changing thermal insulation requirements for walls and the evolution of wall design
- Appreciate the different aesthetic qualities that different cladding systems can achieve

This chapter contains the following sections:

INFO POINT

- BS 5427, Part 1: Code of practice for the use of profiled sheet for roof and wall cladding on buildings. Design (1996)
- BS 8297: Code of practice for the design and installation of non-loadbearing precast concrete cladding [no longer current, but cited in the Building Regulations] (1995)
- BS 8298: Code of practice for the design and installation of stone claddings (2010)
- BRE Paper 18/98: Stone cladding panels: *in-situ* weathering information (1998)
- *Best practice for the specification and installation of metal cladding and secondary steelwork*. Publication 346. Steel Construction Institute (2006)
- Bevan, R. and Woolley T. (2008) *Hemp Lime Construction, A guide to building with hemp lime composites*, IHS BRE Press, Garston, Watford [978–1–84806–033–3]
- Brookes, A. J. (2008) *Cladding of Building*, 4th edn, Taylor & Francis [978–0–415–38387–5]
- *CIRIA Special Publication SP87F Wall technology*. Volume F: Glazing, curtain walling and overcladding (1992)
- *Cladding and curtain walling*. AJ Focus 10.99, Architect Journal (1999)
- Halliday, S. (2008) *Sustainable Construction*, Butterworth-Heinemann [978–0–750–66394–6]

10.1 | Functions of external walls and claddings

Introduction

- After studying this section you will have developed an understanding of the criteria that any choice of external wall system should satisfy.
- You should be aware of functional requirements of external walls and be able to understand their implications when choosing cladding forms.
- In addition, you should appreciate the basis of all of these and the drivers behind the development of increasing levels of performance requirements.
- You should also be able to set out a series of criteria to enable comparisons of different cladding systems to be undertaken.

Overview

The most commonly used **claddings** are non-loadbearing. The independent supporting frame takes the load of the weight of the cladding and any loads imposed on the cladding.

The weatherproofing of high-rise buildings may take a variety of forms, the most common being the use of external **cladding** and curtain walling, or the adoption of infill walling. Infill walling is possibly the simplest form to create, the walling being constructed traditionally, generally of block and brick, off the floor or edge beam section of the frame. All these systems are non-loadbearing and the dead weight of the walling system is carried, along with any weather loadings, through the frame to the foundations. The choice of external walling system is usually due to appearance and the structural engineer needs to know the system of cladding that the architect has specified before the frame is designed.

Performance criteria and main forms of external walls

External walls provide for a number of specific performance requirements or functional needs, and the different cladding options available today have evolved to meet these needs.

The performance requirements of the building fabric have been discussed earlier. However, it is appropriate to summarise them as they relate directly to the construction of external walls. In general they may be considered to include the following:

- Strength and stability
- Exclusion of moisture/weather protection
- Thermal insulation
- Sound insulation
- Fire resistance
- Ventilation
- Durability and freedom from maintenance
- Aesthetics.

In addition, the level of buildability is important since this has a direct effect on time, cost and quality in construction. The design and construction of external walls takes into account these requirements with features incorporated to satisfy each.

PART 3

Strength and stability

The achievement of required levels of structural stability is essential if the building is to withstand the loads that are imposed upon it during its life. Vertical, oblique and lateral loadings must be safely transmitted through the structure to the loadbearing strata; the external walls may or may not take an active role in this transmission. It is possible to consider external walls in categories, related to the extent to which they act as load-bearing elements of the building.

To allow for differential movements between the structural frame and the wall structure there has to be adequate support to carry the weight of the wall structure, and also restraint fixings that will maintain the wall in position and at the same time allow differential movements without damage to either the fixings or the wall material.

Exclusion of moisture/weather protection

In the opinion of many building users this may be the main purpose of external wall construction. The ability to exclude wind, rain, snow and excessive heat or glare from the Sun is paramount in the list of user requirements, yet this must be achieved while still allowing best use to be made of natural light and ventilation. The users must, of course, also be able to enter and leave the building, thus creating the need for numerous openings to be formed in the building enclosure. In these areas, particular care must be taken to exclude moisture, although the nature of the details will naturally vary with differing wall forms. In modern construction three differing approaches may be taken to achieving the exclusion of moisture, the choice depending on the nature of the building, its use and its location.

Traditional walling materials, such as brick, stone and block, exclude rain from the inside of the buildings by absorbing rainwater that evaporates to the outside air during dry periods. It is common practice to construct solid walls as a cavity wall with an outer leaf of brick as a rain screen, a cavity and an inner leaf of lightweight block that acts as a thermal barrier and solid inside wall surface. Non-absorbent materials are vulnerable to rain penetration and therefore the joints in the wall should be watertight and flexible and serve as a protective seal against rain penetration.

Thermal insulation

U values are measured in W/m^2K. A U value of 0.25 means that one-quarter of a watt of heat is lost through each square metre of the element when a one-degree temperature difference exists between the inside and the outside of the element.

Some additional material or materials have to be used to improve the thermal properties of wall structures built with solid, panel or thin sheet materials. Insulating materials are constructed or formed as an insulating inner leaf or lining behind solid or panel walls or as a lining to panels. External walls should be constructed to reduce heat loss to acceptable levels (**U values**). Materials used to enhance the thermal properties of wall structures should be a continuous lining behind or in the wall structure, and cold bridging must be avoided.

Sound insulation

The need to minimise the level of sound transmission through external walls can arise for a variety of reasons, but in general it could be considered necessary when sound levels differ greatly from the inside to the outside of a building. Sound travels in two distinct ways, via a solid material (impact sound transmission) or via the air (airborne sound transmission). Airborne sound transmission requires a massive construction to reduce it, whilst physical breaks in the structure may stop impact sound. The creation of physical breaks, however, allows airborne sound to pass, hence there is an inherent problem here. In the case of masonry construction there is seldom a problem in respect of sound transmission, since it is sufficiently massive. Thus it is common to use masonry at low level in industrial buildings which are otherwise clad with lightweight materials, in an attempt to arrest sound transmission from industrial plant.

In built-up areas loud noises are intrusive and this has to be reduced to acceptable levels. It is possible to use external methods to reduce sound from reaching the building in the first place, but these methods can be costly. The various methods for reducing noise are shown in Figures 10.1 to 10.4.

If these methods are not suitable, then insulation has to be provided by the external wall itself. Impact sounds (when solid material transfers the energy of impact by vibrations) can be most disturbing and the use of resilient fixings in door frames and resilient brushes help eliminate this problem. However, the dead load that the frame will have to carry due to a heavy mass walling may require a larger frame and larger foundations.

Figure 10.1
Mechanisms for reducing sound penetration.

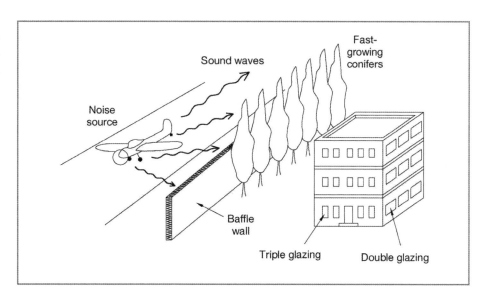

Figure 10.2
The effects of embankments on noise transmission.

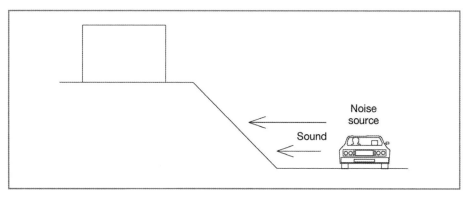

Another method of reducing sound transmission is through the use of a 'layered' wall construction. The advantage of this method is that each 'layer' can be light in weight, therefore overloading of the frame and foundations does not occur, and sound transmission is reduced.

Fire resistance

External walls are required by the Building Regulations to limit external fire spread for a set period of time. The materials that make up the elements of a building are required to have properties that will protect the structure from collapse and must not support spread of flame or fire from one part of the building to another and between adjacent buildings.

Ventilation

Sealed weathertight walls restrict natural ventilation. This can be overcome by the use of air conditioning, where hot air rises causing the movement of air; an alternative is a stack system where open-plan areas around an atrium are used as an open central core. However, in areas where there is little pollution it is sometimes preferable to allow for opening windows, but this is not suitable in all the cladding forms available.

Figure 10.3
The effects of cuttings on noise transmission.

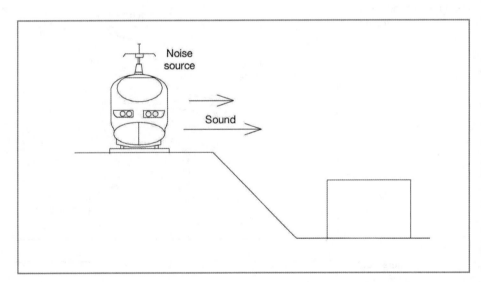

Figure 10.4
Landscaping to reduce sound transmission.

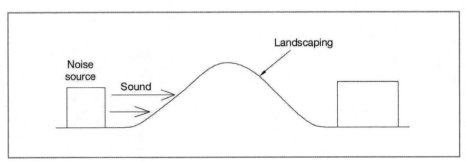

Freedom from maintenance and durability

Walls of brick and stone facing require very little maintenance over the expected life of most buildings. Precast concrete wall panels, which weather gradually, may become dirt-stained due to the slow run-off of water from horizontal joints. This irregular and often unsightly staining is a consequence of the panel form of this type of cladding. There must also be access to maintenance. The level of freedom from maintenance is a measure of the frequency and extent of maintenance work required to preserve functional requirements.

Aesthetics

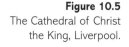

Cladding for buildings are most susceptible to damage at lower levels. It is common therefore to clad these levels in materials that are more durable than those at the higher levels.

The external wall **cladding** gives a building its identity, and the style the building takes is entirely due to the designer. People have very different preferences with regard to wall cladding: many prefer the traditional brickwork finish, whilst others prefer more futuristic glass finishes. Potentially this aspect is probably the most important, because designers will know what they want the finished building to look like and will specify the external wall system accordingly, whilst still ensuring that all the other criteria are fulfilled by the system chosen. The use of new systems is popular with designers and this can potentially create problems for the future, because installation procedures and maintenance requirements may not be fully understood.

An example of this is the Cathedral of Christ the King in Liverpool (Figure 10.5). The cladding to the section between the main structure and the glass tower section was originally specified as aluminium to allow for a shiny finish. However, during installation,

Figure 10.5
The Cathedral of Christ the King, Liverpool.

PART 3

tape was stuck to the panels to protect it. On completion of the work the tape could not be removed simply by peeling it away, and solvents were used. The reaction caused the shiny finish to be dulled and the overall aesthetics of the building to alter dramatically. This section has required a tremendous amount of maintenance over the years at great expense, and recently the panels have had to be removed and replaced with zinc panels.

REFLECTIVE SUMMARY

With reference to the performance of external walls, remember:

- The main performance criteria are strength and stability, exclusion of moisture/weather protection, thermal insulation, sound insulation, fire resistance, ventilation, durability, freedom from maintenance and aesthetics.
- Most designers will decide what they want the building to look like, specify it and then try to ensure that it satisfies all the criteria, which is where problems with the choice of system can occur. The system may simply not be suitable for the type and size of building and/or location.
- Very heavy external walls will mean that a bigger frame and foundations will be required, which will increase the overall cost of the building. This extra cost needs to be offset against the cost of the cladding.

REVIEW TASK

- For what reasons could the location of a building affect the choice of cladding system for multi-storey framed buildings?
- Visit the companion website at www.palgrave.com/engineering/riley2 to view sample outline answers to the review task.

10.2 | Options for walls and claddings to industrial and commercial buildings

Introduction

- After studying this section you will be familiar with the main systems that are used for the external cladding of commercial and industrial buildings.
- You should be able to understand the reasons for the development of specific design features within each form and to relate these to generic performance requirements.
- You should be able to compare and contrast the advantages and disadvantages of each system and their differing features, and make informed choices of the most appropriate type of cladding in different given scenarios.
- In addition, you should have an understanding of the implications of selection of external wall types for other elements of the building, such as the frame, floors, roof and foundations.

Overview

Loadbearing walls may be used in framed construction, but it is unusual. Generally the frame is designed to take the dead load of the external wall and transmit weather loadings from the walls through the frame down to the foundations. The most common forms of non-loadbearing claddings fall into three main categories: infill walling/panels, cladding and curtain walling. These broad categories can be subdivided further, and the specific features of each type will be examined, along with the different materials that can be used for each system.

Loadbearing walls

In modern domestic construction, and in many older buildings, most external walls are designed to be loadbearing, in that they carry their own self-weight together with some element of loading from the rest of the building, such as floor or roof loadings (Figure 10.6). Masonry is the most common material utilised, with brick or blockwork being almost ubiquitous in this form of construction. The loadings transmitted through the wall are transferred to foundations below ground level. In order to ensure that stability is maintained, certain restrictions are made within the Building Regulations relating to the height and thickness of walls, the number and positions of openings and so on. The provision of lateral restraint is also essential to resist the lateral loadings applied to build-

Figure 10.6
Loadbearing walls.

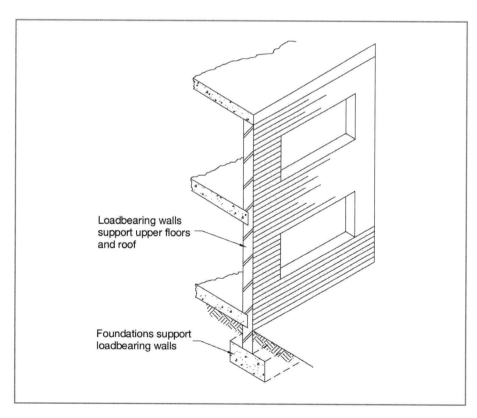

Loadbearing walls
support upper floors
and roof

Foundations support
loadbearing walls

ings from wind etc. However, for multi-storey and large-span buildings constructed these days loadbearing walls are rarely used, as the frame will have been designed to support all dead, live and wind loadings which include the loadings incurred by the cladding material. There are examples of loadbearing brickwork being used for multi-storey buildings, but these are rare. One advantage of using precast concrete loadbearing external walls is that the cladding panels, being structural, eliminate the need for perimeter columns and beams and provide an internal surface ready to receive insulation, attached services and decorations.

Infill walling

This is a form of external wall that is used in large-scale construction. The use of infill walling is common in framed buildings, where the wall sections are non-loadbearing and act only to exclude weather, the thickness being calculated on the basis of the small panel sizes used (Figures 10.7 and 10.8). An important point to note is that the wall section must remain stable, and hence be secured to the structural frame, but at the same time must allow for differential movement between dissimilar materials. Special movement accommodation fittings and details must therefore be incorporated into the construction. Problems of **cold bridging** must also be considered and have resulted in this mode of construction being used less frequently in recent years.

Cold bridging occurs when a component of relatively high thermal conductivity extends partly or completely through the thickness of an element of a building

Differential movement is another problem that can occur if infill panels are used, since the frame, normally concrete, and the infill panel, normally clay brick, will be subject to differential thermal movement with environmental temperature variation. This must be catered for, whilst at the same time the infill panel must be securely tied to the frame to prevent collapse. Hence sliding anchors and wall ties are utilised, which accommodate such movement whilst effecting a restraint on the panel (Figure 10.9).

Figure 10.7
Infill walling section.

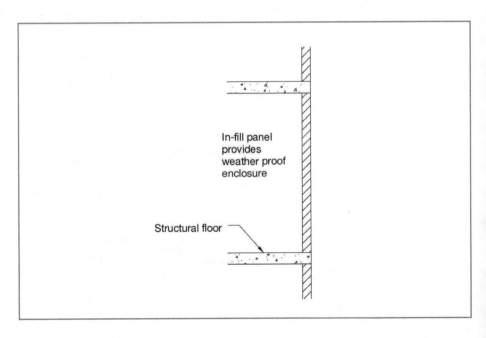

In-fill panel provides weather proof enclosure

Structural floor

Figure 10.8
Infill walling elevation.

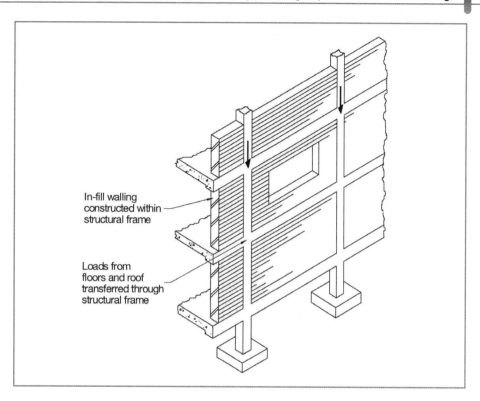

In-fill walling
constructed within
structural frame

Loads from
floors and roof
transferred through
structural frame

Figure 10.9
Differential movement
in cladding.

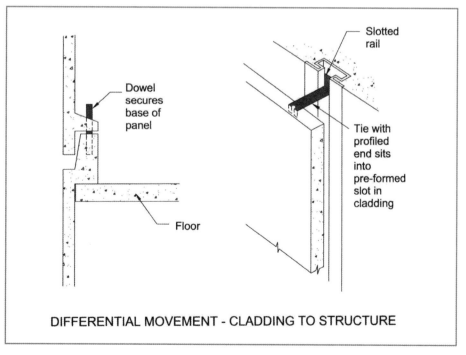

Slotted
rail

Dowel
secures
base of
panel

Tie with
profiled
end sits
into
pre-formed
slot in
cladding

Floor

DIFFERENTIAL MOVEMENT - CLADDING TO STRUCTURE

Non-loadbearing cladding

Another common form of external enclosure to framed buildings is that of cladding, in its strictest sense (Figure 10.10). Pre-formed cladding panels, incorporating window openings, exterior features and surface finishes, may be secured to the building frame in a number of ways. Figure 10.11 illustrates one possible solution. However, a wide variety of options is available in practice.

As in the case of infill walling, allowance must be made for **differential movement** between the frame and panels and between adjacent panels, whilst still maintaining a weathertight enclosure. This imposes demands on the design of the jointing between the panels, and it is common for baffles, drained joints or flexible seals to be incorporated to ensure moisture exclusion. It is essential that all panels are securely fixed to the frame, yet at the same time are free to move independently of each other – the panel and fixing design must take this into account. One of the most common forms of cladding is precast reinforced concrete, although steel, aluminium and plastics are also available and can provide excellent exterior cladding options.

The use of external brickwork cladding has increased in popularity over recent years as designers have tried to produce a 'traditional' appearance in framed buildings. The treatment of this form falls somewhere between infill walling and true cladding. The brickwork is constructed in panels, supported off appropriately spaced supports which are concealed by the cladding, often taking the form of steel angle sections fixed to the structural frame. As in the cases previously described, tying mechanisms which allow for movement must be included, and this is generally in the form of sliding anchor bars as illustrated in Figure 10.12.

One of the obvious features of infill walling is the fact that the main structural frame of the building is exposed, but in the case of cladding, the frame is concealed. The non-loadbearing covering to the building acts only to present a pleasant appearance and to resist external weather. As with infill panelling the effects of differential movement must be considered and allowed for between individual panels and between the cladding and

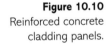

Differential movement occurs when elements of a building that butt up to one another move in different ways and to different extents.

Figure 10.10
Reinforced concrete cladding panels.

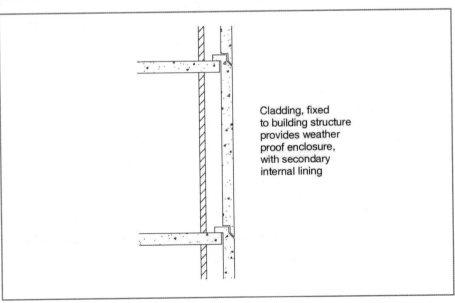

Cladding, fixed to building structure provides weather proof enclosure, with secondary internal lining

Figure 10.11
Concrete cladding
system.

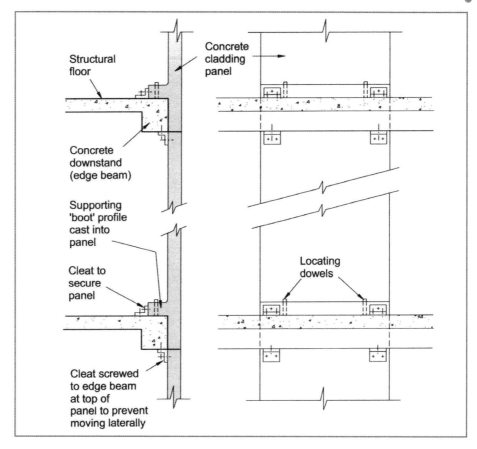

the building structure. The fixings used to secure panels must be very strong and durable; hence stainless steel is normally used, although bronze is also quite common.

Curtain walling

Curtain walling is, perhaps, the most advanced of the cladding forms, consisting of a self-supporting framework tied laterally to the structural building frame. This self-supporting framework carries the loadings from the external cladding, which acts independently of the main building frame whilst totally concealing it. This independence results in the ability to change the external enclosure with relative ease, leaving the building unscathed. Curtain walling often includes large areas of glazing and can result in a very 'high-tech' building appearance.

Figure 10.13 shows a curved building with curtain walling as cladding. The internal cladding frame is not attached to the columns except for a small number of restraining bars. The glass panels are mainly supported by an independent frame.

The fundamental difference between cladding and curtain walling is that in the case of cladding the weatherproof panels are fixed to the main structural frame of the building; with curtain walling they are fixed to their own independent supporting frame

PART 3

Figure 10.12
Brickwork restraint by
sliding anchors.

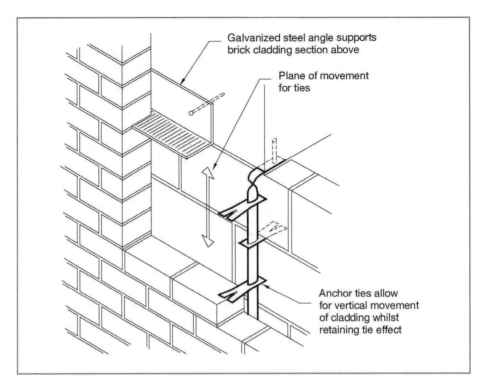

Galvanized steel angle supports
brick cladding section above

Plane of movement
for ties

Anchor ties allow
for vertical movement
of cladding whilst
retaining tie effect

Figure 10.13
Curtain walling.

erected around the building. In most other respects they are similar, with cladding panels of relatively large size, often full-storey height, fixed around the building to form a weatherproof enclosure. Figures 10.14 to 10.18 show various aspects of curtain walling.

Figure 10.14
Curtain walling section.

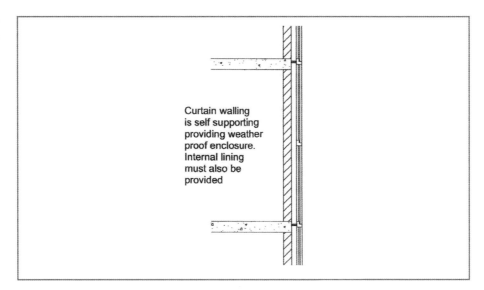

Curtain walling is self supporting providing weather proof enclosure. Internal lining must also be provided

Figure 10.15
Curtain walling elevation.

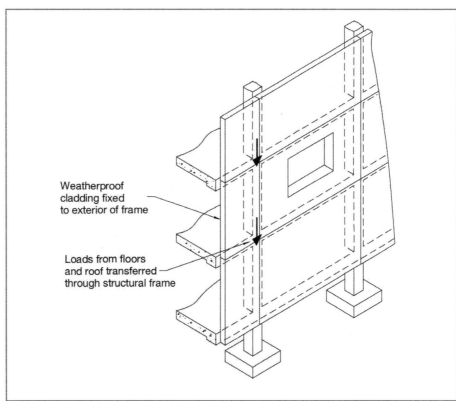

Weatherproof cladding fixed to exterior of frame

Loads from floors and roof transferred through structural frame

PART 3

Figure 10.16
Curtain walling –
typical arrangement of
components.

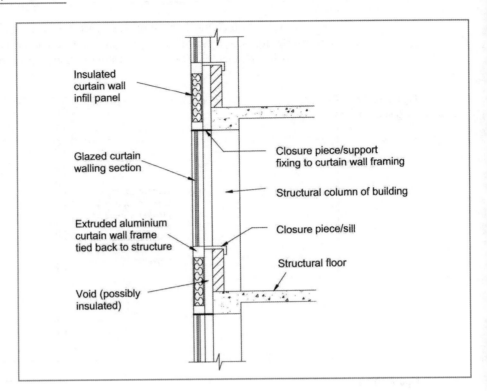

Insulated
curtain wall
infill panel

Glazed curtain
walling section

Extruded aluminium
curtain wall frame
tied back to structure

Void (possibly
insulated)

Closure piece/support
fixing to curtain wall framing

Structural column of building

Closure piece/sill

Structural floor

Figure 10.17
Curtain walling system
details.

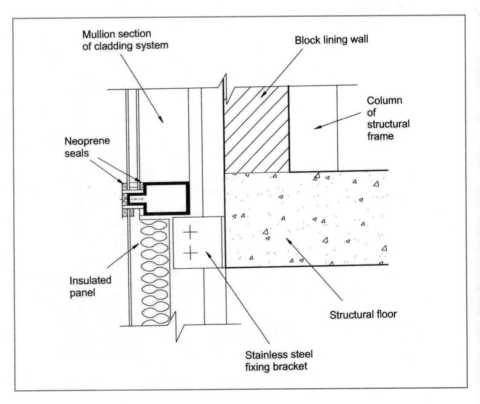

Mullion section
of cladding system

Block lining wall

Column
of
structural
frame

Neoprene
seals

Insulated
panel

Structural floor

Stainless steel
fixing bracket

Figure 10.18
Curtain walling fixing process.

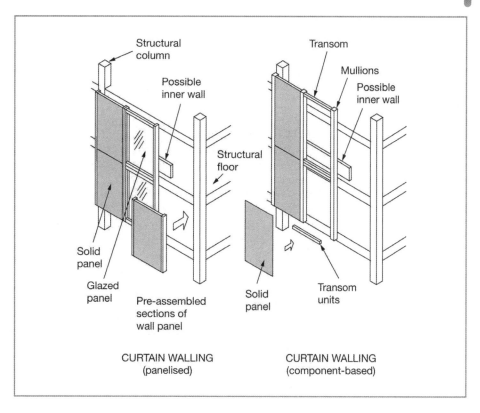

Rain screen cladding

A very simple form of wall construction is the use of blockwork panels built between columns which give stability and a surface to internal plaster or board. In order to protect the blockwork from the elements a rain screen can then be fixed to blockwork externally using a series of brackets.

Figure 10.19
Blockwork with rain screen cladding wall construction.

PART 3

Figure 10.19 shows the brackets attached to the blockwork and the panels attached to the brackets.

A major benefit of this form of construction is that if the rain screen panels become damaged or the building owners wish to overhaul the appearance of the building, the panels can be easily removed and replaced.

REFLECTIVE SUMMARY

- Loadbearing cladding is not commonly used, but has the advantage that pre-formed units can simply be lifted into place, whereupon work can proceed inside the building immediately.
- Non-loadbearing cladding can take one of three forms: infill panelling, non-load-bearing cladding and curtain walling.
- With infill walling the beams and columns of the structural frame are exposed; with cladding and curtain walling they are concealed.
- When using non-loadbearing cladding the structural frame takes all the loadings.

REVIEW TASK

- Look at buildings where you live and work. Try to identify which types of cladding have been used on a range of multi-storey buildings with which you are familiar.
- Visit the companion website at www.palgrave.com/engineering/riley2 to view sample outline answers to the review task.

10.3 | Functions and selection criteria for claddings to low-rise buildings

Introduction

- After studying this section you will have developed an understanding of the criteria that any choice of external wall system should satisfy.
- You should be aware of the functional requirements of external walls and be able to understand their implications when choosing cladding forms.
- In addition, you should appreciate the basis of all of these and the drivers behind the development of increasing levels of performance requirements.
- You should also be able to set out a series of criteria to enable comparisons of different cladding systems to be undertaken.

Overview

The use of structural frames for large single-storey buildings, such as warehouses and commercial and industrial units, allows the creation of buildings with large uninterrupted floor areas, with the possibility of providing great floor-to-ceiling heights. The most common form of single-storey framed building is that which follows the form of a large

Figure 10.20
A building in which brickwork cladding has been used at the lower level and coated aluminium cladding for the upper level.

Figure 10.21
Section through the wall assembly of a steel-framed building.

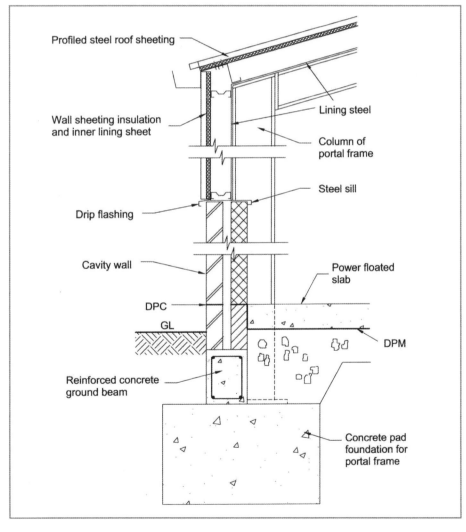

shed and is often referred to as a 'long-span' building. Such buildings are often erected from a standard selection of parts, being then fitted out and adapted to suit specific purposes. The frames are designed to carry cladding of lightweight form, such as profiled steel or aluminium sheeting, although at low level brickwork is also a common form of cladding, with its inherent advantages of security, sound insulation and protection from accidental or criminal damage (Figure 10.20). Figure 10.21 shows the construction detail for the building shown in the photograph of Figure 10.19.

Performance requirements of claddings to long-span single-storey buildings

External walls provide for a number of specific performance requirements or functional needs, and the different cladding options available today have evolved to meet these needs.

The performance requirements for the external walls of such buildings are described below.

Strength and stability

It is not normally expected that the cladding to long-span buildings will be required to carry significant applied loadings. The nature of the external fabric is such that it is intended primarily as a weatherproof enclosure; if major loadings are anticipated, measures must be taken to provide suitable support at the relevant fixing points, possibly by extending the frame of the building. A degree of rigidity is essential however, to ensure that the cladding sheets are able to span between fixing purlins without deformation. This is aided by the inclusion of a regular raised profile in the cladding sections, which increases the strength over that of a simple flat sheet (Figure 10.22).

Exclusion of moisture

The cladding options for long-span buildings are impervious by nature and should exclude moisture as a consequence. Particular care must be taken, however, where cladding sections are joined and at construction details where moisture may penetrate. The creation of details and flashings at appropriate points is essential to prevent moisture ingress, together with the tiling of the cladding profile at exposed sections to prevent wind-driven rain from passing along the raised sections of the sheeting. One of the major advantages of this form of cladding is that joints may be kept to an absolute minimum, since the length of sheets available is limited only by transportation constraints.

Thermal insulation

The nature of lightweight cladding is such that it has little inherent thermal resistance; hence in order to achieve the required levels of insulation it must be subject to the addition of an insulant. Traditionally, this may be in the form of fibre quilt or ridged insulation material laid between the outer cladding and an internal lining profile to upgrade the thermal resistance to the required level. A vapour barrier would also be incorporated to reduce the risk of interstitial condensation. Although effective, this method is ineffi-

Figure 10.22
Fixing of profiled sheet
cladding.

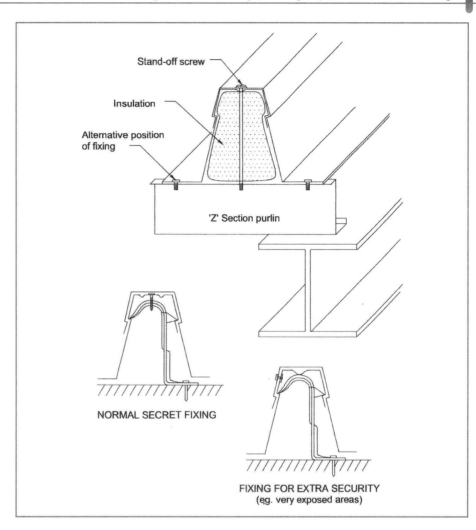

cient in terms of time and cost-effectiveness. Hence it has largely been superseded by the use of **composite panel** sections comprising the outer cladding profile and inner lining profile, bonded together with an expanded foam insulation material (Figure 10.23). This form of cladding requires far less fixing time than the traditional method, and has become very common in long-span building construction.

Composite panels
comprise a number of
layers which usually
include an internal and
external leaf that
sandwich an insulating
leaf.

Durability

The resistance of cladding to degradation is aided by the adoption of techniques for applying moisture-resistant coatings in the case of steel. Problems may occur where the material is cut or trimmed and the protective coating is broken. In such areas localised treatment is necessary to prevent corrosion. In the case of aluminium, of course, the material is naturally resistant to corrosion, but may receive a decorative application for aesthetic reasons.

Figure 10.23
Cladding panel with
bonded insulation.

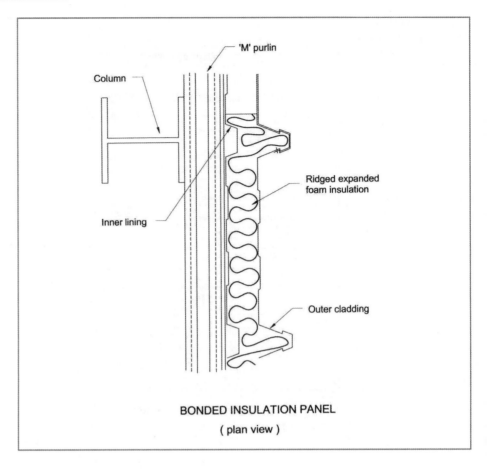

Column

'M' purlin

Inner lining

Ridged expanded
foam insulation

Outer cladding

BONDED INSULATION PANEL
(plan view)

REFLECTIVE SUMMARY

■ The main functional requirements of external walls to long-span single-storey buildings are strength and stability, exclusion of moisture, thermal insulation and durability.

■ This type of building form is most commonly used to build offices. The use of cladding for the top half of the building and brickwork for the bottom half is very popular.

REVIEW TASK

■ What are the advantages of using brickwork to clad some part of a low-rise single-storey commercial building?

■ Visit the companion website at www.palgrave.com/engineering/riley2 to view sample outline answers to the review task.

Sustainability

The use of truly sustainable buildings for multi-storey cladding is still fairly unusual, although there are examples of where more sustainable techniques than the traditional have been used. Figure 10.24 shows an example of where plants have been used to form the external screen to a multi-storey building at the lower levels.

Figure 10.24
Green plant wall.

There are however some very sustainable methods of cladding low-rise buildings and in fact it is both the structure and the cladding that could be classed as 'green'.

Clay wall construction

Clay wall construction is not new and is used extensively around the world as a form of construction. Nevertheless it is an extremely green method of construction and there are examples of its use more recently, especially in the south-east of England which sits on a large clay cap, and this means that the materials required are naturally and abundantly available. A foundation/slab is formed as with traditional construction but a kicker is formed along the lines of the proposed wall. Wall formwork is then erected and clay of layers of 100–150 mm is placed in the formwork. This is then mechanically rammed to ensure it is compact. The process is repeated until the wall is the required height. The wall is cast in panels and expansion joints are incorporated between the panels. Once all the panels are complete, the wall is then topped with a concrete ring beam onto which the roof structure is fixed. Figure 10.25 shows the foundation to wall configuration used in clay wall construction.

Figure 10.26 shows sliding formwork being used to help form the walls.

Straw bale wall construction

Straw bale wall construction is a system that is becoming more common in the UK. For the building shown in Figure 10.27, a timber frame has been constructed on a traditional

PART 3

Figure 10.25
Foundation/slab/wall
details for clay wall
construction.

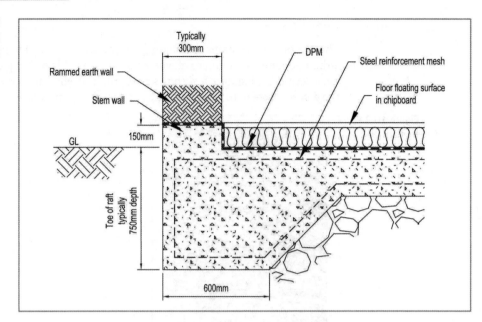

Figure 10.26
Clay wall formwork.

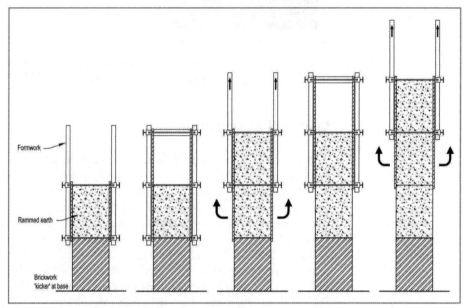

slab and then straw bales used to fill in between the frame elements. The straw has then been rendered over to form the finished wall. Straw is an abundant and highly renewable material and therefore this type of construction is extremely green. Figure 10.27 shows the timber frame, the straw contained in the wall and the first coat of render. The inclusion of this detail has become traditional in straw bale construction and it is known as a 'honesty window'.

Alternatively, the straw bales may be outside the timber frame rather than between the elements and they may be clad with timber or other materials rather than rendered.

Figure 10.27
Straw bale wall
construction.

They can also be used with alternative foundation and flooring systems which are more sustainable than traditional wet concrete systems.

Hemp lime/hempcrete

Hemp lime is a 'novel' construction material. The composite material combines fast-growing renewable and carbon-acquiring plant-based aggregates (hemp shiv) with a lime-based binder to form a lightweight material that is suited to various construction applications, including solid walls, roof insulation and under-floor insulation and as part of timber-framed building. It also offers good thermal and acoustic performance and the ability to regulate internal relative humidity through hygroscopic material behaviour, contributing to healthier building spaces and providing effective thermal mass.

The hemp shiv is mixed with a lime-based binder which binds the hemp aggregates together, giving structural strength and stiffness. The lime also protects the shiv from biological decay. Interestingly, this material can be used to manufacture solid non-load-bearing panels, which can be installed as the cladding to timber (or other) framed buildings. A number of studies have shown that hemp lime has good thermal and acoustic insulation properties. The lightweight hemp lime absorbs sound, dampening transmission through walls and other elements.

Figure 10.28
External walls sprayed
with hempcrete.

PART 3

Figure 10.29
Traditional wall with
hemp lime inner leaf.

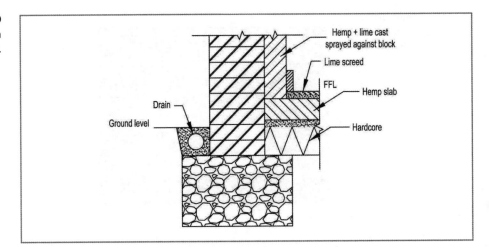

Figure 10.30
Hemp lime loadbearing
wall with rain screen.

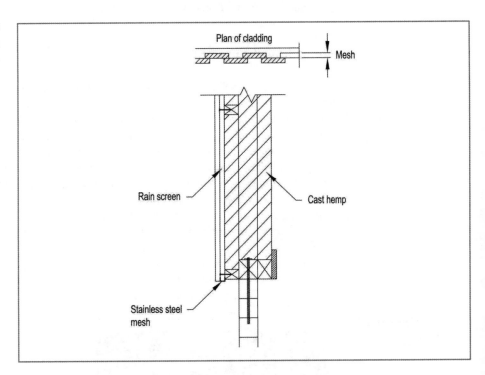

Hempcrete is a mixture of hemp hurds (woody core) and a hydroscopic (breathing) binder, usually lime based, which produces an insulating and heat-storing material that ideally is used to build the walls and floor and to infill the roof structure to provide a very-low-energy building which is very healthy to work in. Figure 10.28 shows the finish to the external walls when sprayed hempcrete has been applied.

Cladding using hemp lime/hempcrete can be constructed by means of techniques similar to those utilised in clay wall construction using formwork or it can be formed into blocks and the cladding constructed using techniques similar to those utilised for traditional blockwork. The hemp lime can also be used for roofing when formed into sheets.

The main advantages of hemp lime are:

- It is a means of achieving energy efficiency due to its insulating properties
- It is a breathable material which can help to create healthy buildings
- It is made from renewable sources and creates little environmental damage
- There is no pollution created at end-of-life disposal
- Hemp is a crop-based material which helps to benefit farmers and is a good use of land
- It offers the possibility of acquiring carbon into a building fabric as opposed to emitting it
- Hemp lime has the ability to make an impact on the future of sustainable building by reversing the damaging effects of greenhouse gases. It is claimed that hemp lime can lock up approximately 110 kg of carbon dioxide per cubic metre of wall.

Figure 10.29 shows a traditional wall with an inner leaf of hemp lime cast onto it and Figure 10.30 shows the main loadbearing wall constructed using hemp lime with a rain screen attached to it to prevent damage. These are two potential situations where the use of hemp lime would be recognised.

Formation of openings

The creation of openings in the external envelope of the building is essential to allow access and egress to the occupants and to allow for the provision of natural light and ventilation or the passage of services. Openings require the provision of a supporting beam or lintel if brick/blockwork is being used above, and materials commonly used for the manufacture of lintels are steel (treated to prevent corrosion) and reinforced concrete.

In addition to the loadbearing capabilities of the lintel, a number of other factors must be noted. Of particular importance are thermal transmittance and cold bridging. The design and installation of lintels is aimed at addressing the issue of the passage of moisture and heat. However, when using non-loadbearing cladding the openings can be formed in the panels themselves by cutting the opening into the panel or by fixing the sheets in such a way as to form the opening. As the cladding is non-loadbearing no lintel is required. However, if blockwork is used to form an inner leaf this will require lintels where openings are to be formed.

Movement accommodation

All building materials change their volumes to some extent when affected by changes in temperature and moisture, and masonry is no exception. If restrained, the effects of such movement can be considerable; hence movement accommodation joints must be incorporated to allow for such occurrences.

It is quite common to fill the joints between solid cladding panels with mastic as this acts as a movement joint.

REFLECTIVE SUMMARY

- Clay wall, straw bale wall and hemp wall construction are examples of technologies that are not only labelled as green, but as natural green because they use natural materials and systems in their construction.
- The use of these systems for construction the walls/cladding of low-rise large-span buildings is increasing in popularity but there is still some reluctance from designers to specify them.
- With all materials used, some form of expansion joint will need to be incorporated to allow for differential settlement.
- If using brickwork cladding, lintels will need to be used to form openings. When using other materials the opening can be formed by cutting into the cladding panel or by forming the opening during manufacture.

REVIEW TASKS

- Sketch a typical detail through the cladding of a low-rise building and identify the function of each component.
- Why do you think there is a reluctance on the part of designers to specify the use of greener technologies for cladding?
- Referring to the comparative study table below, identify criteria that you may want to use to compare the different systems.
- Visit the companion website at www.palgrave.com/engineering/riley2 to view sample outline answers to the review tasks.

COMPARATIVE STUDY: CLADDING

Option	Advantages	Disadvantages	When to use
Loadbearing brickwork	Familiar technology Good loadbearing performance No need for a frame Good loadbearing capacity	Slow construction Expensive Height of building limited	Unusual in industrial and commercial buildings, but could be used for low-rise forms
Infill panels	Cheap Familiar technology Supported off structural frame	Potential for cold bridging Old-fashioned appearance Exposed frame requires maintenance	Technically possible for most concrete-framed buildings, but unusual in modern construction
Cladding	Cheap if mass produced Fast Frame is covered Variety of materials can be used	Maintenance of joints may be required Heavy if concrete used	Suitable for use with all types of frame and for any height of building
Curtain walling	Modern aesthetic appearance Independent of the frame Renewable Upgradable	Expensive Non-traditional appearance	Suitable for use with all types of frame and for any height of building

CASE STUDY: CLADDING

The photograph shows a multi-storey hotel building that is being clad with concrete panels. The panels are pre-finished and need no further treatment. You will note that you cannot see the internal beams and columns as you would if infill panels had been used.

This photograph shows a close-up detail for the cladding panels.

This photograph shows the front elevation with one panel of the 7th floor left to fix in place. The cross-walls for the 8th floor are in position and the panels will be fixed shortly.

This photograph shows the cladding panels on the back of the lorry waiting to be lifted into place.

PART 3

(*continued*)

CASE STUDY: CLADDING (*continued*)

These two photographs show the cladding panel connection details at the rear face of the panel.

At the bottom a corbel has been formed into the back of the panel, and this rests on the concrete slab and is bolted into place.

At the top a steel angle plate has been bolted onto the back of the panel and this then bolts onto the underside of the slab.

The metal studs that you can see running vertically up the inside of the cladding panels are used to fix plasterboard to. This will give a good finish to the interior of the building.

The photographs below show a different type of cladding that is being used here in conjunction with a multi-storey steel frame. The steel posts shown are brickwork sliding anchors, but are being used additionally as wind braces.

Each post is connected to the underside and top plate of the steel beams.

(*continued*)

CASE STUDY: CLADDING (*continued*)

The slots in the posts are oval and therefore give some tolerance for movement of the block ties to correspond to the horizontal joints of the blockwork.

The blockwork forms the internal wall and you can see in the photograph that blockwork ties have been built into the blockwork as it is built. Another skin of blockwork will be tied into these to form the external wall, which is to have a paint finish.

PART 3

AIMS

After studying this section you should be able to:

- Appreciate the functional requirements for upper floors in multi-storey buildings
- Outline the various components of upper floors
- Determine the suitability of particular types of floor construction in given building scenarios
- Appreciate the form and function of stairs in multi-storey commercial buildings
- Determine the most efficient method of achieving internal access in multi-storey buildings

This chapter contains the following sections:

INFO POINT

- Building Regulations Approved Documents B, K and M (2010)
  - AD B (fire safety), concerned with the means of escape in case of fire
  - AD K, concerned with the minimum standards of safety for able bodied persons for normal access in buildings and for protection from falling
  - AD M (access and facilities for disabled people), concerned with making buildings accessible by disabled people
- BS 5268: Structural use of timber. Code of practice for trussed rafter roofs (2002)
- BS 5395, Parts 1, 2 and 3: Stairs, ladders and walkways (2010, 1985, 1985)
- BS 5578, Part 2: Building construction. Stairs. Modular co-ordination. Specification for co-ordinating dimensions for stairs and stair openings (1978)
- BS 5655, Part 1: Lifts, escalators, passenger conveyors and escalators (for lifts installed before 18 March 1994) (1986)
- BS 5655, Part 10: Lifts and service lifts. Testing and examination of lifts and service lifts (1986)
- BS 5950, Parts 1–9: Structural use of steelwork in building (2000, 2001)
- BS 8110, Parts 1–3: Structural use of concrete (1987)
- BS EN 13015: Maintenance for lifts and escalators. Rules for maintenance instructions (2001, 2013)
- CIBSE Guide D, Transportation systems in buildings, 3rd edn, 2005

11.1 | Upper floor construction in multi-storey buildings

Introduction

- After studying this section you should appreciate that the construction of upper floors for multi-storey buildings depends largely on the form of construction used for the structural frame.
- You should also appreciate that the advantages and disadvantages of using different materials are also largely the same as for the frame itself, which were summarised in the Case study in Chapter 7.
- You should understand, however, that the functional requirements of floors are the same regardless of the materials and method of construction used.

Overview

The primary function of upper floors must be to provide a sound, level surface that is capable of supporting all loadings over a given span. In addition, floors must provide adequate fire protection, durability, thermal and sound insulation, and provision for services incorporation. Speed and safety of construction are also important considerations, and increasingly the environmental impact of the use of materials is assessed.

Functional requirements of upper floors

Strength and stability

Floors can be classified by the form of construction. The two most common forms of floor are those supported by arches and those supported by beams. In arch construction all the structural elements are in compression, and in beam construction the top of the beam will be in compression and the bottom in tension.

Both of these forms allow for uninterrupted clear floor spans, but the most versatile for a whole range of floor spans is beam construction. Upper floors are different from ground floors in that ground floors tend to be supported in their entirety by the solid ground beneath them. Upper floors are supported intermittently using beams and columns, with large areas suspended.

The floor needs to carry all of the dead loads, i.e. the weight of the floor structure, plus any imposed loads from the people who will be using the building and furniture etc. Deflection is allowed, but must not be excessive, and consideration needs to be given for any potential change of use. Overdesigning the floor will lead to excessive cost of construction, and therefore a realistic evaluation of any future loads needs to be undertaken.

Dimensional stability

Movement in floors is generally less than in roofs as floors are not exposed to the elements. However, movement can occur due to differential movement of different materials in the floor itself, e.g. in reinforced concrete.

Prevention of the effects of vibration

Long-span lightweight floor structures are especially prone to movement through vibration. Movement joints need to be incorporated to prevent damage due to vibration.

Vibration can be particularly troublesome where a change of use occurs and vibrating equipment is installed.

Deflection

Excessive deflection in floors can create problems on two counts:

■ If a material is stressed so much that the materials used in construction reach their elastic limit, the floor is at risk of collapse. It is relatively easy to design the floor to prevent this problem occurring at the initial design stage, as the dead and intended imposed loads can be calculated and a factor of safety applied. However, if there is a change of use and the imposed loads are dramatically increased, overloading will occur that can cause a collapse.
■ A certain amount of deflection is allowed, usually 1/360th of the span in a beam-constructed floor. The allowable deflection is calculated along with the actual deflection based on the known dead loads and assumed imposed loads. If the actual deflection is less than the allowable deflection then the level of deflection is acceptable. Any greater deflection can lead to cracks in the underside of the floor, which can damage finishes and become unsightly, even if the stability of the floor is unaffected.

Thermal properties

A major problem with the **insulation** of floors is that holes need to be cut to allow service runs to pass from floor to floor. These holes form a natural passage for heat loss between floors.

Cold bridges in floors must be avoided, and this can be achieved by ensuring consistency in the **insulation** provided. The shape and thickness of the floor affect the thermal properties: generally floors of greater mass will provide greater thermal insulation, but greater mass leads to greater weight, which requires bigger foundations and also increased cost. The main problem in upper floors with regard to thermal insulation is ensuring that the holes left in the slab for services access are made good to the same standard as the rest of the floor area.

Moisture

Moisture in upper floors tends to be less of a problem than in ground floors, as there is no contact with ground water. However, moisture can be a problem and its presence occurs for a number of reasons:

■ Excess moisture during construction – an example of this is where a screed finish is applied to a concrete slab. Screed is made up of sharp sand, cement and water, and the water evaporates as the screed cures. If the screed is covered with an impermeable floor finish before it is sufficiently cured, the water remains trapped in the screed and will ultimately damage the integrity of the floor. One solution is to drill screed holes in a concrete slab which go right through the slab: any remaining water will drain through and potential damage is reduced. Other solutions include avoiding the need for screed by either applying a power float finish to the concrete, or by using a raised floor system which is fixed directly to concrete.

■ Surface and interstitial condensation can occur if cold bridges are present and the heating system is only turned on intermittently. If this is the case, then the internal air temperature will be greater than the temperature of the structure and condensation will form on the floor. Both types of condensation can damage the integrity of the floor.

■ Water spillages during construction or during building use can create problems if insufficient drying time is allowed. During construction water can be spilled (for example when sprinkler systems are being tested) and during use excessive water can be used for cleaning purposes. Although these are accidental, whenever they occur they must be treated correctly to avoid damage to the floor structure. Concrete floors can be sealed using sealing paints that are waterproof and can limit damage, although this is a costly process.

Options for upper floors to industrial and commercial buildings

In-situ concrete suspended slabs

In-situ concrete floor slabs (Figure 11.1) are generally used when in-situ concrete is chosen as the structural framing material. The concrete floor slab is poured at the same time as the beams to form an integrated system. The floor slab may be solid, contain hollow ceramic pots that remain in the slab or be a form of waffle construction where waffles are formed in the slab by using glass-reinforced plastic formers that can be removed once the concrete has cured.

Solid concrete floors are usually only used in buildings of up to four storeys, because for buildings of this size it would be more economical than a hollow floor. Solid floors can be used for any floor layout plan and are therefore highly flexible, as the thickness can be increased if and when required due to increased loadings. However, this form of construction is very heavy and the structure required to support it would need to take this into account.

Concrete composite floors

These are based on the tee-beam principle, and they are much lighter than the solid floor systems. A flat surface is created under the beam, and there is space in the slab for services installations (Figure 11.2), which remove the need for a suspended ceiling. The blocks are laid in place on the formwork, supported by small concrete wedges, and then the in-situ concrete is poured. The topping, which should not be less than 25 mm, provides a finish on which a floor finish can be applied.

Waffle floors are the most common type of in-situ concrete floor used in in-situ concrete frame structures. The waffles, usually made from glass-reinforced plastic (GRP), are nailed to the formwork, the reinforcement is fixed over the waffles and the in-situ concrete poured (Figure 11.3). Once the concrete is cured the formwork is struck and the waffles will release at the same time. Thus the underside of the slab will look like a waffle, and the gaps can be used to fix services. Although a suspended ceiling is generally required, this system will be the lightest form of floor construction, and therefore the size of the frame members may be reduced. Figure 11.4 shows the waffles fixed

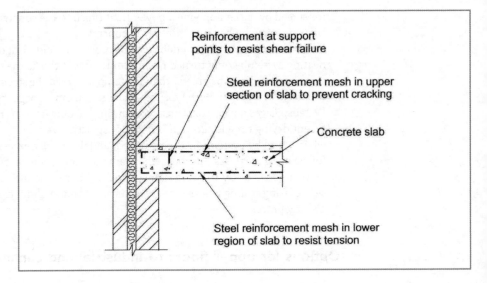

Reinforcement at support
points to resist shear failure

Steel reinforcement mesh in upper
section of slab to prevent cracking

Concrete slab

Steel reinforcement mesh in lower
region of slab to resist tension

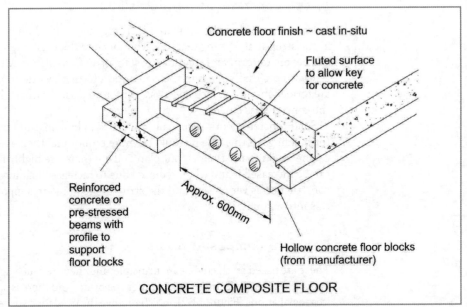

Concrete floor finish ~ cast in-situ

Fluted surface
to allow key
for concrete

Approx. 600mm

Reinforced
concrete or
pre-stressed
beams with
profile to
support
floor blocks

Hollow concrete floor blocks
(from manufacturer)

CONCRETE COMPOSITE FLOOR

in position awaiting the reinforcement to be fixed and the concrete to be poured. These are smaller waffles than are usually used but the principles of construction are the same.

The main advantages of using the hollow ceramic blocks and waffle floor constructions are:

- The hollow blocks and waffles can be used to house services such as electrical cables
- The slab reduces in weight but has the same strength and stability
- Slab-to-slab heights can be reduced because the space required for services is lower, and therefore the cost of construction is less.

Figure 11.3
In-situ waffle floor construction.

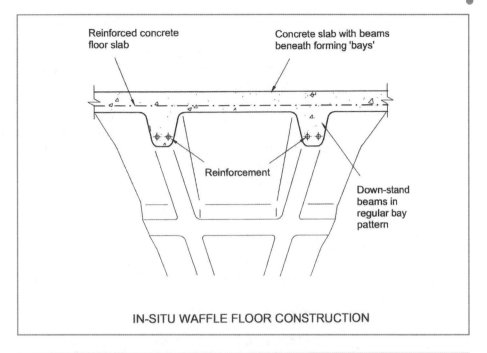

Figure 11.4
Waffles in position awaiting reinforcement and concrete to be poured.

Method of construction for solid, hollow block and waffle *in-situ* concrete floors

1. After the columns on the previous floor have been concreted, the formwork is often left in place and the beam and slab formwork connected to it.
2. The beam and slab formwork is constructed using either timber or a proprietary system, to the required shape and dimensions.

3. The formwork is supported using adjustable props.
4. Mould oil is painted onto the formwork.
5. Waffles or ceramic pots can then be fixed in place if being used.
6. The reinforcement is then fixed in the correct position on the formwork for the beams and slab.
7. If the floor beams butt up to an existing building, a form of slip joint will need to be included to take account of any differential settlement.
8. The formwork is checked for horizontal accuracy (correct to gridlines).
9. The formwork is checked for vertical accuracy. The dumpy level is set up on the slab below and the inverted staff used to check the level of the underside of the supporting timbers.

The abbreviation **SSL** is commonly used on construction drawings. It means Structural Slab Level, which is the level of the structural floor before any coverings have been placed.

The method for checking the level of concrete slabs is shown in Figure 11.5.

The correct level for this point is calculated by deducting the depth of the slab and the thickness of all the supports from the SSL. In this case it would be SSL – 20 – 150 – 200. This is checked by using the inverted staff on the floor below.

Example: the **SSL** is 35.200; therefore the level of the underside of the supports is

$$35.200 - 0.370 = 34.830 \text{ m}$$

A **TBM** is a temporary benchmark used to establish the correct vertical position of all elements of a building.

The **TBM** is fixed at 33.000 and the backsight reading onto it is 0.356. The Height of Instrument is therefore 33.000 + 0.356 = 33.356. The reading on the staff should therefore be 34.830 – 33.356 = 1.474, which will be inverted (Figure 11.6).

10. If fine adjustments are required, these can be achieved by raising or lowering the props. (Note: it is good practice to slightly raise the level of the slab before pouring (5 mm), as any settlement of the formwork will reduce the slab level and therefore reduce the amount of space for services.)
11. The edge of the formwork will be set at the correct structural slab level, but at the centre of the slab adjustable supports can be used to ensure accuracy of the level of

Figure 11.5
Checking the level of concrete slabs.

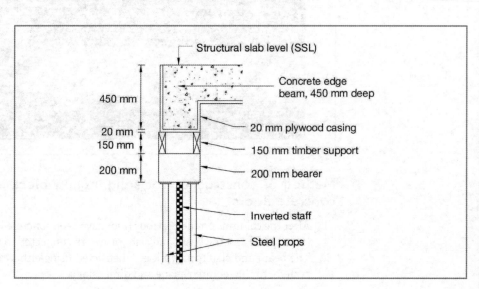

Figure 11.6
Level measurement for floors.

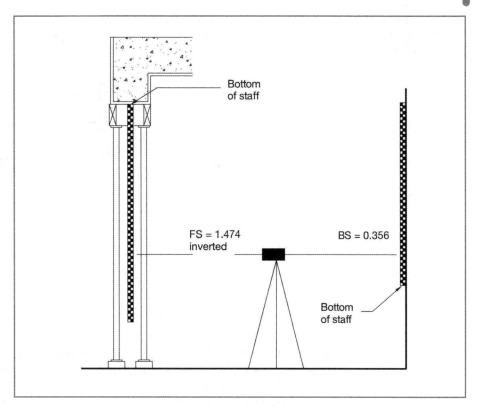

Bottom of staff

FS = 1.474 inverted

BS = 0.356

Bottom of staff

the concrete (Figure 11.7). The adjustable supports are set to the correct level and scaffold tubes used between them. The concrete will be poured to the underside of the scaffold tube level.

12. The formwork must be cleared of all rubbish.
13. The concrete is then poured. A concrete pump can be used or concrete skips attached to a crane. Care must be taken to ensure that segregation of the concrete components does not occur, as this affects the quality of the concrete, and the concrete must be vibrated sufficiently to avoid unwanted voids and to produce a good finish to the underside of the concrete.
14. Concrete cubes must be produced from each batch of concrete to assure concrete quality.
15. The concrete can be power float finished if required, just before the concrete is cured.

The advantages and disadvantages of using *in-situ* concrete for floor construction are as detailed for using *in-situ* concrete frames for the main structural elements in Chapter 7. The main advantage for the client is that, as construction is relatively slow, the pressure on the client's cash flow is reduced at the start of the contract. Conversely this can also be a disadvantage if the client requires the building to be constructed quickly. The other major requirement of using *in-situ* concrete is the level of skill required by the formwork joiners and steel fixers. It is argued that both of these trades are semi-skilled, but many, the authors included, believe that both of these trades require high skill levels that may

PART 3

Figure 11.7
Ensuring the correct level for the top of a floor slab.

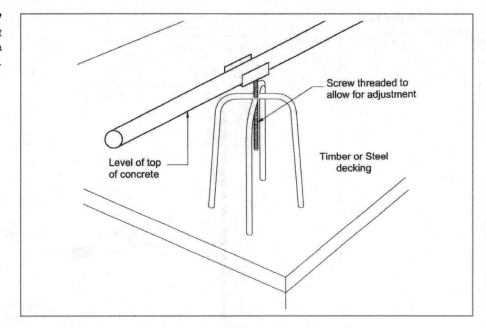

Screw threaded to allow for adjustment

Level of top of concrete

Timber or Steel decking

not be readily available. From a health and safety perspective, the use of formwork can be dangerous, with large sections being hauled into position which can cause back problems. Further, during striking it is not easy to predict where the section will fall, increasing the risk of accidents due to falling formwork. If formwork is allowed to lie around the site it can be hazardous, as it can have large nails protruding from it that can be stood on. Possibly the biggest hazard is the use of nails and reinforcing steel instead of the correct steel pins in the supporting props. This can lead to collapse of the formwork and serious eye injuries if operatives walk into them.

Reinforced concrete could not be considered a sustainable form of construction, but there are no real alternatives other than a steel frame encased in concrete, which has an equivalent environmental impact. However, if the formwork used is timber from renewable sources for the beams and recycled card for the columns, then it is sustainable as much as is possible.

Precast concrete slabs

Precast reinforced concrete floors are a popular choice in multi-storey buildings. This is mainly due to the potential speed of construction and the lack of the need for formwork.

Precast concrete planks or beams are generally used. They can be dense or aerated concrete, and they can be prestressed and in many cases hollow. Prestressed floors are popular because the thickness and hence dead weight of the sections can be reduced for large spans. Prestressing of the precast units is the easiest way to manufacture the units, as this operation would be carried out in a factory as opposed to on-site. Post-tensioning would be a site-based activity and therefore prone to quality problems.

The main disadvantages of using precast units to form upper floors in multi-storey buildings are the possible length of the lead-in time for delivery, and the difficulty of lifting the sections into position. A crane, either mobile or permanent, would be required, and as with any use of cranes there are health and safety implications.

Beam and block floors

These consist of inverted tee beams that are placed alternately with precast infill blocks.

These can be used in conjunction with *in-situ* concrete frames, **precast concrete** frames or structural steel frames. The tee beams and infill blocks are manufactured off-site and are delivered and lifted into place using a crane. The advantage of using this type of system is that the quality of factory-manufactured systems is generally always better than the quality that can be achieved using *in-situ* methods. However, some *in-situ* work is required when using precast flooring systems, as most systems require a form of concrete topping over the precast units.

Alternative forms of precast floor include:

- Hollow plank floors (Figure 11.8)
- Single or multiple tee beams (Figure 11.9)
- Permanent shuttering in the form of profiled troughs carried on beams
- Solid prestressed planks.

As with *in-situ* concrete frames, the steel is protected from fire by the concrete and therefore no further protection is required. The main advantages of using precast flooring are that the quality of the concrete is usually high and less *in-situ* work is required; indeed, it can be said that the precast concrete panels are in fact permanent formwork. Less propping is required during concreting of the floor slab, and the floor will be produced more quickly than it would be using an *in-situ* option. The main disadvantage is the long potential lead-in time required before delivery of the precast flooring system.

Using **precast concrete** units reduces the working time on-site, and also ensures a better quality of product because precast units are made in factory conditions as opposed to site conditions.

Figure 11.8
Precast plank (narrow slab) floors.

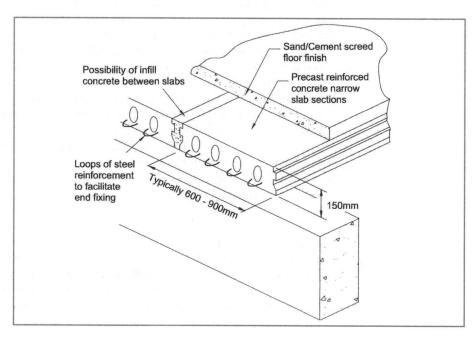

Possibility of infill concrete between slabs

Sand/Cement screed floor finish

Precast reinforced concrete narrow slab sections

Loops of steel reinforcement to facilitate end fixing

Typically 600 - 900mm

150mm

Figure 11.9
Beam and block floors.

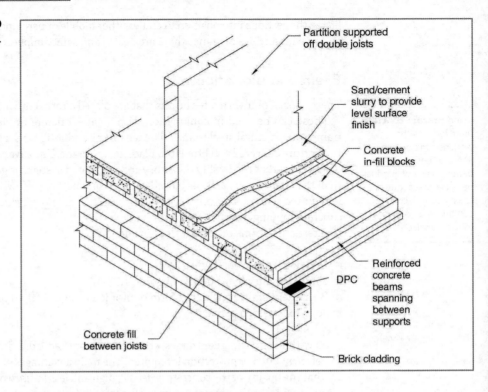

Partition supported
off double joists

Sand/cement
slurry to provide
level surface
finish

Concrete
in-fill blocks

Reinforced
concrete
beams
spanning
between
supports

DPC

Concrete fill
between joists

Brick cladding

Floor systems for steel frame structures

When forming a floor to structural steel multi-storey buildings, *in-situ* concrete floors can be used by fixing formwork to form the beam and floor slab, as in an *in-situ* concrete frame, or by using precast, prestressed concrete planks that span from one steel beam to another. However, the most common method used for forming upper floors in steel frame buildings is through the use of cold rolled steel decks that are then covered in concrete to give fire protection and a smooth or tamped finish to which a raised access floor can be fixed (Figure 11.10).

The steel sheets are easily handled, as they are light, easily fixed using shot-fired fixings, and galvanised on both sides to protect from corrosion. When using this form of construction no floor slab formwork is required as the decking acts as permanent formwork, and is structurally strong enough to support dead and applied loads during concreting works if shot-fired into position. It is also a very quick form of construction. This type of system requires the least amount of *in-situ* concrete when combined with a steel frame.

The methods used for encasing beams and columns in structural steel frames are given below.

- *Columns*: Concrete may be used to allow for fire protection if the columns are exposed and could be accidentally damaged. Concreting of columns will also provide a sound finish for decorating. However, fireline board or intumescent spray may be used if the columns are to be enclosed; therefore the use of *in-situ* concrete is minimal.
- *Beams*: Again, beams may be encased in concrete, but if they are to be covered with a suspended ceiling, fireline board, vermiculite spray or intumescent spray could be

Figure 11.10
Concrete and steel
composite floors.

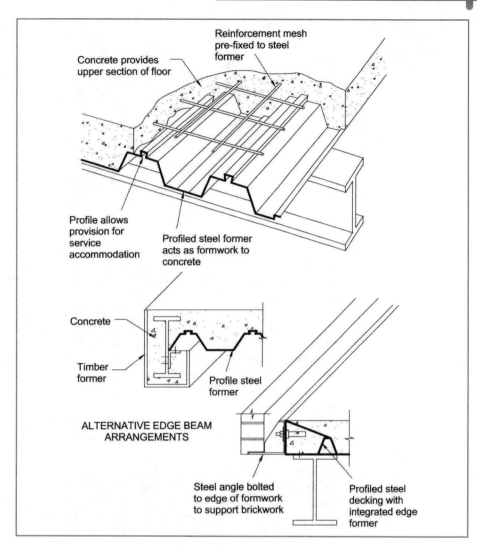

used to protect the beams from fire. Therefore the need for *in-situ* concrete is removed.

- *Floor*: The underside of the floor may be sprayed with vermiculite spray, which is very quick to undertake. However, a suspended ceiling could be used that has sufficient fire-resisting properties to obviate the need for fire protection to the decking. The decking could be manufactured with intumescent properties, and fire barriers could be used to prevent the spread of fire within the suspended ceiling, which would ensure that only limited damage would be caused to the slab in the case of fire. All of these methods require no *in-situ* concrete.

It is therefore common to see only *in-situ* concrete used for casing columns on structural steel frames.

The steel sheets are fixed to the steel floor beams either on the top or on an angle that has been welded to the side of the floor beam or column: reinforcement mats are then

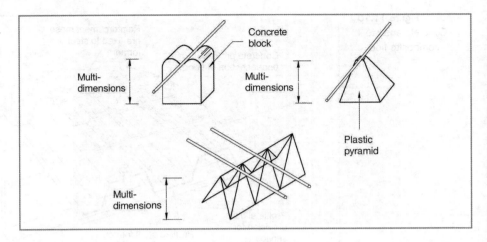

laid over the decking, supported by spacers that are either small concrete blocks or upturned V-shaped mini lattice frames (Figure 11.11). These are used to ensure that the mesh is fixed in the top third of the slab.

Adjustable supports are then fixed with scaffold tubes at the correct level for the concrete, as in the method used for levelling in *in-situ* concrete frame floor finishes, and the floor is then concreted. A major advantage of using this form of construction is that the aggregate used in the concrete may be 'Lytag', which forms concrete that is very wet, with small balls as the aggregate instead of bigger lumps of multi-shaped stones. This type of concrete is very easily placed using a concrete pump, which is very quick and reduces the need for a crane.

Space frames can also be used to form floor slabs where large unbroken spans are required (Figure 11.12). However, they are mainly seen in roof structures as opposed to floor structures.

There are proprietary steel floor systems on the market, one of which is Slimfloor, which is manufactured by Corus. With this system the principles of decking floors are the same, but this system allows space for services. The services can be installed within the depth of the decking itself and a suspended ceiling fitted directly to it. However, this would not accommodate large service runs such as ductwork.

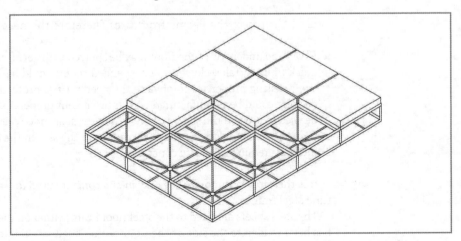

Timber floors

These are generally used in conjunction with timber frame multi-storey buildings. The principles of construction are generally the same as used in domestic construction. However, sometimes additional steel or timber laminated beams are required to support the floor because of the increased loadings due to higher floor areas.

REFLECTIVE SUMMARY

- The functional requirements of upper floors to multi-storey buildings are:
  - Strength and stability
  - Dimensional stability
  - Prevention of the effects of vibration
  - Prevention of excessive deflection
  - Thermal performance
  - Prevention of damage by moisture.
- The type of floor chosen is generally linked to the type of frame being used.
- Composite floors are lighter than solid floors and allow the passage of some services through holes on the sections.
- If using precast or *in-situ* concrete, steel reinforcement is needed in the floor slab to withstand tensile stresses. In steel decking flooring, the only steelwork required in the slab is anti-cracking mesh, as the metal decking will take the tensile stresses.

REVIEW TASK

- Using sketches, illustrate a form of upper floor slab suitable for the following frame types:
  1. *In-situ* concrete
  2. Precast concrete
  3. Structural steel.
- Visit the companion website at www.palgrave.com/engineering/riley2 to view sample outline answers to the review task.

11.2 | Internal access in multi-storey buildings

Introduction

- After studying this section you should have developed an understanding of the performance requirements of staircases in multi-storey commercial, industrial and residential buildings.
- You should also have a knowledge of the different systems that are available and be able to evaluate when and where the different forms of staircases would be used.
- You should also be able to assess the most appropriate finishes that can be used for staircases and floors.
- In addition, you should have a basic understanding as to how mechanical transportation is achieved in buildings.

PART 3

Overview

The movement of people around buildings is extremely important. Safety is of paramount importance and the ability to evacuate a building in the case of emergency needs to be considered carefully at the design stage. In addition to this, there are now stringent requirements for the accommodation and movement of people with disabilities in buildings.

Although safety is the highest priority when considering internal access, the economic viability of a building can also be assessed through the internal access systems. Ease of movement will increase the efficiency of the building, and therefore it will be more attractive to occupiers. Staircases can be designed for purely practical reasons and/or aesthetic purposes, but a combination of both is usually adopted these days. Although the structure of the staircase may be purely functional, the finish can give a very pleasing appearance.

Staircases

The main aims of staircases are to:

- Ensure that movement around a building is as effective as possible for all occupants
- Ensure that the economical space of the building is not compromised by the provision of internal access provision
- Allow for safe and effective means of escape from a building.

Staircases can be built from concrete, metals or timber and can be used to enhance the aesthetics of the interior of a building, as well as allowing for the vital element of permitting movement around a multi-storey building. In addition to this, where possible steps and stairs that are suitable for disabled people should be provided. They should be installed in addition to ramps and lifts. In order to achieve this, **stairs** should:

Stairs that are constructed with uneven pitch depths are very dangerous. Even very small differences can significantly increase the likelihood of somebody tripping.

- Have a shallow pitch and a generous tread width and landing to make them easier for a disabled person to climb
- Include a handrail to hold onto, preferably on both sides to use with either hand
- Incorporate refuges to allow wheelchair users to wait clear of escape routes
- Be well lit
- Ensure that the head and foot of the stairs are in the direction of the handrail, to enable safe use by people who have impaired vision
- Be wide enough to allow people to pass obstructions in case of emergencies
- Have colour contrasts in the decoration so that it is easy to quickly identify which floor a person is on.

Concrete stairs

Concrete stairs are used widely because of:

- Their resistance to damage from fire
- The relative ease with which they can be produced.

In-situ cast concrete stairs

In-situ cast stairs may be designed with or without strings. In the latter case, the stair itself is a reinforced slab and the major supporting element.

In-situ concrete stairs may take one of the following forms:

- String beam stairs
- Inclined slab stairs
- Cranked slab stairs
- Monolithic cantilever stairs
- Continuous slab stairs
- Spiral stairs (a variation of the monolithic cantilever).

String beam stairs

In this type of staircase, the strings may span between landing trimmers or be cranked to span beyond the landings to take a bearing at the perimeter of the stair (Figure 11.13). This type of stair will be thinner than the slab type and lighter in weight. Strings may be upstand or downstand and span between floor and landing, and landing to floor. Upstand strings prevent cleaning water and dropping articles from spilling over the stair. They also give the impression of a slab flowing between floors and are very useful for long spans.

Figure 11.13
In-situ concrete stairs.

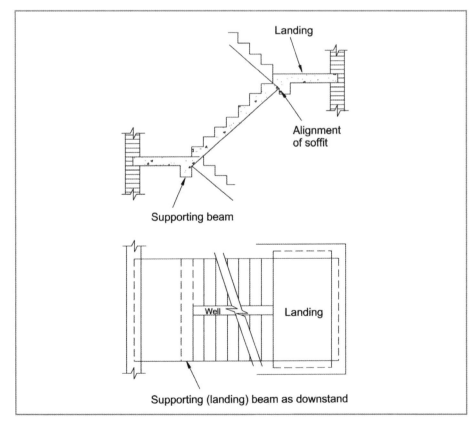

Stair flights span from string beam to bearing wall or from string beam to string beam. A central string may also be cast *in situ* to have precast treads added later.

Inclined slab stairs

In this type of stair, the landings span from well edge to loadbearing wall and the stair flights span from floor to landing and from landing to floor.

Cranked slab stairs

In this type of stair, the stair flights span as a cranked slab from floor to landing edge and from landing edge beam to floor. They therefore have what is sometimes known as a 'half landing' (Figure 11.14). Sometimes they have some form of structural support at landing levels, and sometimes not. If they do not they are sometimes known as 'continuous slab' or 'scissor' stairs.

Monolithic cantilever stairs

This type of stair (Figure 11.15) has landings that cantilever on both sides of a stub wall, which is a cantilever beam out of the rear edge of the spine wall. Stair flights are cantilever flights about the spine wall.

The main disadvantage of using *in-situ* concrete to form staircases is that the formwork required is very complex, and highly skilled labour is required to erect it. It is slower than other forms of construction and the staircases need to be completed as the building is erected floor by floor.

Precast concrete stairs

Precast concrete stairs are widely used for:

■ Simple solid or open riser steps
■ Small spiral stairs.

Using **precast concrete stairs** is only really cost-effective if there is a lot of repetition of the same stair system within a building.

Figure 11.14
Continuous slab stairs.

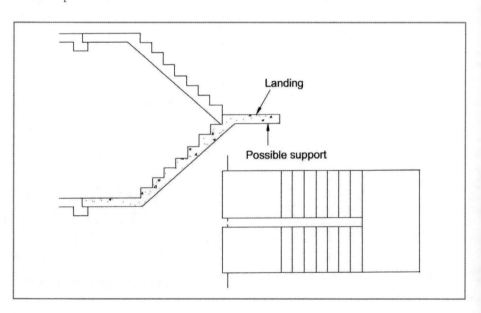

Figure 11.15
Monolithic cantilever
stairs.

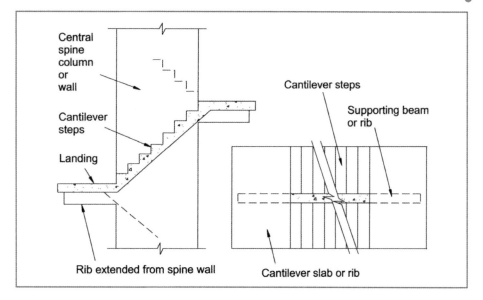

Larger stairs used to be most commonly constructed using *in-situ* concrete, even in a precast concrete frame, but precasting of stairs is far more common these days. Precast concrete stairs can simply be dropped into the building by the crane during the construction phase, even after the majority of work has been completed in the rest of the building. For example, if the stairwell was being used to house the tower crane, after the crane has been used to construct all the main elements of the building, it can be lifted out and a mobile crane used to lift the stair units into place through the top of the staircase that will have been left open at roof level.

Stairs may be precast:

- With strings and steps as separate pieces
- In whole flights
- In whole flights and landings.

Using precast concrete stairs can have the following advantages:

- Good quality control of the finished product
- Rapid construction
- Immediate access for following trades
- Saving in site space, since formwork fabrication and storage will not be required
- No propping required
- Fire resistance of up to 4 hours can be provided
- Can carry very high imposed loads
- The stairs can be installed at any time after the floors have been completed, thus giving full utilisation to the stair shaft as a lifting or hoisting space if required
- Hoisting, positioning and fixing can usually be carried out by semi-skilled labour.

However, there must be a sufficient number of staircases using each mould to make the operation of casting cost-effective. Therefore, for a one-off two-storey building with one staircase using precast concrete would be inefficient, but for a six-storey building with two identical staircases for each floor, i.e. 12 staircases, precast stairs would be efficient.

Typical reinforcement and stair/floor slab connection details are shown in Figures 11.16 to 11.18.

Figure 11.16
Precast concrete stairs and landings.

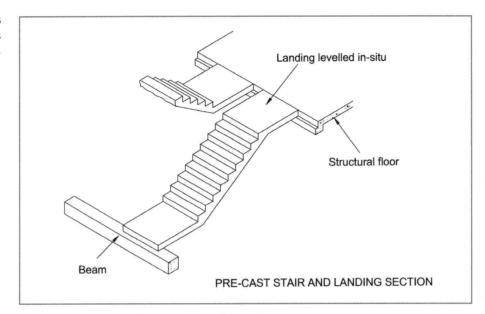

Landing levelled in-situ

Structural floor

Beam

PRE-CAST STAIR AND LANDING SECTION

Figure 11.17
Precast stairs being used for a multi-storey building. Here we see the stair at the junction with the landing. Note the difference in surface level due to the need to accommodate a screed finish to the landing.

Loadings

The loadings that the staircase needs to be able to withstand are laid out in the Building Regulations. They are summarised in Table 11.1.

Figure 11.18
Precast 'closed string'
stairs.

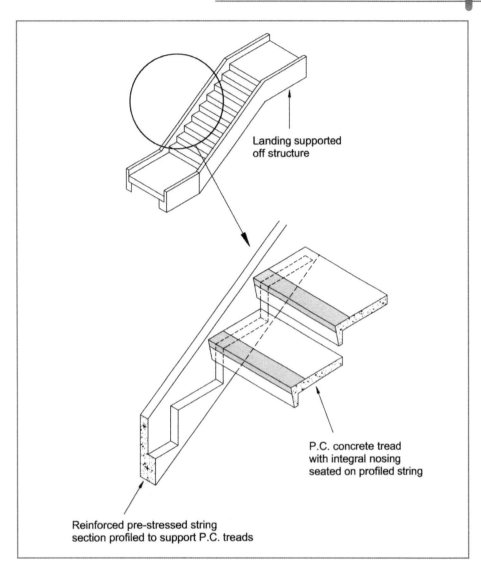

Landing supported
off structure

P.C. concrete tread
with integral nosing
seated on profiled string

Reinforced pre-stressed string
section profiled to support P.C. treads

PART 3

Table 11.1 Loadings that must be withstood by staircases.

Type of building	Loading
Domestic stairs	1.5 kN/m²
Most other occupancies and places of public assembly	4.0 kN/m²
Grandstands	5.0 kN/m²

Cast iron, metal and aluminium alloy stairs

These stairs (Figure 11.19) can be used as a means of escape or for internal accommodation stairs. Metal stairs tend to be purpose-made for a particular building and are therefore quite expensive. The major advantages of metal stairs are:

- No formwork is required
- They are very light and easy to lift into position.

However, the main disadvantage is that they may require a high level of maintenance due to damage to the paint finish.

They tend to be used predominantly for escape stairs, either internally or externally, but can be used internally in the main areas of the building for aesthetic purposes.

Figure 11.19
Metal staircase.

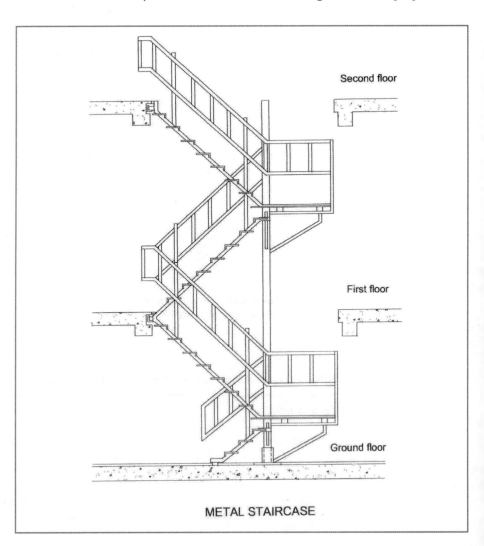

Second floor

First floor

Ground floor

METAL STAIRCASE

Spiral stairs

These are a variation of monolithic cantilever stairs. Spiral stairs are usually formed as a helix around a central column, whereas helical stairs have an open well, and may be circular or elliptical in plan. Figure 11.20 shows an elliptical metal staircase.

Figure 11.20
Elliptical metal
staircase.

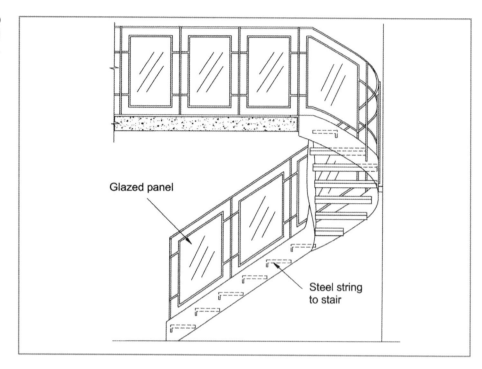

Glazed panel

Steel string
to stair

11.3 | Floor finishes

Introduction

- After studying this section you should have gained an understanding of the different types of screed floor finish that are used in commercial and industrial buildings.
- You should also have developed a knowledge of the different types of screed that are commonly used and how they are laid.

Overview

Ultimately the choice of **flooring system** will depend on the durability required in a particular building and aesthetics.

In the majority of commercial buildings the use of raised access floors is commonplace as they allow for services to be installed in the void, which are hidden from view and allow for flexibility of office layouts. This type of flooring is covered in detail in Chapter 13. However, there are instances where raised access floors are not required, certainly in industrial buildings. In this type of building the **flooring** may need to be very heavy duty and/or very level to allow for the smooth running of equipment. In this case the use of a power float concrete finish is sometimes used, where the concrete is finished to a smooth and level surface just before it has cured.

An alternative to this is the use of screed. Screeds are usually laid to receive flooring on a concrete base for the following reasons:

- To provide a degree of level and smoothness to suit a particular flooring
- To raise the level of the floor
- To provide falls such as in kitchen or bathroom areas
- To accommodate services, as shown in Figure 11.21
- To accommodate floor warming installations
- To provide thermal insulation (Figure 11.22)
- To provide impact sound insulation in the form of floating screeds
- To provide a nailable base for some types of flooring
- To level off an existing floor that has an uneven finish without raising the overall level of the floor unduly by using self-levelling screeds.

Figure 11.21
Ducting provision for floors.

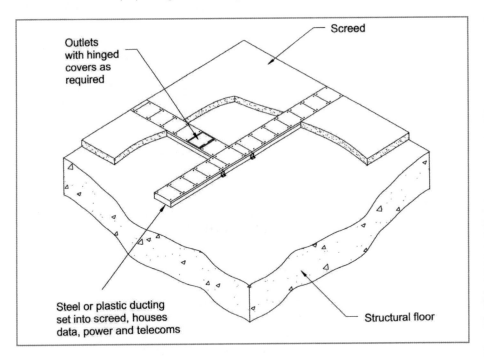

Screeds can be termed 'monolithic' if they are poured on green concrete (usually within three hours of the concrete being placed) or unbonded if they are laid on hardened concrete.

Screeds are usually made up of a mix of 1:3 cement/sharp sand, but the actual mix design may be determined by the finish that is to be applied to the floor.

The concrete floor has a tamped finish, which gives a rough surface to allow for adherence of the screed. The screeder places some small piles of screed at the correct level, and then uses a timber batten to level in the screed between piles.

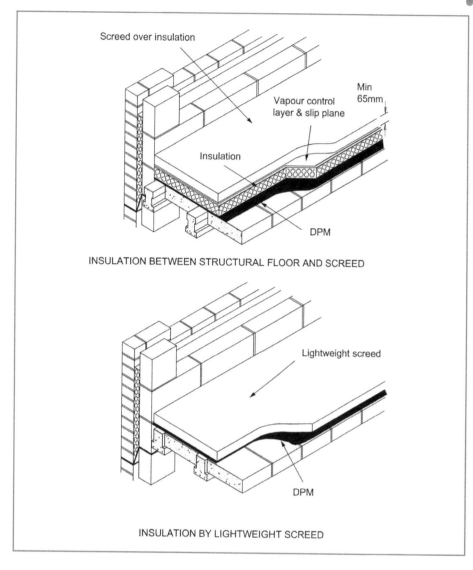

Figure 11.22
Screed and insulation
arrangements.

Floor finishes include:

- 'Plastic' floor coverings that are applied as a liquid, such as mastic asphalt or epoxide resins
- Concrete floorings, such as terrazzo
- Sheet-supplied rolled flooring, such as linoleum
- Boards, such as plywood
- Timber flooring, such as strip or parquet
- Clay floorings, such as brick paviors, ceramic floor tiles and mosaic
- Composition blocks, such as marble, granites and slates
- Carpets and felts in either rolled form or tiles.

All of these are suitable for large floor areas and also for use on concrete staircases.

PART 3

COMPARATIVE STUDY: UPPER FLOORS

Option	Advantages	Disadvantages	When to use
In-situ concrete	Flexibility in design If used with 'waffles', voids for the services are incorporated into the slab Good fire protection	Highly skilled labour required Slow form of construction Formwork can be very expensive Requires extensive propping Possible quality problems as cast on-site	Usually in conjunction with an *in-situ* concrete frame, where the floor and beams are cast as one pour
Beam and block	Precast, therefore quick to install Good quality assurance Good fire protection Semi-skilled labour only required	Heavy sections require lifting No voids for services in the slab Inflexible in design	Can be used in conjunction with all frame types, and are increasing in popularity
Concrete composite	Can be very strong and large spans achievable, especially if concrete beams are prestressed Good fire protection Partially precast therefore reasonably quick to install Semi-skilled labour only required Some voids in the floor for services	Some heavy sections require lifting Fairly inflexible in design Use of some *in-situ* concrete, therefore potential for quality issues Formwork required	Can be used in conjunction with all frame types, especially *in-situ* and precast concrete
Steel composite	Quick to install, easy to cut to the correct shape Light sections to lift Ease of handling Semi-skilled labour required Relatively thin slabs can span large distances	Requires fire protection Care needs to be taken during installation Steel sheets must be fixed securely to the frame before any loading out occurs	Used extensively in conjunction with steel frames Almost ubiquitous

CASE STUDY: FLOORS AND STAIRCASES

This Case study is based on the construction of an eight-storey student accommodation block. The details of the steel frame erection were shown in Chapter 7.

This photograph shows the steel decking loaded out on the frame waiting to be fixed in place.

The hollow rib decking fixed to the steel floor

beams.

The steel decking from the underside.
After the decking sheets have been shot-fire-fixed to the steel frame, the steel mesh reinforcement is

laid on the decking supported by spacers.
The concrete being placed using a vehicular

(continued)

CASE STUDY: FLOORS AND STAIRCASES (*continued*)

concrete pump.
This is a power floater that is used to create a smooth surface to the concrete. The concrete is poured in the morning and will be ready for power

floating by late afternoon.
The power floater has been used on the top floor and the concrete left to fully cure while the floor

below is being concreted.
This is a steel frame building that also incorporates steel staircases. The staircases are fabricated off-site and then bolted in place on-site, in the same way as other steel frame elements. The staircases have been treated with intumescent coatings to provide protection from possible fire damage. The only finish required to the risers is for aesthetic and sound-reduction purposes. The landings are to be grouted before the final covering

11.4 | Overview of mechanical transportation in buildings

Introduction

- After studying this section you should have developed a basic understanding of the systems used to achieve movement mechanically within buildings.
- You should be able to identify which system is most suitable for use in given scenarios, and be able to evaluate the advantages and disadvantages of using each system.

Escalators and lifts

Although realistically out of the scope of this book, it is impossible to discuss internal access without mentioning methods used for the mechanical movement of people around buildings. Internal access is an overall scheme and staircases need to be considered at the same time as mechanical systems if an efficient internal access design is to be achieved. The very form of modern buildings has been liberated and developed as a consequence of advances in the mechanical assistance of vertical and horizontal transportation. Prior to the development of lifts in buildings the realistic maximum height of buildings was around six storeys. Following the development of lifts, however, the height of buildings has continued to increase and the limits of height are now associated with the technological ability to construct tall buildings rather than the ability to move people and goods through them.

Mechanical movement of people in buildings

Mechanical movement of people in buildings is achieved by one of four means:

- Lifts
- Escalators
- Paternosters (although these are now very unusual)
- Travelators.

Lifts

Lifts allow buildings above four or five floors to be used, which would otherwise be the practical limit for buildings that only have stairs. They are a quick and efficient system for vertical movement. Lift installations require liaison at the earliest stage of design between the building designer and the specialist engineer.

The designer will need to determine a series of issues that will affect the selection of an appropriate design solution. These will include:

- The number of floors to be served
- The height between floors
- The types of users
- The quantity of users.

The functional requirements of the lift installation will be affected by various factors which could broadly be categorised as:

- User requirements
- Quality standards.

Some of these issues will be absolute and will be applicable to all buildings within which a lift installation is required. However, many aspects will be bespoke to the specific installation and will be affected by the building form and design, the nature of the building population and the context of their use of the building.

The notion of 'quality' in the context of lift installations relates to the various factors that might be taken into account when taking a view of the 'fitness for purpose' of the particular installation. Clearly this will vary depending upon the particular scenario in which the lift installation features. Typically the following may be considered as factors affecting perceptions of quality and definitions of appropriate quality for the installation:

- The level of prestige of the building
- The accuracy of the installation, including factors such as the levelling at the floor landings
- The required speed of the lifts and the speed and nature of door openings
- The level of use and probability of wear and tear resulting in the need for maintenance.

From an engineer's perspective there is a need for the lift installation to:

- Accommodate basic components for the type of lift
- Ensure safety in operation
- Facilitate appropriate maintenance
- Meet the functional needs required to be deemed fit for purpose.

Types of lift

There are two fundamental types of lift that are in common use. They can be described by the mechanism by which they operate which are:

Electrical lifts are commonly used for people, and **hydraulic** lifts for goods. This is not a general rule but is commonly seen in buildings today.

- **Electrical** traction lifts
- **Hydraulic** lifts.

Figures 11.23 and 11.24 illustrate typical lift shaft and car details for a simple traction lift arrangement. Whilst the details of location and installation of lifts will vary from building to building, the general requirements for all installations will be essentially the same. With a few exceptions, all lifts operate within a vertical shaft that links the floors of the building and which provides a structure within which the lift car and its operating mechanisms are housed. This shaft often forms the structural core of a building and in many cases is the first structural element to be formed. (This can be seen in the earlier sections dealing with slip form and jump form building techniques.)

General factors

The lift shaft or shafts will be located so that they provide a solution to the functional needs of the lift operation, but often also to provide a structural core to the building. This is particularly so in the case of *in-situ* concrete buildings. In addition to the structural constraints that may affect the building design, the location of the lift shaft will depend on the:

- Location of the entrance or entrances of the building
- Design of circulation spaces – these will seek to optimise pedestrian flow.

Figure 11.23
Typical lift shaft arrangement.

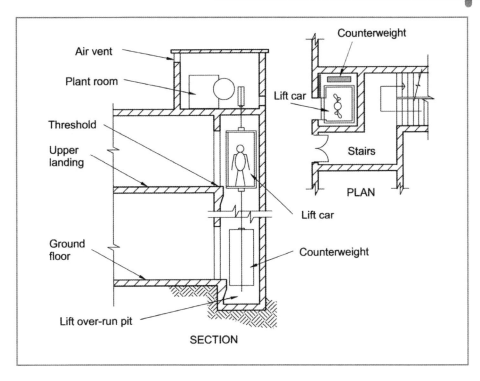

SECTION

Figure 11.24
Compact lift shaft detail.

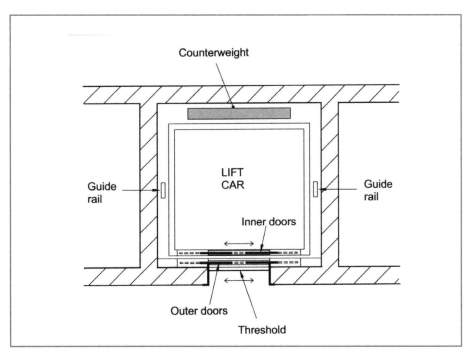

Lifts will be grouped together in large buildings to optimise pedestrian flow and to minimise the need to duplicate the construction of lift shafts and cores. The location will be such that effective management of 'tidal flow' of building users can be achieved. In most smaller buildings there will be one main core which contains the lift and stairs. In larger buildings there may be several of these cores positioned so as to maximise the efficiency of the movement of people through the space.

The velocity of the lift will depend on the function of the building, and the acceptable level of lift speed will vary between different types and scales of installation. For example a lift in a small nursing home may function perfectly adequately with slow speeds due to the low level of user demand, whilst a lift within a busy, modern commercial building will need to be considerably faster to cope with user demand and expectation. The following points need to be considered:

- Speed of circulation required by the user group
- Reliability requirements
- High velocity means high cost – and may only be justifiable in taller buildings. Although velocities of more than 5 m/s are possible the more typical operating range is between 0.5 and 1 m/s.

As the height of buildings increases, with some buildings now approaching 900 m in height, the speed of passenger lifts is becoming a key element in their effective use. Indeed the limiting factor on lift design has become the level of strain that the human body can withstand rather than the technical considerations limiting travel speeds.

Factors affecting system design

A range of economic, functional and technical factors will affect the design of any given lift installation. These include:

- The function of the building
- The number of floors to be served
- The total distance of travel
- The total number of people and/or objects to be conveyed
- Peak concentrations of use, i.e. start of work time and end of work time
- The acceptable waiting time for the arrival of the lift car
- The time delay in loading and unloading.

Relatively simple decisions, such as whether the doors of the lift open to one side or to both sides from the centre will affect the overall performance potential and the subsequent design decisions.

Electrical lifts

Early lift installations were hydraulically powered, but the introduction of electric power allowed lifts to have greater speed, convenience and the ability to service greater heights. Hydraulic lifts are now only used in limited circumstances and usually as goods lifts where greater lifting capacities are required.

Electric lifts suspend a moving car from a driving sheave with one end of the suspension cable attached to a counter balance weight, which is normally half the weight of the car. It also takes load off the driving sheave. A pulley system reduces the load of the driving system further. The use of the counterweight allows the driving force of the motor to be minimised and also reduces the need for excessive cable lengths. The early lift designs utilised a drum around which a cable extending to the full height of the building would be wound. This was inefficient and prone to problems in operation and maintenance.

Modern systems use a design by which the length of cable wrapped around the 'drum' is greatly reduced.

A single wrap system may be used for lightly loaded cars and a 3:1 wrap system for heavier cars.

With a single wrap system the machine room will be at the top of the lift shaft, whereas it will be at the bottom with a 3:1 wrap. The single wrap system provides a compact installation and small shaft dimensions.

Safety

It is an accepted truth that safety is paramount in the design and operation of lift installations. Accordingly there are components in a lift installation to ensure safety is achieved. These include the use of:

■ The overspeed governor
■ Safety gears or brakes.

The governor is connected to the car by a cable and pulley and when excessive velocity is detected, the brakes are applied. Inertia-activated brakes are also utilised that operate friction brakes when the lift accelerates aggressively. These systems are the basis of regular safety testing on lift installations and in the UK there is a statutory requirement to test lifts and assess safety regularly.

Hydraulic lifts

Hydraulic lifts operate on a totally different mechanical principle from traction lifts and are used where:

■ Large and heavy loads need to be carried
■ The building would not be able to carry the loads imposed by conventional lifts
■ Travel-speed and distance requirements are low
■ Lift utilisation levels are relatively low.

Hydraulic lifts are relatively cheap, however:

■ The maximum travel is limited to around 20 metres
■ They are slow compared to traction lifts
■ A limited amount of traffic can be accommodated.

The mode of operation of hydraulic lifts is very simple. Oil supplies the driving force to move the cars by means of a hydraulic ram. Pressurising the oil will extend

PART 3

the ram upwards and the car will ascend. Releasing the pressure will allow the car to descend.

The features of **hydraulic lifts** include:

- The ram, which may be either underneath the car or alongside it
- Controls that are simpler than those of electric lifts and may be remote from the lift shaft
- Hydraulic lifts give a smooth ride and accurate levelling.

Hydraulic lifts offer an affordable solution in instances where larger-scale traction installations may be unnecessary and expensive.

For electric lifts, the machinery is usually held in a roof-level plant room, but for **hydraulic lifts** the machinery is usually at basement or ground-floor level dependent on what is the lowest level of the building.

Lift call systems

The form of lift call system used considerably influences the effectiveness of an installation. The options are:

- *Control-down collective* or *directional directive*
  Down collective – the car ignores all upward calls whilst reacting to a call in upward motion. It will stop at all floors where there is a downward required.
  Directional directive – while in upward motion the car will only call at successive floors that have requested upward movement. The calls for downward movement are recorded and are met in progressive downward motion during which upward calls are recorded but ignored until the lowest level is reached.
- *Automatic control*
  This is a simple form of call system in which no record of calls is made while the car is in motion. The car will respond to a call only after becoming stationary.
- *Group control*
  In large installations where a number of cars are used, a computer is used to mastermind the car movements to ensure efficiency of movements. In very tall buildings lifts may 'home' to particular floors. Individual lifts in group installation may serve a limited number of floors.
- *Attendant control*
  The car stops when an attendant presses a button just before reaching a particular floor. This is almost unheard of in modern lift installation.

The main advantage of using lifts is that they take up relatively small amounts of space compared with escalators. From a constructor's perspective, the installation of lifts and any other mechanical movement of people system is usually carried out by a specialist subcontractor. The builder is responsible for what is termed 'builder's work' which is all the work the builder needs to do before the actual system is installed. For lifts this would be the construction of the lift shaft. Lift shafts are generally used to house a tower crane during the construction phase. The base of the lift pit acts as the crane base, the lift shaft walls are built around the crane, and before the roof to the lift shaft is constructed the tower crane is lifted out of the shaft in sections using a mobile crane that is bigger than the tower crane. In very high buildings this is not feasible and crawling cranes are used that move up the building as the structure increases in size.

Escalators

Escalators are basically moving staircases. The disadvantage they have over lifts is that they require considerable floor space and only move people through a limited height. However, large numbers of people can be moved. For example a 900 mm wide escalator can move 6,000 people per hour.

Figures 11.25 and 11.26 show typical escalator arrangements.

Figure 11.27 shows a system that can be used in open areas in the event of fire. Roller shutters are activated to cover the escalator and seal off each floor effectively.

Figure 11.25
Typical escalator
installation.

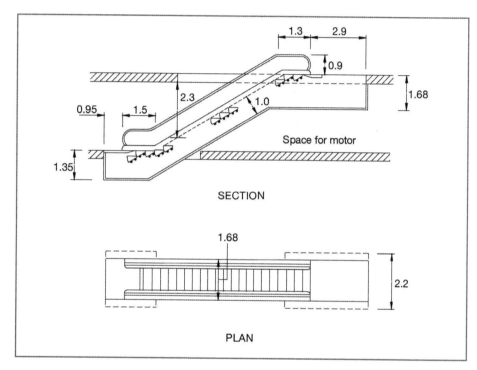

Figure 11.25
Typical escalator
installation.

Figure 11.26
Escalators serving
successive floors.

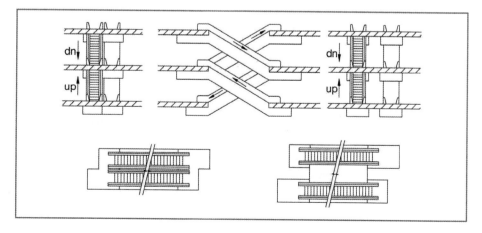

Figure 11.26
Escalators serving
successive floors.

PART 3

Figure 11.27
Roller shutters for fire
compartmentation.

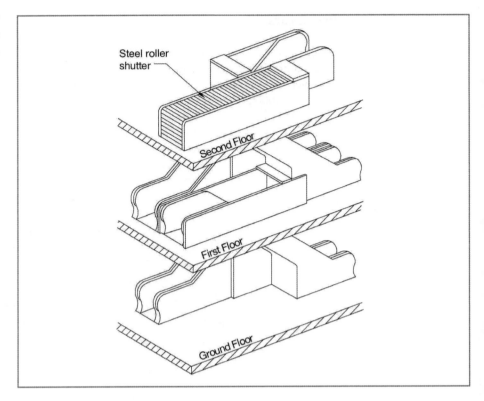

Paternosters

These are rarely used in the UK, but are widely used in mainland Europe. They are appropriate where internal circulation is the critical feature, rather than inward or outward flow. Up to 600 persons per hour can be moved. They use a series of platforms that go around continuously. People step onto each platform as they go around, and then step off on the correct floor (Figure 11.28). The principle is similar to a ski lift.

Travelators

Travelators are basically moving walkways (Figure 11.29). They enable the movement of large numbers of people horizontally or up slight inclines. They are most commonly seen at airports, as they reduce the need for people to carry heavy bags over large distances.

Fire protection of stairs and lift shafts

Flue-like apertures, such as shafts, ducts and deep light wells, should be avoided in the general design of buildings. However, where they are necessary, in the case of lift shafts and staircase enclosures, they should be vented at the top to allow smoke and hot gases to disperse into the atmosphere.

Figure 11.28
Paternosters.

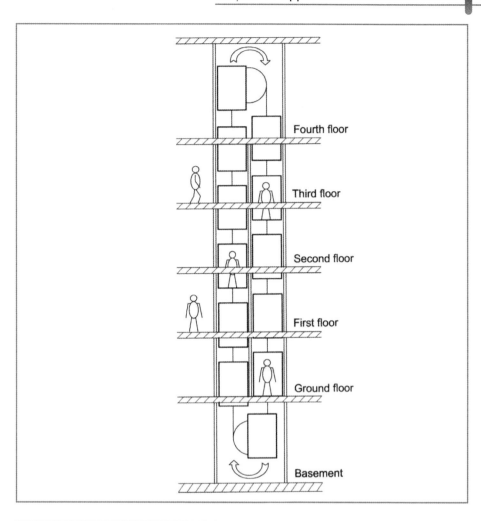

Fourth floor

Third floor

Second floor

First floor

Ground floor

Basement

Figure 11.29
Moving walkways.

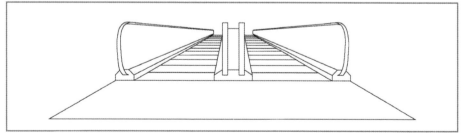

Access for firefighting

External access by ladders is limited to 30 m. Therefore additional requirements are imposed to allow safe internal access for the fire brigade. Providing at least one lobby approach staircase which is regarded as an extension to the street does this. The staircase must be separated from the accommodation on each floor by an enclosed lobby in which all the necessary firefighting equipment is stored.

Both staircase and lobby must be sited next to an internal wall and provided with enough ventilation to prevent smoke accumulation. If this is not possible mechanical ventilation must be used. There must be a secure air supply free of smoke, and a pressure of 50 Pa ± 10 Pa must be achieved. Sometimes natural ventilation, or mechanical ventilation where the staircase is situated away from the external wall, is boosted to the smoke pressurisation level when the smoke alarms are activated.

If there are any protected corridors these should be pressurised to 5 Pa below the protected staircase to prevent smoke entering the staircase.

The enclosing walls to staircase and lobby may be required to have double the standard fire resistance for the building – potentially for up to 4 hours. No openings are permitted

Figure 11.30
Firefighting shafts (based on Building Regulations).

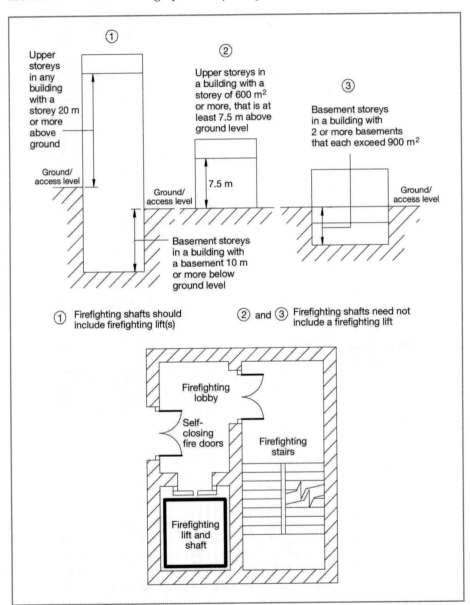

in the enclosure other than for ventilation onto the street or access to the accommodation. Self-closing fire-resisting doors are required and further protection by steel shutters is required for basements and in some other circumstances.

As well as a lobby approach staircase a firefighters' lift must be provided, situated:

- Where possible within the staircase lobby or
- Within the staircase enclosure or
- In a separate adjacent enclosure having the same degree of fire resistance.

This lift need not travel to the top floor.

The term 'firefighting shaft' (Figure 11.30) is used for a protected enclosure containing:

- A firefighting stair
- An eight-person firefighters' lift
- A firefighting lobby.

The lift can be part of the normal circulation of the building, but must be separately powered and capable of being brought under the control of the fire brigade.

REFLECTIVE SUMMARY
- When undertaking a comparative study of the characteristics of metal, *in-situ* concrete, precast concrete and timber floors for use in multi-storey buildings, the criteria used will be the same as for frames constructed of the same materials.
- The choice of floor is *usually* dictated by the choice of framing material and system of construction.
- The functional requirements of upper floors are strength and stability, dimensional stability, prevention of the effects of vibration, prevention of excessive deflection, thermal properties and protection from moisture damage.
- Different types of floor and stair construction will allow for different types of floor finish to be used. Each building must therefore be considered as a separate entity. There is no such thing as a 'standard model' in construction.
- Internal access is concerned with the economical, efficient and safe movement of people within buildings.
- The choice and positioning of stairs, lifts, escalators or travelators will be largely determined by the building use, but may also be chosen for aesthetic purposes.
- Means of escape need to be seriously considered when designing internal access systems.

REVIEW TASKS
- Using a table, compile a list of criteria that you could use to compare timber, precast concrete, *in-situ* and metal/concrete stairs.
- Undertake a comparative study utilising the criteria that you have identified as relevant.
- A proposed steel frame building is in the design phase. Imagine that you are the design and build contractor and recommend a system of internal access for the

PART 3

building that is efficient, economical and safe, as well as aesthetically pleasing. A list of provisions for each floor is required. The building use is planned as follows:
- Ground floor: retail units
- Floors 1–5: offices
- Floor 6: Four luxury apartments.

■ A client has just commissioned the construction of a five-storey steel frame building with precast concrete floors and stairs. Acting in the role of the planning supervisor, identify all potential health and safety risks and determine how these risks can be reduced.

■ Visit the companion website at www.palgrave.com/engineering/riley2 to view sample outline answers to the review tasks.

12 Roof construction

AIMS

After studying this chapter you should be able to:

- Describe the functional performance to be provided by roofs to industrial and commercial buildings
- Appreciate that when dealing with roofs we tend to divide the constructional detailing into roof structure and roof covering
- Understand the different basic methods of forming pitched and flat roof structures
- Describe the different forms of roof covering available for pitched and flat roofs
- Appreciate the various methods used for the collection of rainwater from pitched and flat roofs

This chapter contains the following sections:

INFO POINT

- Building Regulations Approved Document A: Structural strength, Appendix A (2004 including 2010 amendments)
- BS 747: Reinforced bitumen sheets for roofing. Specification [replaced by BS EN 13707: 2004 but remains current] (2000)
- BS 5534: Code of practice for slating and tiling (2003)
- BS 6229: Flat roofs with continuously supported coverings (2003)
- BS 6367: Drainage of roofs and paved areas (1983)
- BS 6399, Part 3: Loading for buildings. Code of practice for imposed roof loads (1988)
- BS 6577: Specification for mastic asphalt for building (1985)
- BS 8217: Code of practice for reinforced bitumen membranes for roofing (2005)
- BS 8218: Code of practice for mastic asphalt roofing (1998)
- BS 8490: Guide to siphonic roof drainage systems [replaces BS 6367: 1983] (2000)
- BS EN 12056, Part 3: Gravity drainage systems inside buildings. Roof drainage, layout and calculation [replaces BS 6367: 1983] (2000)
- BS EN 13707: Flexible sheets for waterproofing. Reinforced bitumen sheets for roof waterproofing. Definitions and characteristics [replaces BS 747: 2000 which remains current] (2004)
- BRE Digest 499: Designing roofs for climate change. Modifications to good practice guidance (2006)
- BRE Report 302: Building elements: roofs and roofing – performance, diagnosis, maintenace, repair and the avoidance of defects (revised 2000)
- *CIBSE Knowledge Series: Green Roofs*, The Chartered Institution of Building Services Engineers, London, September [978–1–903–28787–3] (2007)
- *Flat roofing: a guide to good practice* (Blue Book), 4th edn, Ruberoid plc (2002)
- Garrand, C. (2008) *Roofing, Technical Review*, RIBA [978–1–859–46253–9]

12.1 | Functions of roofs and selection criteria

Introduction

- After studying this section you should have developed an appreciation of the functional requirements of roofs to industrial and commercial buildings.
- You should also be familiar with the more common forms of pitched and flat roof options and you should have a broad overview of some of the other less common variants.
- The nature of the various conditions in which different options are used should be understood and you should be able to select between the available approaches to roof design in given scenarios.

Overview

The functional requirements of roofs to industrial and commercial buildings are essentially the same as the performance criteria required of all external primary elements to the building. The nature and form of framed buildings vary greatly from situation to situation, but the core functional requirements of roofs will remain the same. The methods selected to achieve these requirements will vary and will to a large extent dictate the very shape and form of the building. The decision to select a flat roof rather than a pitched roof, for example, will affect the aesthetics of the building greatly. It should be remembered that when dealing with roofs the normal procedure is to divide this element into two parts: the *structure* and the *covering*. Accordingly we shall consider the various options within these categories. Whichever form is chosen will be based on consideration of the following generic criteria:

- Strength and stability
- Weather resistance
- Durability
- Insulation
- Aesthetics
- Sustainability.

Strength and stability

It is now quite uncommon to find roof structures to simple framed buildings formed in **concrete**. Because **steel** is lighter, easier to form and has a high strength-to-weight ratio, it is by far the most common material for roof structures.

Industrial and commercial buildings are generally framed structures formed on the basis of long-span construction or high-rise skeleton frame construction. As such, the loads from the roof and upper floors are transferred through the building frame through direct application or via beams or purlins. Unlike domestic construction, in which pitched roofs are almost ubiquitous, many commercial and industrial buildings adopt flat roof solutions, whilst others will utilise some form of pitched roof design (Figure 12.1).

The strength of the structure of the roof depends on the strength of the material used to form it. Typically this will be a choice between **steel** or **concrete**, although in some instances timber roof structures may also be found. Whether the structure is of flat or

Figure 12.1
Roof designs.

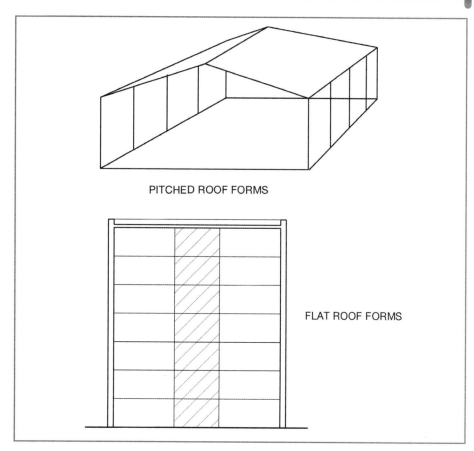

PITCHED ROOF FORMS

FLAT ROOF FORMS

PART 3

pitched type it will still be required to withstand the loads generated by the roof and the actions applied to it, such as wind load. These loads must be safely transferred to the building frame and ultimately to the supporting strata.

Most construction components are designed to withstand the actions of the load that will seek to induce *compression, tension and bending* or *deflection*. In general the components of pitched roofs will be arranged to act in tension or compression, as in the use of props and ties within trusses, for example. Conversely, the structures of flat roofs act to span between supports and will act to resist the forces that seek to induce deflection; hence they must be robust whilst attempting to minimise the self-weight of the roof.

Weather resistance

Weather resistance has always been the main function of a roof, and over the years many coverings have emerged to achieve this function. Generally pitched roofs will utilise overlapping sections of covering to allow water to run off the sloping surface without penetrating to the interior. Flat roofs, in contrast, rely on the water-excluding capacity of a continuous impervious covering.

There are many features of roofs purposely designed to resist moisture entry and each of these will be examined when reviewing the constructional details which apply.

A key issue in resistance to weather entry is the slope or pitch of the roof. Even 'flat' roofs depend upon the creation of a slight pitch to encourage water to run to collection or discharge points.

Durability

The ability of the roof and its covering to be durable has been linked with resistance to rain and to wind action. However, today atmospheric pollution may also be a factor in the equation, particularly in the industrial environments that some framed buildings are located within. This can cause corrosion of metal sheets and fixings, together with premature failure of the coverings that may be applied. Hence the probable exposure to airborne contaminants and corrosives must be taken into account when specifying a design solution. The combination of materials used must also be considered, since the possible reactions between materials and the potential for differential movement to cause premature failure must be taken into account.

Insulation

If we examine the pattern of heat loss from buildings we can see that the roof is of major importance in thermal resistance. As warm air naturally rises, the tendency is for heat to be lost through the roof, and this fact is reflected in placing the greatest need for insulation on the roof element. The large areas that are commonly covered by roofs to framed buildings mean that the effective insulation of the building enclosure is heavily reliant upon the effectiveness of the roof insulation. However, it must also be remembered that some industrial buildings may not be provided with heating to the interior. This is often the case in warehouses, agricultural buildings and so on. In such cases there is no need to provide insulation between the interior and the exterior, since the conservation of fuel and power is not an issue. Climate control in large building enclosures can be problematic, and in very large buildings the creation of internal weather systems or microclimates has been experienced. The environmental moderation of such buildings is complex and is beyond the scope of this book.

Insulation of the roof structure not only means insulation against the loss of heat but also against heat gain. Indeed, in modern commercial buildings much of the cost associated with climate control relates to the need to cool the interior rather than to heat it. This is a consequence of the actions of solar gain and the heat that is generated within the building by people, processes and equipment.

Insulation may also be required to reduce the level of acoustic penetration or emission from the building. The dBA (acoustic decibel) scale is the means by which we measure sound levels, and the roof can be quite effective in making dB reductions to create an acceptable internal environment. High noise generators such as aircraft may still prove problematic however, if the building is located on a flight path to or from an airport, or if the property is close to a major transport route such as a motorway. Industrial processes within the building may also need to be insulated against to prevent nuisance to the surrounding area.

Aesthetics

The shape of a roof is generally dictated by the scale and layout of the building and the structure of the walls or frame which support it.

The adoption of pitched roof forms is generally accepted as being more aesthetically pleasing, although in large buildings they are costly and can prove to be expensive. One solution to the problem of aesthetics is to utilise a combination of pitched and flat roof forms that allows the flexibility of form afforded by the flat roof whilst creating the appearance of a pitched roof. This is often adopted in the construction of modern super-stores, for example (Figures 12.2 and 12.3).

The slope or pitch of the roof will also have an impact on aesthetics, as shown in Figure 12.3. Different visual effects will also arise from the use of different roof coverings and their colour, texture, shape, size and laying pattern.

The more complicated the plan layout or footprint of the building, the more complicated becomes the pitched roof structure solution. For this reason, many complicated building shapes rely on the use of flat roof solutions.

The nature of the roof forms utilised in the construction of large buildings differs substantially from those utilised in the construction of smaller buildings, primarily as a result of the scale of the building. Large roof areas dictate the adoption of design solutions which are capable of spanning relatively long distances and, in some cases, carrying substantial loads. There are two distinct groups of roof design that are adopted for industrial and commercial buildings. For the purposes of our consideration here these are conveniently termed 'flat roof' and 'pitched roof' options. Historically, the flat roof forms were utilised most commonly in the construction of high-rise framed buildings. One of the reasons for this was the need to gain ready access to the external enclosure of the building, the flat roof providing a mechanism for housing access gantries and cradles and allowing ready access to the roof area itself. An additional benefit was the ability to provide for housing of mechanical and electrical plant and to gain access to it for maintenance etc. The complexity of building shape or footprint often makes the creation of pitched roofs difficult, and this is another reason why flat roofs may be favoured in the construction of commercial buildings (Figure 12.4). Conversely, pitched forms tended to be adopted for long-span or low-rise building forms, where building footprints have

PART 3

Figure 12.2
Changing appearance through shape.

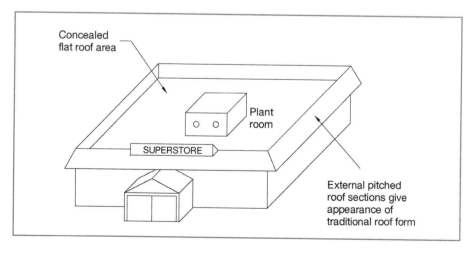

Concealed flat roof area

Plant room

SUPERSTORE

External pitched roof sections give appearance of traditional roof form

Figure 12.3
A typical example of creating the appearance of a pitched roof on a flat-roofed building.

tended to be simple. For buildings of this type it is also common for very large areas to be covered by the roof, requiring careful treatment to ensure that rainwater run-off is facilitated effectively.

Figure 12.4
Roof solutions for complex shapes.

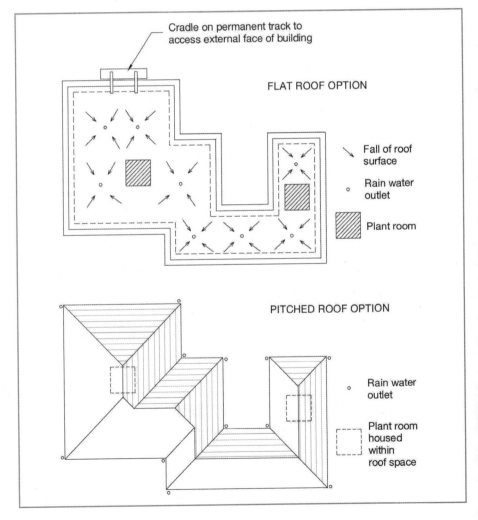

Figure 12.5
Laminated timber roof
construction.

 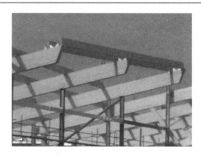

Figure 12.5
Laminated timber roof construction.

Sustainability

The issue of sustainability is dealt with in some detail in Part 4 of this text and we shall not repeat that material here. However, it is axiomatic that the design, form and construction of building elements, including roofs, have evolved significantly and rapidly as a result of the need to develop sustainable solutions. The increased use of sustainably managed timber products is a visible illustration of this. The need to create large structural members to provide for reasonable roof spans in commercial and industrial buildings once limited their construction to steel and concrete forms. The advances in technology associated with laminated timber, for example, now allow for large sections of structures to be formed from timber, whilst adhering to sustainable construction practices. Figure 12.5 shows a long-span building that utilises laminated timber beams and purlins to achieve large open-floor areas.

REFLECTIVE SUMMARY

When considering the selection of appropriate roof options for commercial and industrial buildings, remember:

- The selection will be driven by
  - The structural form of the building
  - Aesthetic requirements
  - Functional requirements, such as thermal insulation and durability.
- The long spans and/or heavy loads that may be encountered demand a robust structure and a relatively light covering.
- The spans and loads encountered preclude the use of domestic-type roof formations.
- Pitched roof forms tend to be used on long-span buildings, whilst high-rise buildings may use pitched or flat forms.

REVIEW TASKS

- Identify a range of industrial and commercial buildings in your area and try to consider why the roof forms that they have were selected.
- Prepare a matrix of criteria for the selection of pitched or flat roof forms and apply it to your chosen buildings.
- Visit the companion website at www.palgrave.com/engineering/riley2 to view sample outline answers to the review tasks.

PART 3

12.2 | Pitched roof forms

Introduction

■ After studying this section you should have gained an appreciation of the various forms of pitched roof that can be utilised in the construction of framed buildings.
■ You should be aware of the various loadings that the roof structure and covering are subjected to, and their implications for the selection of an appropriate design solution.
■ You should understand the technology involved in the fixing of roof coverings and be familiar with the function and form of the individual components that make up the roof assembly.

Overview

The subject of pitched roofs to framed buildings was touched upon in the chapters dealing with structural frames and cladding. Indeed, the technology of pitched roof construction to industrial and commercial buildings is based around the interaction of some form of roof cladding with the roof members of the building's structural frame. Unlike domestic construction, the structural form of the roof tends to be considered as part of the totality of the building frame or structure.

As noted in an earlier section the use of trussed, lattice and portal roof forms for long-span buildings is common. The selection of the frame form for the building will inevitably result in a tendency to favour certain roof options. Portals and trussed forms that have a natural incline in the roof members make the selection of pitched roof options obligatory, and in these instances the pitch of the structural members dictates the pitch of the finished roof covering. Roofs of this type are often constructed at very low pitches and are generally clad with lightweight profiled sheeting formed from steel, aluminium or some form of fibre-reinforced material. This is by far the most common method of providing long-span pitched roofs, but other forms are also available. One of the major developments in the technology of roof design has been that of the space frame roof, the principles of which were considered earlier. Unlike most other forms of long-span roof structure, space frames are capable of spanning very great distances in two directions, often resulting in a large roof area supported only at the four corners. The space frame is basically a three-dimensional grid of standard components, linked to form a rigid framework typically less than 1 m deep. The advantages of this form of roof are great, in that:

■ It can be readily assembled at ground level and then be craned into place as a complete unit
■ It is light in weight, thus reducing the overall load in the building structure
■ It allows easy incorporation of building services
■ Various roof designs can be made up from the range of small, standard components.

The covering to space frame roofs will typically be of the same form as that for long-span buildings.

Pitched roof structures

The pitched roof forms that are found in industrial and commercial buildings are generally integral with the framed structure of the building as a whole. As such, they have been considered in some depth in Chapter 8. There is a clear distinction between the roof forms adopted for long-span buildings and those for high-rise buildings, which will tend to have shorter spans. Historically the use of flat roofs was most common in buildings with a complex 'footprint', owing to the ease of creating the roof shape. The problems associated with the durability of older forms of flat roof covering have resulted in the growing use of pitched roof construction. Hence it is now common to find pitched roofs on long-span and high-rise building forms. Where this is the case the roof must be carefully designed to allow for the removal of rainwater using concealed gutters in order to avoid the need for externally mounted rainwater pipes on the face of the building.

Pitched roof structures can be considered to fall into broad categories as follows:

■ *Integral with the structural frame*, such as in portal frame construction. This form of roof is generally found in the construction of long-span buildings. The roof members are important structural elements in the building frame.
■ *Independent of the structural frame*, such as in lattice or truss construction. This form of construction may be found in low-rise buildings of relatively modest spans or in high-rise buildings. In both instances the roof structure, although connected to the main structural frame of the building, does not contribute to the overall strength of the frame.

In both of these roof types the most commonly used material is steel. Indeed, even in situations where the structural frame is formed on concrete, it is common to find independent steel roof structures. This is because the steel structure is lighter in weight and the pitched form is easier to create with steel because of its inherent flexibility of design and ease of assembly. A further benefit is the ability to fix purlins and cladding to steel simply and securely with the use of a range of self-drilling and self-tapping fixings.

Since the various options for pitched roof frames have been dealt with in Chapter 8, we shall not consider them further in detail here.

Pitched roof coverings

The scale of roof construction for industrial and commercial buildings is such that the materials used must be capable of providing large areas of impervious covering economically. By far the most common design solution is the use of profiled metal sheeting. Historically asbestos cement and other fibre-reinforced cement sheets were also used for roofing. However, these have now been largely superseded by the use of steel and aluminium sheeting of various types. Fibre-reinforced sheeting may still be found in specialised situations, such as when chemical attack may be an issue.

Whether formed in steel or aluminium, the broad principles of performance and assembly are the same. The metal sheets are extruded into a ridged profile and are cut to the desired length at the factory. In the case of steel sheeting, a plastic coating is often applied to the sheet during manufacture to provide suitable protection from corrosion. Aluminium is naturally more resistant to corrosion, and as such it is possible to use the sheets in an unprotected state, often termed 'mill-finish'. In most instances, however, aesthetic considerations will require some form of coating. The ridged profile of the sheets introduces longitudinal strength and allows them to be fixed such that they span between supports in the form of rolled metal purlins. Individual manufacturers will produce load/span tables for their own products and different profiles will differ in their ability to span between supports. Typically spans between purlins will be in the region of 1.5 to 2.0 metres, depending on the profile and the thickness of the metal sheet material. The cold rolled steel purlins are fixed directly to the main structural members of the roof: raking roof members in the case of portals and pitched rafter sections in the case of trusses. Two types of purlin are in common usage, the 'Z' section purlin and the 'M' section purlin (Figure 12.6), so called because of their shape in cross-section. The fixing of these to the structure of the roof is facilitated by the provision of metal cleats, allowing the purlins to

Figure 12.6
Purlin and cladding
configuration.

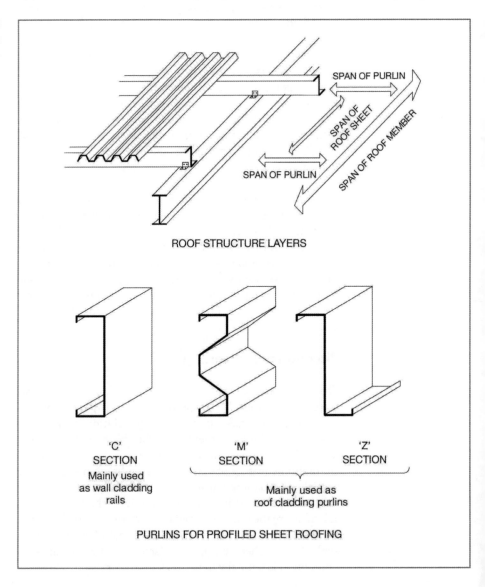

be securely located using self-tapping screws or bolts. The roofing sheets are then fixed directly to the purlins using self-drilling, self-tapping screws to complete the roof assembly.

Alternative systems

Depending on the nature of the building, its location, its intended use, its structural form and the budget available, a specific system will be selected. There are many such systems available in the marketplace, each with their own advantages and disadvantages. However, they all fall into one of the broad definitions set out below (Figure 12.7):

- *Single skin systems*: These are often utilised in industrial buildings and sheds which are not heated internally. As such they do not require the inclusion of thermal insulation and are formed by fixing a single sheet layer directly to the roof purlins.
- *Double skin, site-assembled systems*: These systems allow for the provision of insulation, typically in the form of fibre quilt insulation, set between the outer cladding sheet and an inner lining sheet which has a shallower profile. These are assembled from the individual components on-site and require the use of spacer rails (mini-purlins) between the two layers to create the void in which the insulation is positioned.

An alternative to this is the use of a profiled, rigid insulation panel which sits between the inner and outer layer and does not require the use of spacer rails. This is sometimes termed a 'site-assembled composite'.

Both single and double skin systems are available as 'secret-fix' systems which utilise a series of clips and special fasteners to secure the panels without the need for the use of penetrative fixing:

- *Factory-assembled composite systems*: These systems are pre-assembled at the factory and comprise the outer cladding sheet bonded to the inner liner with a rigid insulation material. They are produced to provide a desired thickness of insulation and are fixed as a single thick panel. These remove the need to deal with the assembly of components on-site, but are more costly and can be unwieldy when dealing with large panel sizes.
- *Linerless or boarded systems*: These systems adopt the use of rigid, boarded insulation positioned beneath the outer cladding and do not utilise a profiled lining sheet. Instead, the underside of the insulation board is pre-finished, often using a white plastic or aluminium foil coating.

The principles of assembly and fixing of these systems are similar to those discussed in Chapter 10, dealing with cladding systems, and will not be repeated here.

Moisture exclusion

Although the principles of design and assembly of profiled metal roofing systems are generally the same as for wall cladding systems, as described earlier in the book, there are some significant elements that must be considered when dealing with roofs. The exposure to external weather conditions and the risk of moisture penetration are greater

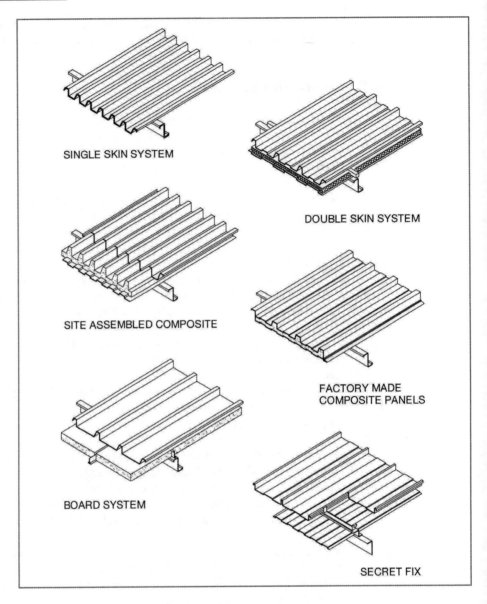

SINGLE SKIN SYSTEM

DOUBLE SKIN SYSTEM

SITE ASSEMBLED COMPOSITE

FACTORY MADE
COMPOSITE PANELS

BOARD SYSTEM

SECRET FIX

for roofs than for walls. The level of exposure and the form of the roof must be considered in detail when arriving at a suitable design solution. The pitch of the roof will greatly affect the degree to which rainwater is directed to gutters from the main roof area. In recent years there has been a desire to make roofs as shallow in pitch as possible to reduce the amount of wasted space internally. In particular, this has been driven by a desire to reduce the cost of heating the internal air volume in the building. This has resulted in some secret-fix roof systems being constructed with pitches as low as 1°. In most cases, however, the typical minimum pitch is likely to be in the region of 4° to 5°. These low pitches are achievable with suitable overlapping to the end and sides of adjacent sheets. Indeed, it is possible to provide sheets that are of sufficient length to cover the entire roof span without the need for any end laps. The practicality of handling such long sheets makes this a rare occurrence, and in most cases end laps are inevitable. A range of typical end lap and side lap details is illustrated in Figure 12.8.

As with any pitched roof construction the shape of the building will often dictate the formation of ridges, hips and valleys. These must also be considered in great detail when attempting to ensure that the roof remains weathertight. A vast range of standard flashings, cover trims and profiles is available to allow for these features to be formed using standard components on-site. A selection of typical roof details is shown in Figure 12.8.

Alternative roof forms

In situations where there is a wish to display architectural innovation or where functional needs dictate a different solution, the use of pitched or flat roofs may be inappropriate. We have previously considered roofs with reference to structures, decks and coverings.

Figure 12.8
Pitched roof covering
details.

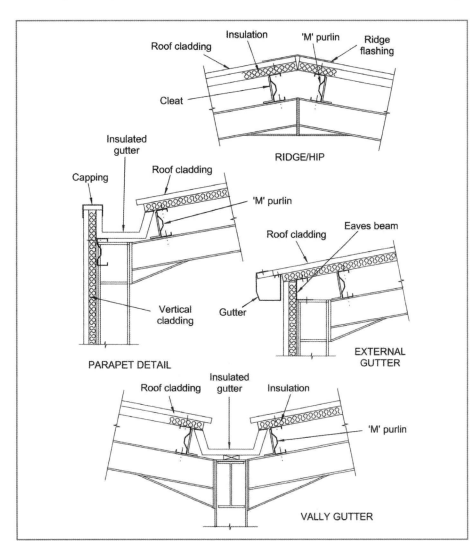

PART 3

Figure 12.9
Vaulted roof formation.

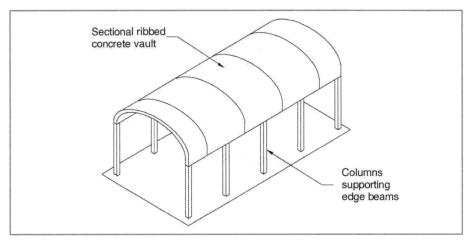

Figure 12.10
Vaulted roof detail.

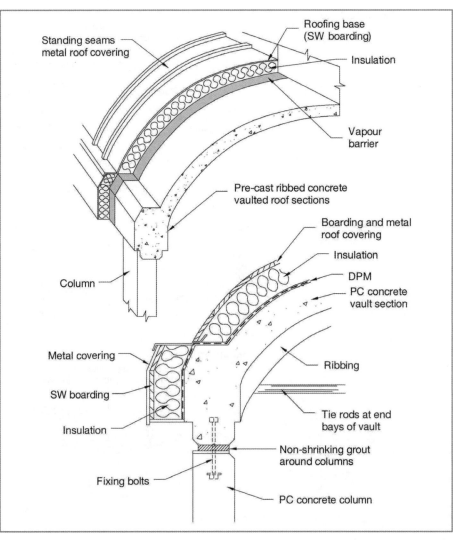

However, a form of roof also exists which combines the structure and deck in the form of a shell or vault. These roofs utilise a curved roof profile which acts as both structure and deck, and in some cases also incorporates the waterproofing function. These are termed 'shell roofs' or 'vaulted roofs', depending on the shape of the roof. Shell roofs tend to adopt a form that resembles a section sliced from a sphere, whilst vaulted roofs tend to resemble a section sliced from a cylinder (Figure 12.9). They are not used as a common alternative to pitched or flat roofs; rather they are selected in situations in which the aesthetic needs require such an approach.

By far the most common form is the simple barrel vault, which generally consists of a reinforced concrete panel constructed to form an arched profile in one direction. This acts in the same way as a structural arch to provide the structure and deck for the roof in a combined section. As with structural arches, the strength of the section relies on the lateral restraint of the bottom sections of the member. This is generally facilitated by the use of reinforced concrete edge beams, which allow the connection of the vault section to the main structural frame of the building. One of the benefits of these roof forms is that the arched profile permits a reduction in thickness of the concrete structure/deck, allowing a slender section to be used. This may be assisted by the use of ribbing where loads are high.

As with all roofs, there is a need to provide an adequate degree of resistance to the passage of heat, and the use of insulation materials must be considered. Figure 12.10 indicates an example detail for a vaulted roof using a warm deck configuration.

REFLECTIVE SUMMARY
- Pitched roof structures may be an integral part of the main building structure or they may be separate structural elements.
- Even though termed pitched roofs, it is possible to create roofs of very low pitch: well below 5°.
- The profile of pitched roof coverings assists in directing rainwater to gutters and also introduces longitudinal strength into the sheets.
- Sheets span between rolled metal purlins.
- Insulation is important and its position dictates the way in which the building responds to thermal changes.
- Vaulted roofs offer an alternative when aesthetic requirements dictate an innovative approach. They act in the same way as a structural arch.

REVIEW TASKS
- What is the implication of very low pitches on roofs of industrial and commercial buildings? How does this affect the detailing of the roof covering?
- Explain the difference between a roof that is an integrated part of the building structure and one that is separate. Give examples with reference to buildings that you know of.
- What is the link between the thermal response time of an industrial building and the position of the insulation in the roof assembly?
- Compare and contrast *three* alternative roof covering materials with reference to the performance requirements for roofs.
- Visit the companion website at www.palgrave.com/engineering/riley2 to view sample outline answers to the review tasks.

12.3 | Flat roof forms

Introduction

■ After studying this section you should be able to understand the different forms that a flat roof may take.
■ You should also appreciate the importance of details to prevent condensation within the roof and the principles of cold deck, warm deck and inverted roofs.
■ The implications of dealing with large roof areas should be understood and the potential issues arising as a result of the adoption of differing sub-bases for the roof structure will be recognised.

Overview

The term 'flat roof' is rather misleading, in that all roofs require an inclination or pitch in order to ensure the effective discharge of rainwater. 'Flat' roofs are considered to be those with a pitch of less than 10°. This presents a potential confusion in the case of long-span roof constructions, since many pitched roof forms to industrial buildings can operate effectively at pitches well below this, but are still considered to be pitched roofs. This is because they adopt the design principles of pitched roofs and share the same approach to construction detailing. To avoid confusion these are referred to as 'low-pitched' roofs, and it is possible to find examples constructed using pitches as low as 2°.

We will restrict our consideration in this section to the more commonly accepted forms of flat roof. Although many different construction forms are possible for flat roofs, they all adhere to the same basic design principles and typically incorporate the same generic elements. The main elements of flat roofs as constructed in industrial and commercial buildings are as follows:

■ The roof structure, provided by the building frame
■ The roof deck, which acts as the base on which the covering will be applied
■ The roof covering.

In addition there will be other associated elements such as insulation, vapour barriers and ventilation fittings.

Roof structure

The roof structure provides support to the roof and transfers the dead and applied loadings to the main building structure. In domestic forms the structure tends to be fabricated in timber, resembling a suspended timber floor in design. Although this is a rare form in industrial and commercial buildings it is sometimes used if the scale of the building is appropriate. Far more common for large roofs of this type is a range of concrete and steel roof structures, depending upon the general form of the building and, if appropriate, its frame. The main options of these were examined when considering building frames, such as plain beams, space frames and lattice structures. Hence we shall not consider them again here – instead, we shall focus on the potential options for roof decks.

In many commercial buildings the roof structure and deck may be combined, as when constructed in concrete, either by casting *in situ* or by using precast sectional units. Concrete plank or pot and beam systems are available of exactly the same type as those used for the construction of suspended floors. Increasingly, however, systems are being used that incorporate profiled steel decking as a base onto which insulation or boarding is laid, or acting as permanent formwork for concrete that is cast *in situ*. Indeed, there is little difference between the roof structure for a flat roof to a framed building and the upper floor structure to the building. Naturally, the roof structure will receive additional treatment to resist the passage of moisture and to allow for thermal insulation, but the structure is essentially the same. Such structures are able to carry heavy loads and span large distances, which are essential in industrial and commercial buildings, often being required to support roof-level plant rooms for example.

The roof deck

The roof deck provides a surface onto which the impervious roof covering is laid, and in the case of the concrete and steel forms described previously is integral with the structure. Steel decks are also available, where the roof structure is formed by a framework or series of supporting beams. In these cases profiled steel sheet, woodwool slabs, plywood or resin-bonded chipboard spanning between the beams of the main structure are commonly used options, although the use of steel decking is becoming most common. This is partly a result of its durability and partly due to its inherent strength and ability to span between supports without the need for secondary structural elements.

Options for flat roof structures and decks

Since the structures and decks of many buildings are essentially combined we shall consider them as such within this section. The selection of the specific form chosen for an individual building is restricted by the structural form of the building and the position of intermediate supports for the flat roof. There are a number of possible options (Figure 12.11), and the more common of these are considered below.

Concrete forms

Concrete flat roof elements include both precast forms and forms that are cast *in situ*. The selection of an appropriate form will largely be affected by the structure of the main building. If the frame of the building is formed *in situ* then the most probable option for the roof slab will also be an *in-situ* formation. In most other situations it is more likely that a precast variant will be adopted.

In-situ concrete
The formation of a reinforced concrete slab *in situ* relies on the use of formwork and temporary support, which can interfere with the construction process underneath. The

principles of construction and the technical detailing are very similar to those described earlier for the formation of *in-situ* floors.

The presence of construction moisture in concrete structures and decks cast *in situ* can result in some problems in the period following construction, as entrapped moisture expands and vaporises beneath the impervious covering. The formation of bubbles in the covering can lead to premature failure and must be avoided. To this end, vents are incorporated to allow the release of entrapped moisture (Figure 12.12).

Precast concrete

As with *in-situ* formed slabs, the structure of precast concrete roof decks shares much of its technical detailing with that for the formation of precast upper floors. The options available include beam and block floors, plank floor panels and prestressed floor units for use where heavy loads or long spans are anticipated (Figure 12.13).

Figure 12.11
Overview of flat roof deck options.

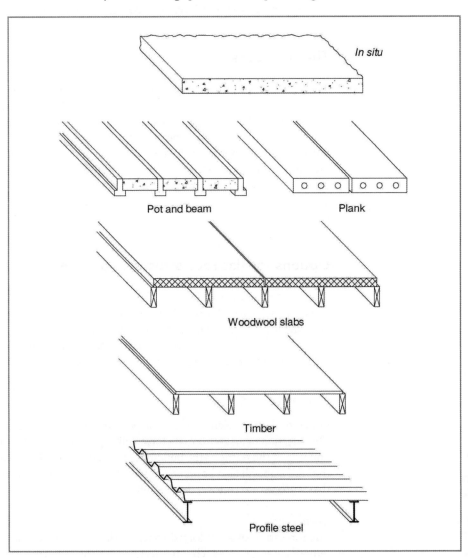

Figure 12.12
Air vent to release vapour from concrete roof decks.

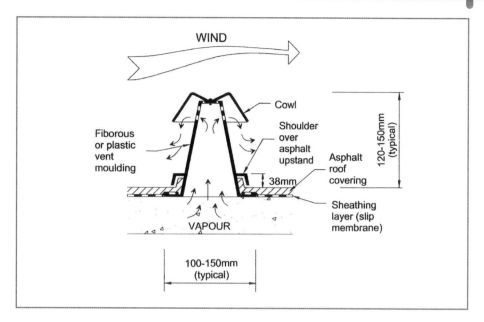

Metal decks

Profiled metal decking (Figure 12.14), usually steel, is used to span between supports, but does not itself provide a fully supporting substrate or base. The addition of a concrete upper section, as with hollow rib floors, is often used to provide a robust and smooth structure capable of accepting insulation and the waterproof membrane. Alternatively timber decking or rigid insulation boarding may be laid over the profiled sheets to provide a smooth surface onto which the membrane is laid.

Woodwool slabs

The use of woodwool slabs was once common in the formation of roofs to large buildings. However, the advent of fast-track systems using steel decking, as described above, has largely overtaken this type of construction. We shall, however, consider the use of woodwool slabs here for the purpose of completeness. 'Woodwool' is a term used to refer to the material formed when long shreds of timber are bound together by a cementitious slurry. The resultant material is compressed into a variety of forms (Figure 12.15), including flat slabs of between 25 mm and 100 mm thickness. The slabs have an open textured finish and were used extensively for the formation of roof decks and as permanent formwork for *in-situ* structures. Their use as decks for industrial and commercial buildings is limited by the need to provide support for the slabs at around 600 mm centres. The development of slabs with metal channels along the sides, and the use of deep channel forms which incorporate a downstand beam profile, allow greater spans to be achieved. Typically, spans up to 3.5–4.0 m are achievable with these channel units, although the risk of cold bridging through the metal channel must be avoided.

PART 3

Figure 12.13
Options for concrete
roof decks.

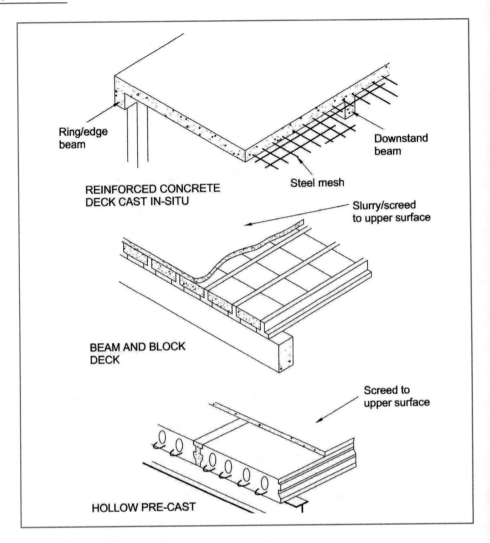

Ring/edge beam

Downstand beam

Steel mesh

REINFORCED CONCRETE
DECK CAST IN-SITU

Slurry/screed
to upper surface

BEAM AND BLOCK
DECK

Screed to
upper surface

HOLLOW PRE-CAST

Flat roof coverings

In flat roof construction, unlike pitched roof forms, the weather resistance is provided by a continuous, impervious covering. Such coverings take three principal forms: built-up felt, mastic asphalt and liquid-applied membranes.

Built-up felt roofing

Roofing felts are classified by BS 747, which includes felts based on asbestos, glass fibre and polyester. In addition, felts may be classed as base, intermediate or surface types, depending upon their structure and surface treatment. Figure 12.16 summarises the classification of felts under BS 747.

The nature of felt coverings is such that a series of adjacent sections are laid, with overlaps to the sides and ends, to attempt to create a continuous, impervious covering

Figure 12.14
Options for metal flat
roof decks.

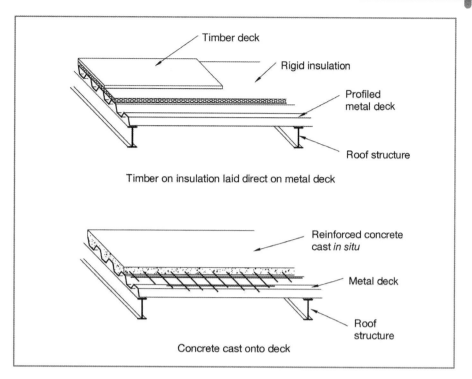

Timber deck

Rigid insulation

Profiled
metal deck

Roof structure

Timber on insulation laid direct on metal deck

Reinforced concrete
cast *in situ*

Metal deck

Roof
structure

Concrete cast onto deck

(Figure 12.17). Differential thermal movement between the structure and the felt must be accommodated. If this is not done there is risk of the felt creasing and stretching respectively when the structure is subjected to shrinkage and expansion. This is of particular concern in industrial and commercial buildings, where the proportions of the roof can be significant. For this reason it is usual to allow some slippage between the roof deck and the covering by laying a partially bonded base layer rather than a fully bonded layer. Hence the covering is bonded to the deck only at localised spots, allowing some differential slippage whilst ensuring a secure fixing. Partial bonding may be effected in a series of ways, such as spot nailing or more commonly by laying loose a perforated base layer over which bitumen is poured, allowing bonding only where it passes through the perforations. Subsequent layers are fully bonded using hot bitumen to provide a two- or three-layer system. This traditional method has been developed further by the introduction of 'torch on' felts, which incorporate their bonding bitumen within their structure, allowing the felt to be laid quickly by applying a blow torch on-site.

In recent years the development of elastomeric or high tensile roofing felts has been a major advance in the technology of flat roofing. The nature of these materials is such that thermal movement in roof structures is accommodated without damage to the covering, and hence defects are drastically reduced. This is of particular benefit in large roof areas, where the effects of differential movement are greatest. These materials are generally accepted to be far more durable than the older types of felt roofing, but they share essentially the same details of application and construction. One development that does differ substantially from the traditional approach to applying felt roofing is the use of single-ply membranes.

Figure 12.18 shows cross-sectional details of concrete and metal felted flat roofs.

PART 3

Figure 12.15
Woodwool slabs for flat
roof decks.

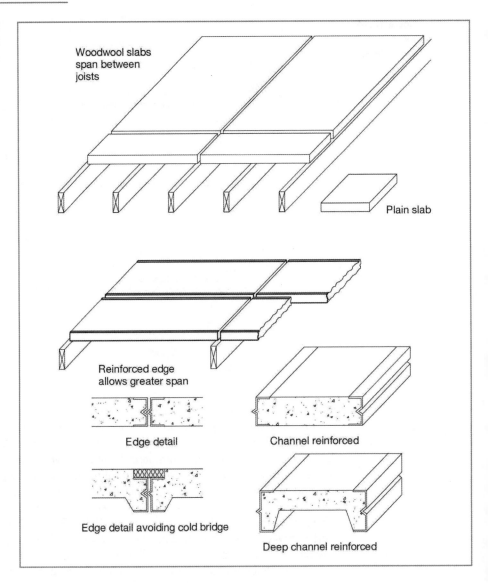

Figure 12.15
Woodwool slabs for flat
roof decks.

Woodwool slabs
span between
joists

Plain slab

Reinforced edge
allows greater span

Edge detail

Channel reinforced

Edge detail avoiding cold bridge

Deep channel reinforced

Single-ply roofing

Single-ply membranes are not derivatives of the bituminous-based felts previously described. Developments in the rubber and plastics industries have resulted in the emergence of sophisticated, highly engineered polymer-based membranes; for our purposes these are termed collectively 'single-ply' roofing systems.

Single-ply membranes are manufactured as sheets which, when applied as a single layer, offer a reliable and durable method of waterproofing. They are often used in refurbishment projects when they can be laid over the existing roofing to provide a long-term upgrade or repair of the previous roof covering. They are also being used increasingly in the construction of new industrial and commercial buildings. Two broad categories of polymeric single-ply membranes exist:

Figure 12.16
Built-up felt
classification.

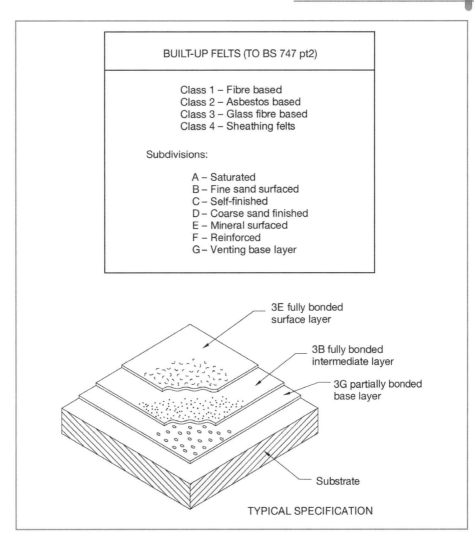

BUILT-UP FELTS (TO BS 747 pt2)

Class 1 – Fibre based
Class 2 – Asbestos based
Class 3 – Glass fibre based
Class 4 – Sheathing felts

Subdivisions:

A – Saturated
B – Fine sand surfaced
C – Self-finished
D – Coarse sand finished
E – Mineral surfaced
F – Reinforced
G – Venting base layer

3E fully bonded
surface layer

3B fully bonded
intermediate layer

3G partially bonded
base layer

Substrate

TYPICAL SPECIFICATION

PART 3

Figure 12.17
Laying of built-up felt.

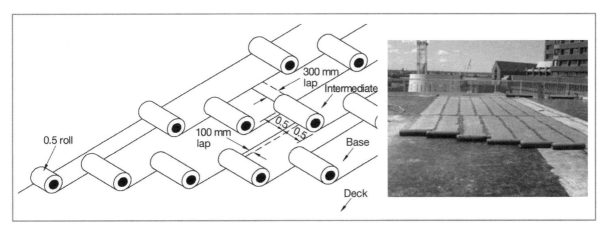

300 mm
lap Intermediate

0.5 roll

100 mm
lap

0.5/0.5

Base

Deck

- *Thermoplastic*: these tend to soften with increased temperature, allowing thermo-welding of the edges of the sheets to create a continuous membrane. They have poor resistance to chemicals and organic solvents and thus may have durability problems in some installations.
- *Elastomeric*: these are less temperature-sensitive, but are more elastic in their behaviour, readily accommodating thermal movement in the roof structure without rippling or tearing. They also have enhanced resistance to chemical attack. They are laid in the same manner as traditional roofing felts with side and end laps bonded to resist moisture penetration.

The membranes are commonly adopted for the creation of warm or cold roofs and they are favoured in situations where a degree of natural thermal movement in the structure is likely to occur. As with all flat roof designs it is essential the membrane is seen as part of the composite roof assembly rather than simply a covering layer to the structure. The various components must be matched to ensure that the deck, insulation

Figure 12.18
Felted flat roof details.

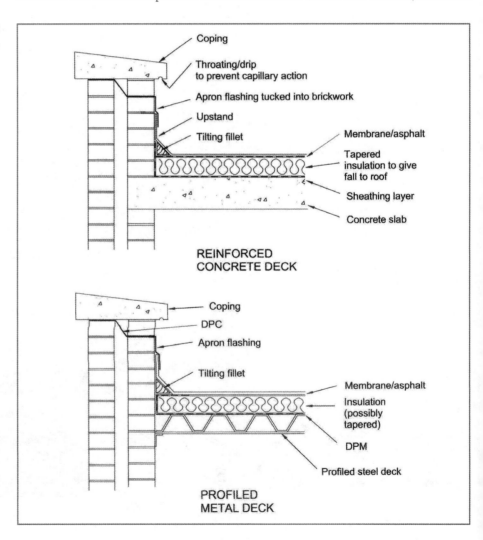

and membrane act together as an effective composite. Three common methods are adopted to place the membrane on the roof deck: adhesion, mechanical fastening and loose laying.

When using adhesion the membrane is fully or partially bonded to the deck using a range of prescribed adhesives. When constructing warm deck roofs, the adhesion is to the insulation material and special care must be taken to ensure that the adhesive and the insulant are compatible.

Mechanical fixing (Figure 12.19) relies on the use of fasteners with flat bars or large-section washers screwed to the roof deck and securing the membrane. This approach may be adopted when there is great risk of uplift on the membrane, requiring the use of an effective holding-down system for the covering.

Loose laying relies on the use of mechanical fixing or adhesion around the roof edges and at penetrations, with the main part of the covering laid without bonding to the deck. Alternatively, the membrane may be weighted down or ballasted to prevent uplift. In the case of inverted roof construction the membrane would be weighted down by the insulation material and some form of ballast.

Asphalt roofing

Asphalt for roof coverings (Figure 12.20) differs from felt in that it is a homogeneous coating, applied in liquid form to create a continuous, jointless coating. As with felt, it is essential to cater for differential movement between the coating and its substrate, and this is done by introducing a separating or sheathing layer, which allows for slippage. Two layers of asphalt will normally be applied by trowelling to produce a finished thickness of approximately 20 mm. One advantage of this form of impervious covering is that patch repairs can be easily carried out, since localised areas may be cut out and replaced with new material, which can be fully bonded to the existing by the application of heat. A disadvantage of asphalt is that it has little inherent strength when subject to

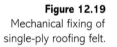

Figure 12.19
Mechanical fixing of single-ply roofing felt.

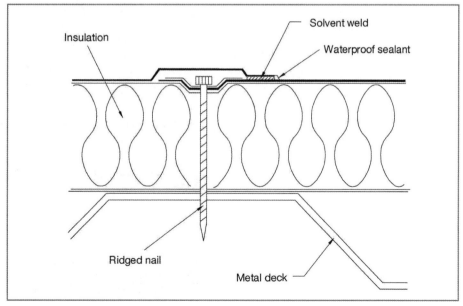

PART 3

Figure 12.20
Asphalt roof coverings.

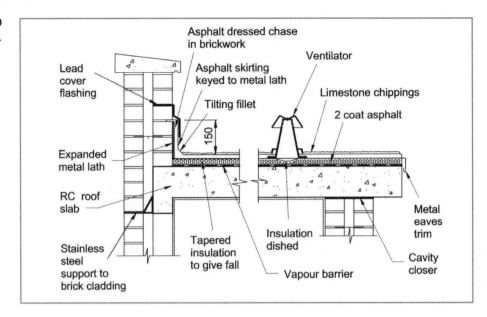

heat; hence at details which require stability, support must be provided to prevent sagging in direct sunlight. Such support may take the form of raked-out brickwork joints or expanded metal lath fixed to the backing of the detail.

Asphalt is a naturally durable material with good resistance to abrasion and excellent wear characteristics. For this reason it is often favoured for situations in which the roof

Figure 12.21
Soft metal roof covering.

is likely to be trafficked, as is the case in many commercial buildings where access to rooftop plant is required.

'Soft' metal coverings

A range of what we may think of as 'soft' metal roof-covering materials are available, such as lead, zinc and copper. However, although once common, these are costly and as a result tend to be specified only where architectural prominence is required. The materials can be used for flat or pitched roofs and they tend to be recognisable by the characteristic 'standing seam' formation. These are really a specialist form of roof covering, and as such we shall not deal with them in detail here. Figure 12.21 illustrates a finished roof covering using a soft metal standing seam formation, in this case based upon the use of rigid sheet units combined with a patent glazing system.

Green coverings

There is an increasing realisation that the landscaping aspect of any development can contribute significantly to its sustainability rating. If building on previously used land then the landscaping aspect needs to be designed in and can be difficult to achieve due to lack of space around the proposed building. The temperature experienced in city centres is generally higher than in park and suburb areas. This proves that landscaping is essential in preventing heat islands in heavily built-up areas. However, land costs in city centres are usually high and land is at a premium. Therefore, innovative ways to landscape in city centres need to be developed.

One way of increasing the amount of landscaping without increasing the footprint of a development is by the inclusion of what is known as a 'green roof'. Not only do these types of roof increase the amount of landscaped area in a development, but they can also be developed as relaxation areas for occupants. They also lead to significant cooling in buildings and are becoming increasingly popular in warm climates. These green roofs can be formed using traditional turf but are usually formed using 'sedum' as the top layer. Sedum is a genus of plants which have water-storing leaves and are therefore ideal for green roof construction. Water that falls during periods of high rainfall can be stored by these plants to irrigate the roof during dry periods.

The most common types of green roof are:

- *Extensive*, which means they cover the whole roof and the plants are delivered to the site from out of the area, thus implying they will have a consistent appearance as shown in Figure 12.22.
- *Extensive but biodiverse*, which means that the plants that form the top surface will not be planted at the time of construction but will seed naturally. This may lead to an inconsistent appearance but have a high level of biodiversity. As can be seen in Figure 12.23, seeds have been dropped on the roof by birds or have blown on the roof, leading to different plant types growing after the original green roof was laid.
- *Intensive*, which are sometimes called roof gardens. The plants are used to form a 'park' on the roof, which can be used as an amenity as well as providing the other benefits of turf roofs. Figure 12.24 shows the beds being prepared to take the plants. You will

note that there is space to walk in between the beds and this will be a roof park which is accessible by the building occupiers.

Figure 12.25 shows the typical elements of a green roof build-up. A slab is constructed and then a layer of insulation applied. As there will be significant moisture held in the top layer, a waterproof membrane is required. To prevent the plant roots damaging the waterproof membrane, a root barrier layer is required. In order to prevent the top layer from drying out, a water-retention layer needs to be included that complements the water-retention properties of sedum plants and stores water

Figure 12.22
Extensive green roof

Figure 12.23
Extensive biodiverse roof.

Figure 12.24
Preparing for a roof park.

Figure 12.25
Eco green roof
build-up.

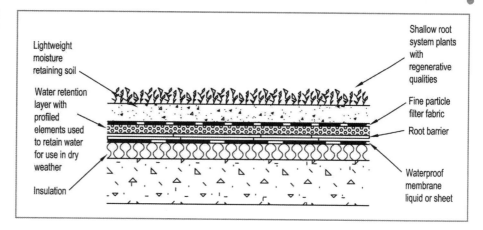

during wet periods for use when the weather is drier. Lightweight soil is used to reduce the load on the structure and shallow root system plants are used to form the top layer. These plants need to be durable and easily regenerated to ease the maintenance requirements of the roof.

Solar protection

Bitumen felt and asphalt suffer degradation when subjected to ultraviolet radiation. Hence, in order to provide satisfactory durability, they must be provided with solar protection. This commonly takes the form of solar reflective paint applied to the surface, white mineral chippings bonded to the surface or, in the case of some felts, an application of mineral chippings applied to surface felts at the factory. By far the most cost-effective and hence the most common on large roofs is the application of solar reflective paint. The use of mineral chippings may, however, be favoured in some situations because it provides a natural protective coating against burning embers that might land on the roof in the event of a fire in another building. The degree to which this is significant depends on the overall approach taken to fire protection of the building.

Insulation

Like all elements of the external fabric of buildings, roofs must provide a satisfactory degree of thermal insulation. It is therefore essential that roofs are provided with thermal insulation material to maximise their thermal insulation properties. The positioning of the material within the roof can take one of a number of alternatives, as illustrated in Figure 12.26.

The position of the insulation provides a convenient classification method for roof types as follows:

- *Cold deck*: the insulation material is below the roof deck
- *Warm deck*: the insulation material is above the roof deck
- *Inverted*: the insulation material is above the weatherproof covering. In this case the

choice of insulation material must be carefully considered to ensure satisfactory performance, even when wet, and durability in the hostile external environment.

Figure 12.26
Alternative insulation
positions within roofs.

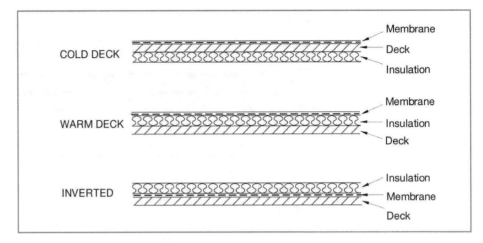

It should be noted that when insulation is included within the roof structure a significant temperature drop will occur across it. This may pose problems of interstitial condensation, where condensation occurs within the fabric. In order to minimise the risk of this the moist air from the building interior is inhibited from entering the roof structure by the inclusion of a vapour check on the warm side of the insulation. In addition, any moisture which does enter the structure should be removed by effective ventilation within the roof. The risk of condensation is significantly greater with cold deck roofs than with the other forms; hence it is uncommon in industrial and commercial buildings.

As we know, the thermal performance of the roof is largely dictated by the installation of a suitable insulation material within the roof assembly. The structural form of the roof will pose some limitations on this, as will the desire to create a building that is either 'thermally light' or 'thermally heavy'. These concepts were discussed in *Construction Technology 1*.

There are three main forms of flat roof, which are distinguishable by the position of the thermal insulation relative to other parts of the roof assembly, particularly the deck. Figure 12.26 illustrates the three main forms of flat roof in common use. It should be remembered that in most cases the cold deck detail tends to be avoided due to long-term durability issues and problems associated with interstitial condensation. The warm deck and inverted roof forms tend to perform better in these areas.

Cold deck roofs have been used extensively in domestic construction. However, they are relatively rare in industrial and commercial forms of building due to the nature of the roof structures used.

Cold deck roof

Cold deck roofs are so called because the thermal insulation is located below the deck. This is a good position for intercepting the heat which is rising from the building but it does isolate the deck, leaving it cold and providing a surface against which moist warm air may condense. A vapour barrier such as Visqueen sheet tends to be used, and this is always positioned on the warm side of the insulation to prevent moisture-laden air reaching the cold deck. However, this barrier is quite difficult to achieve successfully. In domestic construction, where timber is commonly used for the structure and deck, this is problematic. In industrial and commercial buildings this is not the case since the structure and deck will tend to be formed in concrete or steel. There are instances, however, where timber may be used, and the implications of interstitial condensation must always be considered in designing a roof assembly. The main shortfall of this roof configuration in large buildings is the tendency for the roof deck to react to changes of temperature by expanding and contracting relative to the structure.

Warm deck roof

Perhaps the most common form adopted for industrial and commercial buildings is the warm deck roof. In this configuration the insulation is placed on top of the deck and is then protected by the waterproof membrane that is laid directly on top of it. This form of construction allows the deck to be insulated against external temperature variation and the dimensional changes that can result in defects arising in the roof are more readily avoided. The insulation material tends to be laid over a vapour barrier that acts to resist the passage of moisture from the deck or the building interior to the insulation material. It also acts as a slip plane to allow the effects of thermal movement to be accommodated within the roof assembly. Insulation in the form of rigid panels or batts is then laid over the membrane to provide a base for the waterproof roof membrane in the form of felt or some other appropriate covering. One feature that has become common in this form of roof is the use of tapered insulation to create the required fall or incline on the 'flat' roof to facilitate rainwater collection and disposal. This innovation allows the design and construction of complex roof shapes with relative ease and without imposing the need to create a structure or deck which incorporates the required fall.

Inverted roof

The inverted roof is based on the idea of positioning the roof insulation outside the waterproof membrane, and when constructed appropriately these roofs perform very well. The positioning of a durable water-resisting insulation material above the roof felt or other covering ensures that the membrane is protected both physically and from the effects of thermal variations. This form also allows for the upgrading of the thermal performance of the roof by increasing insulation thickness without the need to remove and replace the roof covering. However, the insulation must be securely positioned, and this is generally achieved by the laying of ballast over the insulation or the use of tiles to allow the roof to be trafficked.

PART 3

REFLECTIVE SUMMARY

With reference to flat roof forms, remember:

- Flat roofs are never truly flat, as they require a slope to clear rainwater.
- Tapered cement-based screeds can provide the fall for reinforced concrete structures, although tapered insulation is now common.
- The location of the thermal insulating layer defines the roof configuration as *warm deck*, *cold deck* or *inverted*.
- Roof membranes can be multiple- or single-layer felts, liquid-applied membranes or asphalt.

- A range of materials can be used for the construction of roof decks, including concrete, steel and woodwool slabs.
- For the continued performance of the roof it is essential that rainwater-removal features are maintained free from blockage.
- The use of more sustainable roof coverings is increasing in popularity. Turf or sedum roofs are a good choice in built-up locations as they can reduce the formation of heat islands.

REVIEW TASKS

- Compare and contrast the performance of *three* different flat roof coverings.
- Explain the differences between *warm deck*, *cold deck* and *inverted* roofs with reference to their performance characteristics.
- Provide a chronology of the evolution of roofing felts with reference to the key advances in their performance.
- What are the advantages and limitations of the use of soft metal roof coverings such as lead and copper?
- Visit the companion website at www.palgrave.com/engineering/riley2 to view sample outline answers to the review tasks.

COMPARATIVE STUDY: ROOFS

Option	Advantages	Disadvantages	When to use
Pitched roof structure			
Structure separate from building frame, e.g. lattice, truss	Flexible in form Can cope with difficult shapes Light in weight Readily assembled on-site or craned into place	Limitation of roof span Slower to construct Potential for quality problems if site-constructed Care must be taken with connections to main frame and wind bracing	Small-scale roof construction for industrial buildings More common on high-rise buildings or where there are complex roof formations
Structure as part of building frame, e.g. portal	Cheaper for larger jobs Roof designed by frame manufacturer The few components used limit the scope for quality problems Long spans possible	Lead-in time required for manufacture Standard frame shapes may limit flexibility of form Transport, storage and assembly of large components	Portal frame roofs have become almost standard in modern long-span low-rise building

(continued)

COMPARATIVE STUDY: ROOFS (*continued*)

Option	Advantages	Disadvantages	When to use
Pitched roof covering			
Slates/tiles	Aesthetically pleasing Available in a range of sizes Easily trimmed on-site	Expensive for large areas Potential for damage and wind uplift on low pitches Slows process	The selection of slates or tiles on modern framed buildings tends to be avoided due to cost and time issues However, they may be used in conjunction with other coverings in an attempt to give a traditional feel
Profiled sheet	Available in a range of colours and profiles Robust, with inherent longitudinal strength Large sizes of sheet allow fast, economic coverage of roof area Large panels demand fewer overlaps/ junctions and assist in moisture exclusion	Handling of large panels can be difficult Transport of large sheets Some forms prone to high levels of thermal movement which affects fixings	Almost ubiquitous for covering to large pitched roofs
Soft metal coverings	Aesthetically pleasing Highly durable Metals malleable and easy to create special shapes	Difficult to detail Requires many joints to prevent failure due to thermal movement Very expensive Requires highly skilled operatives	Used when there is an aesthetic demand
Flat roof structure			
Cold deck roof	Cheap Easy to construct Relatively shallow depth as insulation is contained within structure	Ventilation required to prevent condensation if using timber deck/structure Potential for differential movement between covering and structure	Often used for small-scale buildings and where the generation of a thermally light structure is desired

(*continued*)

PART 3

COMPARATIVE STUDY: ROOFS (*continued*)

Option	Advantages	Disadvantages	When to use
Flat roof structure (*continued*)			
Warm deck roof	Warm deck resists differential movement of covering Allows use of tapered insulation for creation of falls Can be upgraded when recovering roof	Potential for traffic to damage insulation	Possibly the most common form in commercial and industrial buildings
Inverted roof	Protection of impervious membrane Insulation is upgradeable without need to recover roof Entire structure and covering protected from thermal changes	Difficult to detect leaks in the event of failure Potential for damage to insulation Requires protective surface covering Costly	Technically superior performance but used only where there is minimal risk of damage or heavy traffic to the roof surface
Flat roof covering			
Roofing felt	Cheap Familiar technology	Older forms suffer limited lifespan Multi-layer process subject to quality control problems Solar degradation Problems with differential movement	Selection is largely based on familiarity of designer with products Now losing favour with increased use of polymer-based membranes
Single-layer covering (polymer-based)	Enhanced polymer/ single-layer technology Extended lifespan Able to cope with differential movement	Single-layer covering may suffer from localised damage	Simple installation and improved durability have resulted in an increased use of these forms of covering Now becoming the most popular flat roof coverings for larger buildings
Asphalt	Liquid application is flexible Monolithic application results in absence of joints Ease in formation of seals around fixture etc.	Potential to creep at upstands Solar degradation Large areas are expensive and time-consuming to deal with	Often used where the roof is to be subjected to traffic or for small areas with difficult detailing

12.4 | Roof drainage

Introduction

■ After studying this section you should be able to identify the various details that may be employed to collect rainwater from roofs to industrial and commercial buildings. This will include the use of rainwater gutters, downpipes, roof outlets and special details employed with parapet roof solutions.

Overview

The nature of industrial and commercial roofs is such that the areas covered tend to be large. Hence the extent of the rainwater run-off that must be accommodated by the rainwater goods is considerable. The manner in which rainwater is collected and discharged is essentially the same as for domestic construction. However, the scale of the elements is greatly increased and the difficulties associated with provision of rainwater systems to tall buildings must be taken into account. For this reason, and as a result of the need to deal with numerous outlets across the roof surface, the rainwater downpipes to tall buildings tend to be located internally. With long-span buildings using pitched roof forms the systems are akin to enlarged domestic systems with robust components to ensure satisfactory durability.

Drainage of pitched roofs

The traditional means of rainwater collection from a pitched roof is the rainwater gutter and downpipe system. The design of the rainwater collection system for large buildings may incorporate separate elements sourced from several manufacturers provided that the outlet pipework is of sufficient size to accept the flow from the gutters and that the gutters are adequately sized to accept the run-off from the roof (Figure 12.27).

Generally gutters are formed using steel or aluminium – plastic rainwater goods of the type found in domestic construction do not provide sufficient durability for buildings of the type considered here. There are many different gutter profiles, which can achieve quite different visual effects, but it should be remembered that the prime consideration is in relation to the carrying capacity of the gutter and its ability to handle the run-off from the roof area in question. BS 6367 recommends minimum sizes for gutters of different profiles and positions, typically 500 mm for valley gutters and 300 mm for parapet gutters. These large sizes allow for removal of rainwater at an appropriate rate and also provide for access for cleaning and resistance to blockage.

In situations where the external wall or cladding extends beyond eaves level to form a parapet, a gutter will be formed in the roof structure behind the parapet and connection will be made from this gutter to the rainwater downpipe. Indeed, in most high-rise situations this is the case. The connection to downpipes will be made so as to facilitate rainwater disposal with the pipes positioned within the building enclosure. This approach has a number of benefits in that there is protection from freezing, from deterioration due to exposure to the elements, and from vandalism. In addition, the pipes can be

Figure 12.27
Rainwater goods for
industrial and
commercial buildings.

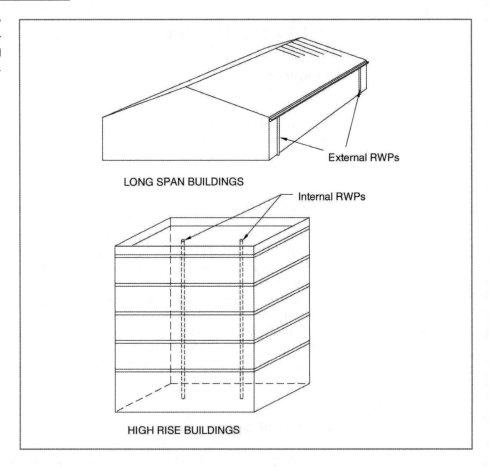

readily accessed for clearing blockages and general maintenance. In most long-span buildings gutters will be mounted at the eaves, although they may often be concealed behind a curved section of cladding to provide protection from vandalism and to produce a pleasing appearance aesthetically.

Drainage of flat roofs

The broad principles of drainage to flat roofs are similar to those of pitched roofs, with the exception that it is uncommon to find gutters utilised in large flat roofs. Although in domestic construction the use of eaves gutters for flat roofs is common, in larger forms the flat roof will tend to be formed within a parapet wall enclosure. The fall of the roof will be designed such that there is a series of internal focus points for the collection of rainwater into a shallow channel leading to rainwater outlets. Alternatively, the roof may be configured so as to drive the water directly to the outlet points. It is important that these outlets are regularly maintained to ensure that they do not become blocked. The use of plastic or metal cages over the outlet helps to prevent the build-up of debris which might cause blockages (Figure 12.28). This is an extremely important issue, since the form of flat roofs is such that blockages will result in ponding of water on the roof, which can result in leakage. In some extreme cases roofs have collapsed due to the excessive loads applied by the build-up of water on the roof surface.

Figure 12.28
Roof outlets for flat roofs.

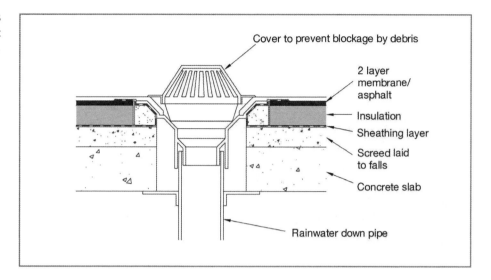

CASE STUDY: ROOF COVERING

The roof shown in the photographs in this Case study is being incorporated into a seven-storey steel frame building. The roof boards are composite, bonded, pre-assembled panels that incorporate the insulation.

The main support for the roof is basically a steel frame which comprises beams bolted to the columns of the previous floor. The only difference is that the beams are fixed with an incline to allow the roof to drain. Bracing has also been incorporated between the beams to give additional support against wind loading.

The underside of the roof structure, showing beams and bracing, the roof support beams, and the underside of the roofing boards.

A section through the roof. The purlins are connected to the steel frame, and then the boarding to the purlins. The void will allow for further insulation of the roof.

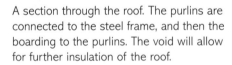

This photograph shows the actual roof finish. Aesthetically this is not very pleasing, but given the height of the building, the low pitch of the roof and the location of the building, the roof finish will not be visible to people other than those undertaking roof maintenance.

In this photograph the ridged effect of the roof finish is more visible. The ridging will allow for rainwater to drain down the roof in channels to the rainwater pipes.

13 Internal division of space and integration of services

AIMS

After studying this chapter you should have developed an understanding of:

- The importance of building services
- The nature of services installations from the perspective of the builder
- The need for and methods of services integration
- The methods used for housing services in buildings, both structural and non-structural
- The methods used for dividing space in buildings
- The performance requirements and different forms of suspended ceilings, raised access floors and partitions

This chapter contains the following sections:

13.1   Requirements of building services and the need for services integration
13.2   Structural and non-structural methods of services integration

INFO POINT

- BS 5234, Parts 1 and 2: Partitions: code of practice for design and installation (1992)
- BS 7083: Guide to the accommodation and operating environment for IT equipment (1983)
- BS 8000, Part 8: Code of practice for plasterboard partitions and dry linings (Workmanship on site series) (1994)
- BS 8204, Part 5: Screeds, bases and *in situ* floorings (2004)
- BS EN 13213: Hollow floors (2001)
- BS EN 13964: Suspended ceilings – requirements and test methods [replaces BS 8290-1: 1991, BS 8290-2: 1991 and BS 8290-3: 1991] (2004)
- BS EN ISO 140, Part 12: Acoustics related to access floors (2000)

13.1 | Requirements of building services and the need for services integration

Introduction

- After studying this section you should have developed an understanding of the nature and complexity of building services installations in modern buildings.
- You will also have gained an appreciation of the role of the building services engineer within the process of building design.
- In addition, you will have gained an awareness of the importance of services integration and the role of the builder in ensuring that this task is undertaken correctly in order to ensure that construction work is carried out as efficiently as possible.
- As services form such an integral part of a building, it is also important that you develop an understanding of the potential problems that can arise if services installations are not installed or maintained correctly.

Overview

Although building services design is outside the scope of this book, services form a major part of the construction of buildings and an appreciation of what is required and of all building work that is related to services needs to be discussed. A building can be described as a shell that houses people in a 'comfortable' environment, and enables people to carry out designated tasks with ease. The process of services integration is a fairly simple concept, but is often not planned in sufficient detail and problems can occur. These problems can range from planned construction activities taking longer than was envisaged and trades working on top of one another, to more serious problems, such as claims by contractors for delays in working and adverse effects on the future health of occupants. For these reasons, services integration should be deemed one of the most important tasks on a construction management project.

Requirements of building services

Buildings must house services to facilitate certain requirements, as discussed below.

Comfort of occupiers

Generally the natural environment does not provide for human comfort. Therefore dwellings are used to provide protection from the elements.

Over time, construction materials that provide the best protection have been developed. Nowadays new environmental problems, together with more exacting demands from occupiers, are forcing more radical designs to evolve. In some cases electronic equipment and manufacturing processes demand more closely controlled conditions than are necessary for human comfort. Mechanical installations were originally used for the control of the environment, and the environment achieved was only credited to these installations – the performance of the building itself in conjunction with the **services**

PART 3

The life of a building structure is far longer than the life of **services installation**, which may need to be upgraded numerous times. This should be taken into account in any building design.

installation was not considered. New materials and consequently possible building forms, plus an increasing awareness of environmental issues relating to the built environment and a wide range of different types of installation, are now changing the nature of commercial buildings. The environmental design must therefore form an integral part of building design at the earliest stage. Not only must environmental considerations be taken into account in the fundamental design of the building, but they must also be balanced against each other, e.g. large windows give better daylighting, but encourage higher heat loss in winter and overheating due to the Sun's rays in summer. In addition, environmental factors interact with other aspects of interior design; and the aesthetics, convenience and maintenance of rooms may be dictated by the positioning of environmental control equipment.

Basic human environmental requirements are for light, air and warmth. In addition, sound must be considered, along with humidity and hygiene. For comfort and efficiency the human body needs to be maintained between certain levels of these environmental conditions.

Throughout history all these factors have been considered to a certain extent when designing buildings. The main differences these days are that with improvements in technology more use can be made of the building shell materials and outside environmental conditions, and there is also a growing awareness of the need for **inclusive design** from the architecture profession.

Inclusive design can be simply defined as design that considers the needs and preferences of potential occupiers from the start of the building design process.

Health of occupants

Developing medical expertise, especially in the 19th century, led to the breakthrough in knowledge that conditions in buildings could give rise to illness. Initially the conditions identified were polluted water supplies, exposed excrement, damp, infestation, fumes and cross-infection due to lack of ventilation. The development in necessary standards in building meant that improved working and housing conditions caused a drastic improvement in public health. The virtual disappearance of some terminal diseases is probably more the result of improved building standards than of medical developments. During the 20th century, Building Regulations changed little with regard to health provision, but over recent years it has become obvious that with the use of new materials and technology in buildings there has been a rise in new health problems. These include radon emission from building components and Legionnaires' Disease.

Illnesses suffered by the occupants of buildings that are linked to the building itself can be classified in two different ways:

- *Sick building syndrome* (SBS) is diagnosed when the affected employees' symptoms disappear once they leave the building.
- *Building-related illnesses* (BRI) are diagnosable illnesses that can be attributed to a defined indoor air-quality problem.

Sick building syndrome (SBS)

Causes of SBS

There is a lack of a definitive answer to the causes of SBS, as it is believed to relate to a complex range of related factors which all have to be present at specific levels for SBS to occur. However, factors believed to contribute include:

- Micro-organisms/bacteria inside ventilation and air-conditioning systems
- Chemical fumes emitted from the contents of buildings, such as furniture, carpets and plastics
- The effects produced by the form of artificial lighting
- Varying levels of air pollution gases such as carbon dioxide
- Inadequately operated air-conditioning systems that use high levels of recycled air
- Heating provision that does not suit the level of activity
- Incorrect levels of relative humidity: for comfort it should be between 40 per cent and 60 per cent
- Level of positive and negative ions in the air
- Noise levels
- Monotony of work activity.

The cocktail of a combination of these factors at specific levels appears to be the formula for SBS to occur.

It needs to be recognised that the cocktail is not a constant, but is changing sometimes by the minute. The *trigger level* is the key that needs to be established, i.e. the level at which occupants appear to suffer. The World Health Organisation has suggested that 10–30 per cent of all buildings are 'sick', and although this has not proved to be a very major problem in the UK, compared with some other **countries**, it must still be considered during the design, construction and occupancy phases of building life. The number of synthetic substances and chemicals used in the manufacture of a building is said to exceed 5000, and therefore care must be taken when considering systems and materials to be used in the building. The Control of Substances Hazardous to Health (COSHH) Regulations (1999) have gone some way in alleviating the use of potentially health-damaging chemicals, but long-term effects are difficult to assess.

Tight buildings are generally those which are artificially ventilated, and these buildings appear to be the ones in which hypersensitivity to chemicals and chemical fumes prevail and environmentally induced illnesses are worst.

Symptoms of SBS include dry eyes, dizziness, lethargy, colds/flu, stress and skin rashes, which are all very unpleasant, and every effort needs to be made to ensure that buildings do not possess an internal environment that allows their development. There are ways to try to avoid the occurrence of SBS in buildings during the design, construction and maintenance phases.

Countries that seem to have most problems are those that are either very hot and need high levels of air conditioning, or very cold and need a lot of artificial heating.

Design

1. Try to use a traditional office plan layout, where some of the offices are ventilated naturally, as opposed to an open office plan layout.
2. Ensure that the relative humidity will be within the correct range through the design of the **HVAC** system.
3. Locate photocopiers away from the workers in separately ventilated rooms.
4. Ensure that the heating and ventilation systems are designed for maximum comfort.
5. Ensure that the design of services is considered at the time of the initial building design.
6. Design in Personal Environment Systems where people have some control over their personal working space with regard to heat, light and sound.
7. Ensure that possible air- and structure-borne sound is designed out as much as possible.

HVAC stands for Heating, Ventilation and Air Conditioning.

8. Carefully consider the type of lighting to be used and ensure that it is suited to the type of work being undertaken.

Construction

HVCA is the Heating and Ventilating Contractors Association.

1. Ensure that the heating and ventilation contractor adheres to the **HVCA** guide with regard to the internal cleanliness of new ductwork.
2. Before construction work is undertaken, check that the coordination of the services installation has been thoroughly planned.

Maintenance

1. Ensure that the relative humidity is always within the correct range. The comfort zone will be different for different people and the provision of freestanding fans may alleviate the problem.
2. Ensure that adequate monies are allocated for checking and planned maintenance of the HVAC system.
3. Ensure that the building is well cleaned and monitor the type of chemicals being used.

A further factor that is out of the scope of responsibility of any of the professionals involved in the construction process, but very relevant to the building occupiers, is that from research it would appear that reducing the monotony of the actual work being undertaken could be advantageous.

It can be seen from the above that the factors which need to be addressed include many parties: the architect, the services engineers, the main contractor, the services subcontractors, the building owner, the building occupier, the facilities manager etc.

This could be the reason why SBS occurs, because the level of coordination required to avoid or implement all these factors makes it untenable.

One also has to consider that in the lifetime of a building it may have several occupiers who have different requirements of personal comfort, and this is rarely considered when a change of use occurs.

Legionnaires' Disease

This is very different from SBS. It comprises two specific diseases, Legionella Pneumonia, which is potentially fatal, and Pontiac Fever, which is generally non-fatal. The general cause is vaporised water from showers and humidifiers that has an aerosol effect. *Legionella* is an aquatic organism, and if you have a low relative humidity of less than 30 per cent, *Legionella* dies. It survives best at between 40 per cent and 65 per cent relative humidity, which coincidentally is the range of humidity that is comfortable to most people. The maintenance and general cleanliness of moist routeways within services need to be carefully considered to avoid the occurrence of Legionnaires' Disease.

Design

Because of the reasons discussed it is essential that the services/environmental engineer must be involved with the building design right from inception. In the past the architect would design the layout of the building and specify materials to be used for construction, and then pass the drawings on to the services engineer to design the services to a client's

PART 3

specification of environmental requirements. From research, it has been proved that many building materials in themselves are inert, and services systems have been checked in laboratory conditions to see whether they produce the desired effect. However, the *whole* system has not been checked together to prove long-term environmental comfort.

Much work is being carried out in this area and services engineers are being encouraged to consider the services side of construction as an integral part of a project.

Duties of the services engineer

At design stage the services engineer will design all the services required for the building. The engineer must coordinate with the architect and design team and the statutory services board to enable the development of a comfortable environment for the occupier, either human or machine.

The engineer produces a set of services drawings that are used by the main contractor to install all the services. The services engineer may also be asked to contribute with information on costings for services, as this is a very specialised area. Throughout the contract the engineer or his or her representative must keep a continual check on every area of the installation of the services, and upon completion of the installation the engineer must oversee the commissioning of the entire system to ensure that it is working safely and effectively.

Integration of services

In modern buildings of all types, the integration of services is of great importance in ensuring that effective utilisation of space is made. In commercial buildings services may occupy 10 per cent of the total building volume; in other types of building this could be even higher (see Figure 13.1). It is therefore essential that their housing be considered at an early stage of building design. If this is undertaken, maximum flexibility and quality of the internal environment (comfort and aesthetics) should be achieved.

To ensure that integration of services is carried out correctly, an understanding of the interaction between services and the building fabric is essential. There are two distinct elements that need to be addressed:

- *Design*: This ensures that the services installation performs to user requirements
- *Planning*: This ensures that the system will fit satisfactorily into the building.

The space required for distribution routes and the positioning of plant rooms are dictated by the size required, weight of plant, noise disturbance issues and access for maintenance and running.

If integration of services is not carried out effectively, this can potentially be disastrous for the smooth running of the construction programme. It needs to be agreed who is responsible for the **integration of the services** before any works are undertaken. The parties who could undertake the works are:

- The architect
- The services engineer

Integration of the services must be carried out by a methodology that is determined before any work starts on-site in order to ensure the efficiency of installation works.

Figure 13.1
Services space
utilisation.

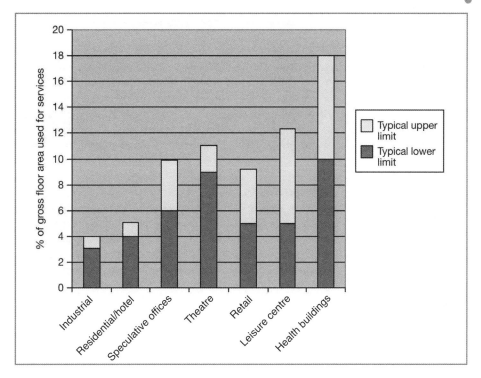

- The main contractor
- The mechanical subcontractor
- The electrical subcontractor.

The organisation that takes responsibility must be allocated sufficient funds to allow for the works to be undertaken.

Services need to be integrated through many elements of the building, and the methods of housing services can be either:

- *Structural*: They pass through the main supporting elements of the building
- *Non-structural*: They are housed in elements that are not part of the main structure and may be fixed in elements that are movable or non-movable.

REFLECTIVE SUMMARY
- A building can be described as a shell that houses people in a 'comfortable' environment, and enables people to carry out designated tasks with ease.
- Buildings must house services to facilitate certain requirements. They include:
  - Comfort of occupiers
  - Health of occupiers.
- Potential problems include:
  - Sick building syndrome: there is no definitive cause of this problem but it is believed to be caused by a cocktail of factors which, when they reach a trigger level, cause the symptoms of SBS

PART 3

- Legionnaires' Disease: this illness is known to be caused by the bacteria passing through 'damp' areas of the building, such as air-conditioning systems.
- At design stage the services engineer will design all the services required for the building. The engineer must coordinate with the architect and design team and the statutory services board to enable the development of a comfortable environment for the occupier, either human or machine.
- In modern buildings of all types, the integration of services is of great importance in ensuring that effective utilisation of space is made. In commercial buildings services may occupy 10 per cent of the total building volume.

REVIEW TASKS

- What aspects of a building are believed to contribute to the existence of *sick building syndrome* in multi-storey commercial buildings?
- At what stage in the design process should the services engineer be consulted about the requirements of the potential building occupier?
- Visit the companion website at www.palgrave.com/engineering/riley2 to view sample outline answers to the review tasks.

13.2 | Structural and non-structural methods of services integration

Introduction

- After studying this section you should have developed an understanding of the methods used to integrate services in buildings, both structurally and non-structurally.
- You should also have an appreciation of the performance requirements of suspended ceilings, raised access floors and partitions, and the different forms these systems can take.
- In addition, you should be able to determine which of these systems is best suited for given building scenarios.

Overview

The need for services to pass from floor to floor in buildings is unavoidable. For this reason, structural methods are required to house services. The term 'structural' relates to an element that passes through the main structural elements of a building, such as the frame, floors and walls. These services are necessary to ensure the physical comfort of occupants and that the intended use of the building can be facilitated.

In addition to physical comfort, people want to live and work in environments that are aesthetically pleasing, and the use of suspended ceilings and raised floors to cover unsightly services can add to the ambience of a room. Internal divisions need to be carefully considered in order for the most efficient use of space to be achieved. If buildings are going to have constantly changing or multi-organisation occupiers, the use of movable or demountable partitions could be the best system to use. These can be moved around to change office layouts very simply and quickly whilst causing little or no damage to surrounding finishes.

Structural methods of services integration

Plant rooms

Plant rooms typically contain:

- Heating boilers
- Electricity distribution and switching gear
- Air-handling plant, standby power supply equipment (generator)
- Lift/escalator machinery.

A number of general principles should be followed when considering the choice and location of plant rooms to ensure efficiency:

- Positioning should ensure that supply and distribution runs are as simple as possible to reduce cost and be less wasteful of space.
- The installation will be most efficient if it is located centrally within the zone it serves.
- The location should take account of the fact that plant is often heavy and noisy and should ensure that as little disturbance to the occupier as possible is achieved.

Roof-level plant rooms

Roof-level plant rooms will have a major effect on the loadings on the structure of a building.

Generally the plant rooms for electric (passenger) lifts are located at roof level, but in addition the provision of general plant rooms is common. The advantages of **roof-level plant rooms** are that valuable internal space is not wasted and heavy pre-assembled plant can be lifted easily using cranes, which avoids the need for fabrication on-site.

However, the additional loading to be placed on the roof must be accounted for in the structural design of the upper floors, which could use lighter construction if no plant room was to be placed on the roof. The incoming mains services will also have to run up the whole height of the building, which is in direct opposition to the idea of using the shortest routes.

Basement plant rooms

For structural and other reasons, plant rooms are often located in the basement. It is sensible to place the heavy plant in this area as the design of the foundations will incorporate any loading and remove the need for heavier construction up the building. The massive construction of basements provides good sound insulation and the length of incoming service routes is minimised.

However, access to the area for installation and replacement of large pieces of equipment will be difficult, and plant may have to be assembled on-site. The high cost of constructing a basement is also a consideration that needs to be taken into account.

In practice, a combination of the two locations for plant rooms may be taken, exploiting the advantages of both systems with vertical service risers linking the rooms (Figure 13.2).

PART 3

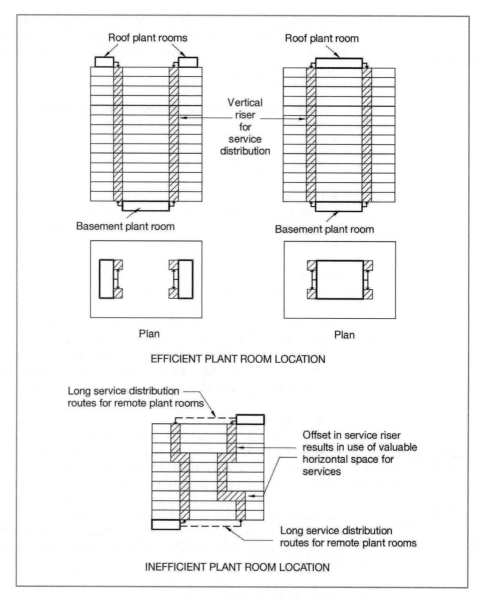

Figure 13.2
Location of plant rooms.

Distribution and accommodation of services

The vertical and horizontal distribution of building services must be accurately coordinated to ensure that all the elements are incorporated within the available spaces, without danger of obstruction or interruption. The forms of distribution services requiring to be accommodated in a modern building may include:

- Air-conditioning/ventilation ductwork
- Water distribution pipework
- Sprinkler system pipework

- Electric cabling and trunking
- Security/fire alarm and detection techniques
- Communication/computer cabling and fibre optics.

With the increasing amounts and complexity of these distribution services it is very important to ensure they are housed with as little wasted space as possible.

Usually service voids are created in the building to allow for vertical access of the services to every floor.

Figure 13.3 shows how a service void is created in a steel frame building. The voids are usually positioned next to the lifts and staircases, as in this case.

The following factors may influence the choice of methods:

- *Accessibility*: Can the services be permanently enclosed or do they require easy access?
- *Volume of services*: The volume and type of services to be enclosed will dictate the nature of the enclosing system.
- *Flexibility*: Can the system be expanded in future due to ongoing technical developments (services installations need replacing a long time before the fabric needs a major overhaul)?
- *Coordination*: Vertical and horizontal distribution must be coordinated to ensure that the most efficient and effective routes are taken through the building fabric.
- *Fire resistance*: Services will need to be encased in fire-resistant enclosures. The main concern is the effect of fire resistance of the building fabric arising from providing pathways for services. Ducts must be fire stopped and ceiling voids included for fire blankets.
- *Durability*: Housing for services must be robust enough to withstand any imposed loadings.
- *Aesthetics*: The visual effects of the services accommodation will vary in importance depending on the nature of the building and its use. In the majority of buildings constructed these days all services are hidden from view as much as is possible unless a feature is being made of them.

Figure 13.3
Service voids in a steel frame building.

■ *Sound insulation*: Steps must be taken to restrict the passage of sound via service routes, e.g. installation of sound insulation in ceiling voids and in openings between areas.

It is impossible to incorporate a services system entirely without some structural aspects being constructed to allow for the installation. The following are some of the aspects that are considered as structural and would be detailed by the Services Engineer on **Builders' Work Drawings**. These drawings are issued to the contractors at the start of the contract and detail all work to be done by them before the services contractors arrive on-site. If the work is not undertaken and/or is not carried out accurately there will be huge problems with the services installation:

> **Builders' Work Drawings** give details of all work to the structure that is required to house services or enable installation of services.

■ Forming holes in foundations for main service supply entry; formwork will be required for this activity
■ Forming holes in slabs for risers; formwork and reinforcing steel details will be required
■ Forming holes in walls for access – if the walls are concrete, then formwork will be required, and reinforcement details not cut out after if possible (if the walls are brickwork, holes must be formed as work progresses)
■ Building of plant rooms and risers etc.
■ Cutting chases in internal walls for conduits and plug sockets
■ Building lift shafts and undertaking structural work in the shaft that is not expected of the lift installer
■ Forming of ducts in slabs.

Figure 13.4 shows how services are installed in structural elements.

Non-structural methods of integrating services

The accommodation of services in ceiling and floor voids is very common and they are easily concealed by the use of suspended ceilings and raised floors.

Services in ceiling voids

It is good practice to separate the void into zones to incorporate wet services, ventilation ductwork and electrical services (Figure 13.5). Wasting this space has a large cost implication, as it results in a large floor-to-ceiling height. For example, in a 40-storey building an increase of 100 mm per floor would give an overall difference in height to the building of 4 m – in effect, an extra floor!

Functions, selection criteria and options of suspended ceilings

The use of a suspended ceiling can effectively create a horizontal planar duct area for the housing of services, thus allowing great flexibility of installation. Depending upon the selected type, the performance requirements set out for a services integration solution may be satisfied to varying degrees. Suspended ceilings can be classified as jointless, jointed or open.

Figure 13.4
Accommodation of
services in structural
elements.

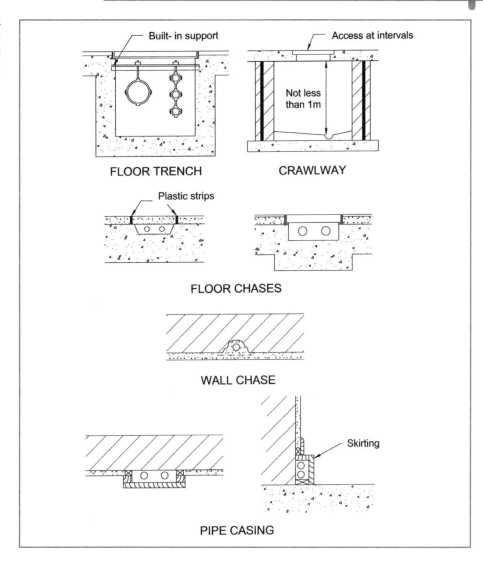

PART 3

Jointless ceilings

These types of ceiling (Figures 13.6 and 13.7) give a conventional appearance, but are suspended beneath the main structure. Figure 13.8 shows the services in the void above the ceiling. Two main construction methods are adopted to form such ceilings, involving either a suspended framework supporting plasterboard, or expanded metal lathing on a framework coated with plaster or some other fibrous coating. The main disadvantage is the restriction in terms of access to the enclosed void. Access hatches need to be formed in the ceiling – these can be difficult to construct to an adequate quality and may get dirty or damaged from use.

One of the main
advantages of
suspended ceilings is
that they allow for ease
of access for
maintenance and
cleaning.

Jointed/frame and tile ceilings

These (Figures 13.9 to 13.11) are the most common form of **suspended ceiling**. They are constructed of a metal grid or frame suspended from the structural floor. Fibreboard,

Figure 13.5
Integrated services accommodation.

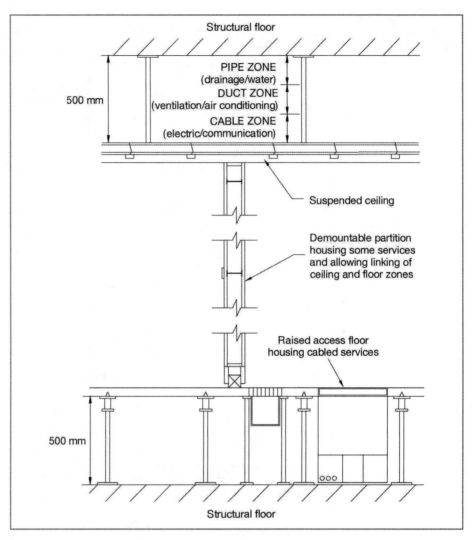

Figure 13.6
Jointless ceiling types (after British Gypsum).

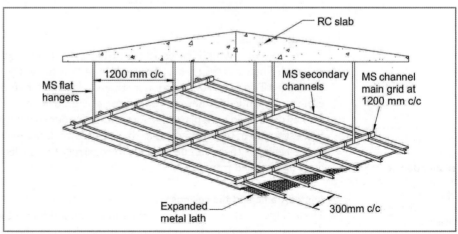

Figure 13.7
Jointless ceiling details.

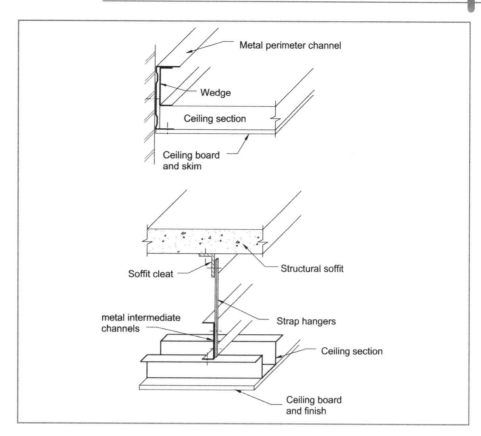

Figure 13.8
Services installed in the ceiling void.

Figure 13.9
Jointed ceiling layout
(after British Gypsum).

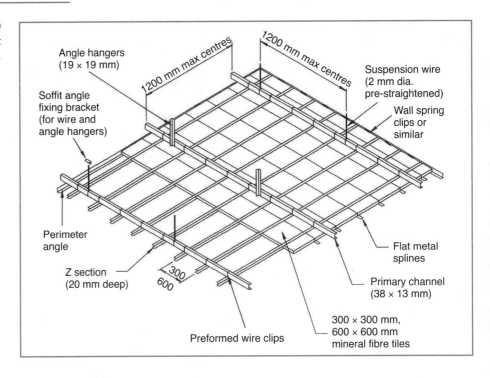

Figure 13.10
Jointed ceiling details.

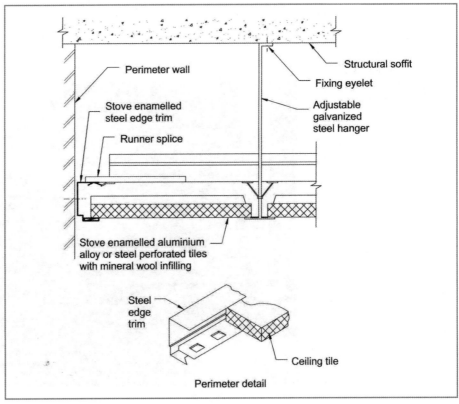

PART 3

Figure 13.11
Ducted services for ventilation/air conditioning housed within the ceiling void. Access is easy using the jointed ceiling.

plaster or metal tiles are then dropped into the frame. The grid may be exposed (termed a 'lay-in grid') or 'hidden' (concealed), and the tiles are modular – usually 300 mm × 300 mm or 600 mm × 600 mm.

These exposed grids give very easy access to the services. The tiles lift out (as many as need to be removed) and the services can then be easily worked on. However, these are generally less aesthetically pleasing than a ceiling with a hidden grid.

Open ceilings

This type of ceiling (Figure 13.12) offers the ability to provide a visual barrier between the room and the services in the ceiling without having a solid physical barrier. By painting the underside of the slab above in dark colours and mounting the light fittings at or below ceiling level the ceiling zone will essentially be hidden, whilst allowing immediate unhindered access to the services.

The performance criteria that need to be considered when choosing a suspended ceiling system are as follows:

- *Services accommodation*: What depth is required (for housing, installation, insulation, maintenance etc.)? Does the ceiling need to be loadbearing? Have the services been integrated properly?
- *Access*: How much and how often is access required?
- *Sound insulation*: Is it required both vertically and along the ceiling void?
- *Sound absorption*: The tile or board will need to satisfy the criteria that have been determined as necessary.
- *Fire*: It can be difficult to achieve fire resistance and precautions can be damaged during maintenance. Great care must be taken to ensure that damage does not occur and that if it does that it is rectified immediately.
- *Finish*: The ceiling finish will depend on the quality required by the client. Prestigious developments will require a high-quality finish.

Figure 13.12
Open ceilings.

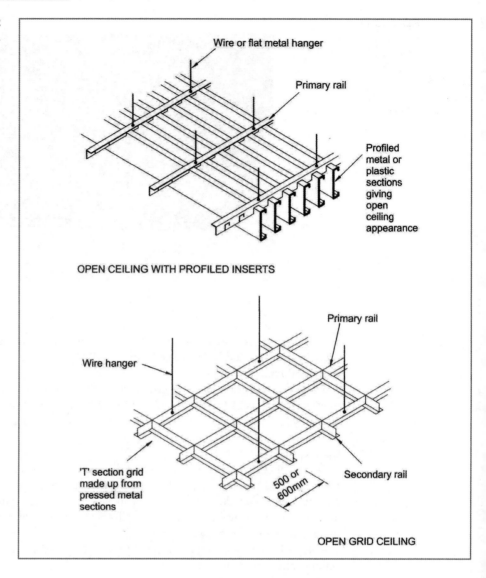

It is becoming increasingly popular not to use a 'suspended' ceiling at all and to leave the underside of the slab of the floor above exposed, and simply fix the services directly to the slab. This is only suitable of course if the quality of the slab finish is to a high specification. It also works better when there is some moulding and then a feature can be made out of the services. Figure 13.13 shows an *in-situ* waffle slab from the underside with the lighting fixed to the waffle profile. In this scenario the lighting is the only service fixed to the ceiling. There is no air conditioning or water supply in the room and heating is provided using a wall perimeter heating system. If a whole range of services was needed at ceiling level then this system would become very unsightly.

Figure 13.13
Waffle slab with
services attached
directly.

PART 3

Functions, selection criteria and options of raised access floors

In most modern commercial buildings these days there is a requirement for large amounts of cabling with ready access. The development of 'raised access' or platform floors aims at addressing these issues. As in the case of suspended ceilings, platform floors provide a flexible horizontal duct space capable of housing a range of services. Although initially developed to house data processing cabling these systems are now being expanded to allow provision of other forms of services, such as ventilation ducting and wet services. The floor is supported on a series of adjustable jacks, allowing a level surface to be created at varying heights above the structural floor. In this way a void is created through which the services can be passed. Typical void depths will be between 50 mm and 1500 mm, depending upon the type and specification of the chosen system.

Cable routes are allowed for by a series of trunkings or trays, fixed to the underside of the raised floor or laid directly on the surface of the structural floor. Outlets such as plug sockets from the cables are provided in the floor to allow for easy connections to items of equipment. Computer network points can also be incorporated into tiles, therefore avoiding 'trailing' cables during the occupancy of the building. Raised access floors are modular, and therefore if the position of a plug or computer network socket needs to be moved to another position, the tile can be lifted out and moved to another location.

Performance requirements of raised access floors

- Raised access floors should be able to take any required loadings
- They must have the required level of durability, which will depend on what the building is to be used for
- They should provide adequate fire protection for the enclosed services
- Sound insulation: a raised access floor creates a layered effect that reduces sound transmission, but if the void is 'full' of services then this benefit is reduced
- They must allow for access to the enclosed services
- They must be flexible enough to allow for movement of plug points
- They must be capable of taking the required floor finish.

Most raised access floors are not pre-finished and therefore a covering is required, usually for aesthetic purposes. The type of finish used must not reduce any of the performance requirements of the flooring system. To ensure maximum accessibility it is usual for the floor finish to be supplied in the same modular size as the floor, so 500 mm × 500 mm carpet and vinyl tiles are ideal for use with a raised access floor where the modules are 500 mm × 500 mm in size. To gain access to services the floor finish and the tiles can be removed where required and replaced without causing any lasting damage to the appearance of the finished floor. Using rolled carpet over a raised access floor defeats the object of the exercise, as the whole carpet would need to be lifted to gain access to the services enclosed in the floor void.

Types of raised floor

Shallow battened

Shallow battened floors are often used in buildings that are being refurbished to modern requirements, but where there is a limited floor-to-ceiling height.

Shallow battened floors are constructed by nailing or screwing shoes to the main floor and then fitting timber battens into the shoes. The floor planks are then nailed down onto the timbers to form the finished floor.

Figure 13.14 shows a system that can be used if the amount of available space for the flooring is limited. Boxes can be fixed in place to take cabling, and then a self-levelling screed can be used to create a smooth finish. Finally, the shoes for the timber battens can be fixed and the floor finish nailed to the battens.

Deep platform

In most modern buildings enough space is left between the slab of one floor and the underside of the slab of the next floor to allow for a raised access floor. These floors are constructed using pedestal legs fixed to the concrete slab, usually with resin. They then have modular floor tiles fixed to them, generally with screws in each corner of the tile. The module size is usually 600 mm × 600 mm and the space under the floor can range from 200 mm to 1200 mm. This space allows for extensive services installation beneath the floor, which allows easy maintenance access.

Method of control for the installation of suspended ceilings and raised floors

These systems are usually installed by specialist subcontractors. All they require is a datum to work to. A datum is given when all the plastering is complete, and this is usually 1 metre above finished floor level (FFL) (Figure 13.15). The datum value is calculated, then a backsight reading taken onto a temporary benchmark (TBM). The value of the required foresight reading is then calculated and transferred to the wall at each corner. The marks are then joined up using a chalkline. (A chalkline is a piece of equipment that contains a ball of string wrapped around a coil, rather like a fishing line, which can be filled with loose coloured chalk. You then hold one end of the string on one point and pull out the string to another point. Then you 'ping' the line and a clear mark is created at the correct level in chalk.) Sometimes ceiling fixers use rotating laser levels to establish the line of the ceiling. The distance to the ceiling level from the datum is calculated and the instrument set up at the correct ceiling line. The laser level then rotates at speed and a 'line' is formed in light on the walls. The ceiling is then fixed at this level.

Figure 13.14
Raised access floors.
SSL denotes the
structural slab level.

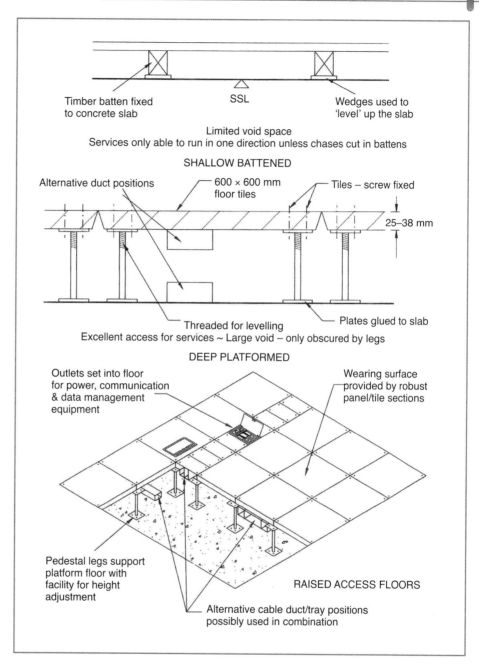

Timber batten fixed
to concrete slab

SSL

Wedges used to
'level' up the slab

Limited void space
Services only able to run in one direction unless chases cut in battens

SHALLOW BATTENED

Alternative duct positions

600 × 600 mm
floor tiles

Tiles – screw fixed

25–38 mm

Threaded for levelling

Plates glued to slab

Excellent access for services ~ Large void – only obscured by legs

DEEP PLATFORMED

Outlets set into floor
for power, communication
& data management
equipment

Wearing surface
provided by robust
panel/tile sections

Pedestal legs support
platform floor with
facility for height
adjustment

RAISED ACCESS FLOORS

Alternative cable duct/tray positions
possibly used in combination

Floor layers sometimes use another type of laser level which is first set up. Then a receiver is attached to a piece of timber or the levelling staff. The receiver is then moved and fixed at the correct level, so that when the floor is in the correct position the receiver bleeps. The receiver will indicate whether the floor needs to be lowered or raised in order to be at the correct position.

Figure 13.15
Installation datum level.

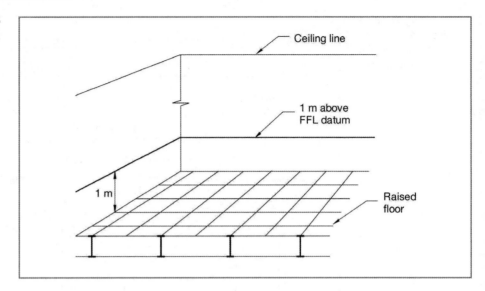

Ceiling line

1 m above
FFL datum

1 m

Raised
floor

Functions, selection criteria and options of partitions

Partitions (the internal walls of a building) are used to create internal divisions of space. They may be permanently loadbearing or non-loadbearing, fixed or movable, solid or hollow.

The design life of a building is usually quoted as 60 years, but in the main the actual structure of a building will last considerably longer. During this lifespan the occupiers of the building could change many times, and each occupier will have different ideas as to how the building should be laid out. Also, changes in technology will mean that the services layout could change dramatically over the whole life of the building. For example, if we think about what the services in a building would have consisted of 60 years ago, we can easily illustrate this point. Open plan offices were (and still are) very popular, but there are occupiers who prefer a more traditional office layout.

Loadbearing partitions

In addition to dividing a space, loadbearing forms also carry some loads from the building, with the division of space a secondary function. Often their location, which is dictated by the loadbearing requirement, is a hindrance to the efficient use of space rather than a benefit. The use of these partitions in commercial and industrial buildings, which are generally of framed construction, is comparatively rare.

Non-loadbearing partitions

Non-loadbearing partitions carry no structural loading other than their own self-weight. The use of these is universal in modern construction.

As with any building element, partitions are subject to a set of performance requirements, which must be met to varying degrees in order that the element fulfils its intended function. The performance requirements for partitions are discussed below.

Division of space

Partitions are utilised primarily to ensure areas of privacy, to provide visual division or simply to allocate areas of activity to individuals or operational activities.

Durability

Partitions must offer suitable levels of durability to ensure that they perform adequately for the duration of their expected life. An element of **flexibility** is essential, since it is normal for the internal layout of a building to be subject to variation several times within the life of a building. Additionally, it is possible that the use of the building may change over time, which will impose its own demands upon flexibility of spatial division.

Flexibility of office layouts is very attractive to potential leaseholders or owners, but allowing for flexibility can lead to problems with fire protection.

Sound insulation

The division of space may in some instances require that sound insulation between spaces is provided in the partition. The prevention of the passage of noise must be taken very seriously. The way in which the partition interacts with the surrounding structure is important in this respect, since there is possibility of sound passing around the unit if the details of connections between the partition and main structure allow (Figures 13.16 and 13.17).

Sound absorption

In some environments the acoustic qualities of a space are very important, so the level of sound which is reflected back into the space is therefore an element which must be considered – the nature of the surface finish and the materials from which the unit is

Figure 13.16
Partitions and sound.

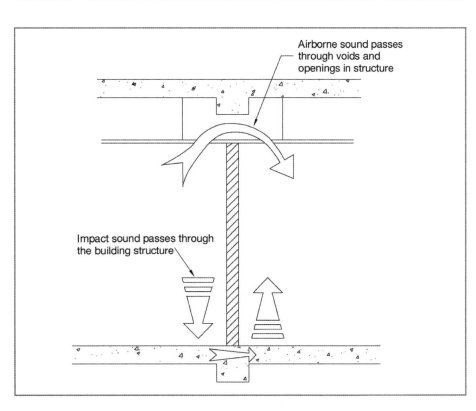

Airborne sound passes through voids and openings in structure

Impact sound passes through the building structure

PART 3

Figure 13.17
Sound bypassing
partitions.

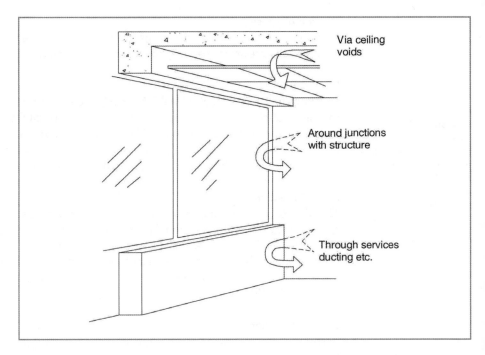

made are important factors in this respect. This will be very relevant in rooms that act as conference facilities or meeting/boardrooms.

Fire

The ability of the partition to resist the passage of fire depends upon the ability of the unit to maintain its integrity and its stability for a requisite period of time. Integrity is the ability of the element to retain its form without allowing the passage of flame or smoke. The level required depends upon the position and function of the partition.

The rate of spread of flame over the surfaces of the partition can be an important factor in the control of a fire. Depending upon the use and type of building the partitions may be required to meet a classification between 0 (spread of flame almost totally inhibited) and 4 (spread of flame unchecked), as detailed in the Building Regulations.

The passage of smoke into spaces of a building in the event of fire can be disastrous; hence fire-resisting partitions are required to resist the passage of smoke as well as flame. It is therefore essential that the integrity is maintained for the required period of time, in addition to stability.

The incorporation of openings and doors into partitions must also be considered carefully with regard to performance in the event of fire, since any openings in fire-resisting partitions can seriously reduce their effectiveness. Junctions of the partition with ceilings and adjoining elements must be treated with care, since the passage of smoke or flame around the partition must be prevented. The use of fire stops in ceiling voids etc. is helpful in this respect (Figure 13.18).

Services

These may be encased within partitions. Access for the purposes of repair and maintenance must be provided for.

Figure 13.18
Sound bypassing
partitions.

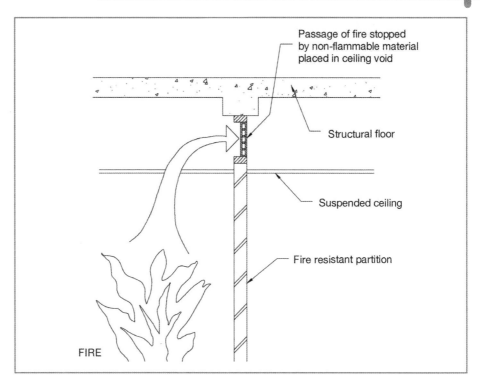

Passage of fire stopped
by non-flammable material
placed in ceiling void

Structural floor

Suspended ceiling

Fire resistant partition

FIRE

Strength/stability

Usually partitions are not loaded heavily because they do not carry any portion of the loads from the main building fabric. However, depending upon their position, they are subject to limited levels of loading and must be capable of resisting them. Forms of loading which may occur are:

- *Lateral loading*, such as that applied by fittings or people leaning on the partition
- *Applied loadings*, such as those resulting from fixing pictures or shelving to the partition
- *Impact loading*, resulting from items colliding with or impacting on the partition.

Appearance

The standard of internal finish must be taken into consideration together with any effects which this choice may have on the provision for maintenance and ease of care during the life of the element.

Types of partition

Partitions can be divided into a number of generic types, each with their own characteristics, advantages and disadvantages. The following forms are the main types which are commonly found in modern buildings.

Solid partitions

This form of partitioning is common in many buildings and is usually constructed from non-loadbearing lightweight blocks or brick (Figure 13.19). Blockwork is by far the most

PART 3

Figure 13.19
Solid partitions (after
British Gypsum).

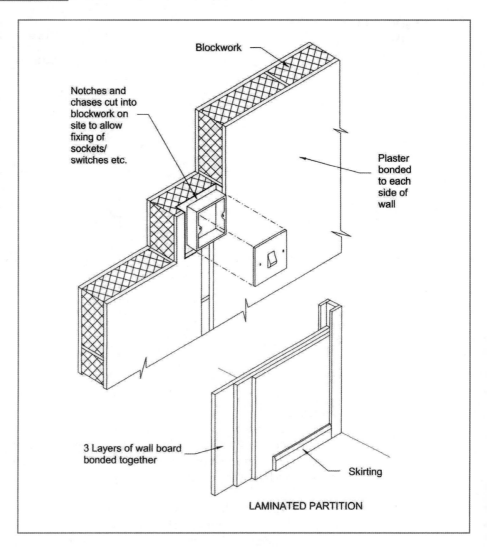

common material of construction, usually 75, 100 or 150 mm thick. Partitions of this type are very stable, but offer limited flexibility in use. A number of proprietary systems are also available however, such as 'Gyproc' laminated partitioning, which is formed by bonding layers of plasterboard to form a solid panel of the requisite thickness. Solid partitions have a number of advantages: they are strong, relatively cheap, easily constructed and allow for flexibility in design. However, the quality of the end product will depend on the skill of the constructor, a problem that is enhanced by the fact that the building of block walls is considered a 'wet trade', and wet trades are always more susceptible to quality issues than dry trades. They are also difficult to incorporate services into, and require the cutting of holes and/or chasing of conduits, which are time-consuming and dirty jobs.

This type of partition is non-movable and therefore after construction allows for very inflexible use of space.

Figure 13.20
Stud and sheet
partitions.

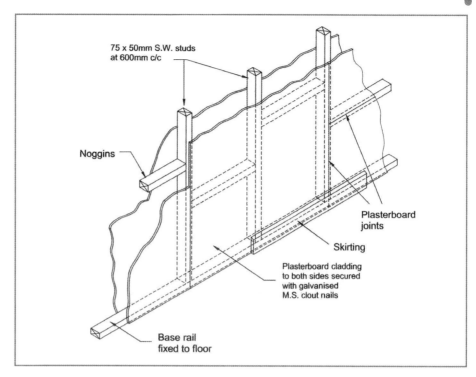

75 x 50mm S.W. studs
at 600mm c/c

Noggins

Plasterboard
joints

Skirting

Plasterboard cladding
to both sides secured
with galvanised
M.S. clout nails

Base rail
fixed to floor

Stud and sheet partitions

This form of partitioning comprises a series of timber or metal studs clad with a sheet material, often plasterboard (Figure 13.20). Pre-finished systems are available, however, which do not require decorating. In their simplest form they are usually constructed using a series of softwood vertical studs with horizontal cross noggins forming a basic framework or carcass. The frame is then clad with plasterboard on each side and may be provided with a skim coat of plaster or an application of tape to the plasterboard joints. More advanced forms consist of galvanised steel studs clad with pre-finished boarding rather than plasterboard.

The advantages of these partitions are that they are easy to construct, the components are readily available, they are cheap, and they are flexible in design. Their disadvantages are, however, that they can be slow to build because they are constructed from sections of material which must be cut to size from large sections. This process can also lead to a lot of wastage, which is not advisable from an environmental perspective. They are also not very flexible in use, as when they are taken down and reinstated elsewhere they are easily damaged.

Frame and sheet partitions

These are similar to the stud and sheet type of partition but they are manufactured from proprietary components, including standard door and window sections. They are a **demountable** form of partitioning, with easily accessible fixings that are usually hidden (Figure 13.21). The partitions can be taken apart and relocated without damaging them.

They have advantages in that there is little material wastage, they are easy to fabricate, they are flexible in design and use, they are reusable and they are manufactured using

Demountable
partitions are easily
movable.

PART 3

Figure 13.21
Frame and sheet
partition.

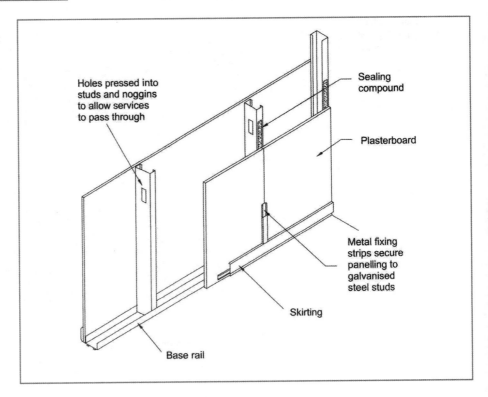

Holes pressed into
studs and noggins
to allow services
to pass through

Sealing
compound

Plasterboard

Metal fixing
strips secure
panelling to
galvanised
steel studs

Skirting

Base rail

totally dry processes. The disadvantages of these partitions are that they can be expensive and can be limited in design by the necessity to use standard components.

Frame and panel partitions

These are very similar to the frame and sheet type of unit, but the structural frame of the partition is left exposed. Typically, pre-finished panels are placed into the profiled forming (Figure 13.22). Again this type of unit is demountable and is commonly found in commercial premises. Their advantages and disadvantages are similar to those of the frame and sheet systems.

Figure 13.23 shows a typical demountable partition. The partitions can be moved around to form pods of the size required. They can then be moved to change the pod sizes, and little making-good work will be needed.

Figure 13.22
Frame and panel partitions.

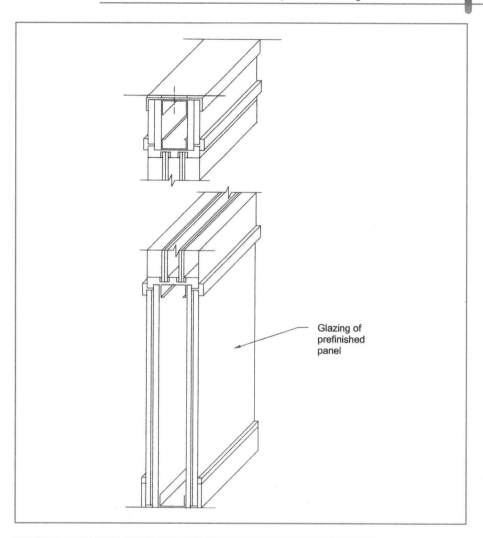

Glazing of prefinished panel

Figure 13.23
A typical demountable partition.

CASE STUDY: PARTITIONS AND CEILINGS

The photographs show how partitions, suspended ceilings and raised access floors can be used to create a comfortable environment whilst being aesthetically pleasing with a minimum of services visible. The main room was not perfectly square and had to be made so because of the design and the quality of the final finish that was required.

To make the room perfectly square, metal studding was fixed to metal sole and soffit plates and plywood sheets fixed to the studs to form partitions that act as false walls (Figure 13.24).

Figure 13.24
Using partitions to create a regular space.

20mm Plywood sheets

600 mm | 600 mm

Existing walls

Suspended ceiling jointed, hidden grid

The suspended ceiling is a 600 mm × 600 mm modular concealed grid system with plaster tiles that have been painted white.

The floor is a shallow batten raised access floor finished with timber strips that can be removed to gain access to the services underneath (mainly electrical wiring). The restricted floor-to-ceiling height limited the floor void depth to 50 mm.

(continued)

CASE STUDY: PARTITIONS AND CEILINGS (*continued*)

The majority of the services are enclosed behind the partitions and within the suspended ceiling.

An interesting feature of this room is that the wall covering is carpet. The reason for the choice of this material as a wall covering is that pictures are to be hung and changed regularly. Using nails will damage virtually every alternative covering, but when nails are removed from the carpet no sign of them is evident.

In the secondary room, the flooring and partitioning use the same methods, but the ceiling is actually a jointless plasterboard ceiling with access panels cut in at strategic positions to allow for access. The detail on the ceiling is merely for aesthetic purposes and is not an integral part of the ceiling itself. The problem with this method is that access is fairly limited, while the ceiling can become dirty and damaged during any maintenance work.

In the corridor areas there is a totally jointless ceiling, and no access panels have been incorporated. Although the finish is very aesthetically pleasing, over the long term, maintenance (either planned or emergency) will be required and the ceiling will ultimately suffer from damage. The walls in the corridor are solid blockwork that has been plastered and painted

---

COMPARATIVE STUDY: SERVICES INTEGRATION

Option	Advantages	Disadvantages	When to use
Suspended ceilings	Ready access if using jointed form Depth of void can accommodate a range of services Enhanced appearance of the room Allow for use of modular light fittings etc. in grid form	Problems with access if enclosed Possible lack of durability in some forms Care needs to be taken to avoid transmission of sound and fire through the void	All buildings where a high level of services provision is envisaged and concealment desirable
Raised access floors	Excellent cable management characteristics Potentially good access Allow for provision of ventilation etc. through floor Cope with need for relocation of electrical/data outlets	Can be heavy Floor covering choices are limited if access is required Rolling loads can be problematic Potential for passage of sound and fire through the void	Almost essential in all modern office buildings, where extensive data and IT provision is required
Partitions	Allow for vertical distribution of services between suspended ceilings and raised access floors Demountable forms allow ready access to services Wide variety available Also act as room dividers Some forms can act as fire compartments	Problems with access in some forms Potential lack of flexibility Regular access to services can lead to damage	Demountable partitions are commonly used in office buildings They are especially useful in loose fit buildings as design evolves

PART 3

REFLECTIVE SUMMARY

■ Integration of services is extremely important in ensuring that the construction process runs smoothly and that as much of the services installation as possible is obscured from the view of the building occupier.

■ Services can be integrated using structural or non-structural methods.

■ Suspended ceilings and raised floors not only act as service void formers, but are also aesthetically pleasing.

■ Partitions can be movable or non-movable. In buildings where changes of occupier are expected, movable or demountable partitions are beneficial.

REVIEW TASKS

■ Who should take responsibility for the integration of services and why is this aspect of the construction process important?

■ Detail the performance requirements of suspended ceilings and raised access floors.

■ Why is the use of demountable partitions beneficial in commercial buildings?

■ Visit the companion website at www.palgrave.com/engineering/riley2 to view sample outline answers to the review tasks.

Sustainable building services

Sustainable engineering systems

Laurie Brady and Derek King

AIMS

After studying this chapter you should be able to:

- Understand why building services are vital in ensuring the comfort of building occupiers
- Appreciate the contribution that building services can make to the overall 'sustainability' of a building
- Understand the basics of mechanical engineering, electrical engineering and public health engineering systems in commercial buildings

This chapter contains the following sections:

14.1   Mechanical engineering systems
14.2   Electrical engineering systems
14.3   Public health engineering systems

INFO POINT

1. CIBSE (2006) *Guide A: Environmental Design*, pp. 1-1 to 1-6, CIBSE, London [978–0–900–95329–3]
2. CIBSE (2005) *Guide B: Heating, Ventilating, Air Conditioning and Refrigeration*, p. 1-5, CIBSE, London [978–1–903–28758–3]
3. BRE (2010) *Part L Explained: The BRE Guide*, IHE BRE Press [978–1–860–81910–0]
4. HM Government (2010) *Non-Domestic Building Services Compliance Guide* (*Compliance Guide for Part L 2010*), RIBA Publishing [978–1–859–46376–5]
5. CIBSE (2005) *Guide B: Heating, Ventilating, Air Conditioning and Refrigeration*, pp. 1-12 to 1-37, CIBSE, London [978–1–903–28758–3]
6. CIBSE (2006) *Guide F: Energy Efficiency in Buildings*, pp. 10-1 to 10-16, CIBSE, London [1–903–28734–0]
7. CIBSE (2006) *Guide A: Environmental Design*, pp. 4-1 to 4-19, CIBSE, London [978–0–900–95329–3]
8. CIBSE (2006) *Guide F: Energy Efficiency in Buildings*, pp. 4-13 and 4-14, CIBSE, London [1–903–28734–0]
9. CIBSE (2005) *Applications Manual 10: Natural Ventilation in Non-Domestic Buildings*, CIBSE, London [978–1–903–28756–9]
10. CIBSE (2005) *Guide B: Heating, Ventilating, Air Conditioning and Refrigeration*, pp. 2-1 to 2-133, CIBSE, London [978–1–903–28758–3]

(continued)

INFO POINT (continued)

11. CIBSE (2004) *Guide K: Electricity in Buildings*, pp. 4-1 to 4-11, CIBSE, London [1–903–28726–X]
12. Society of Light and Lighting (2012) *SLL Code for Lighting*, CIBSE, London [1–906–84621–9]
13. CIBSE (2006) *Guide A: Environmental Design*, pp. 1-9 and 1-10, CIBSE, London [978–0–900–95329–3]
14. CIBSE (2004) *Guide G: Public Health Engineering*, pp. 1-1 and 1-2, CIBSE, London [1–903–28742–1]
15. HMSO (1999) *Water Supply (Water Fittings) Regulations 1999 Statutory Instruments 1999 No. 1148*, London
16. Bicknell, S. (2010) *Rainwater Harvesting: Regulatory Aspects*, RainWaterHarvesting.co.uk, Peterborough, September, http://www.rainwaterharvesting.co.uk/downloads/rainwater-harvesting-regulations-and-incentives.pdf
17. Environment Agency (2011) *Greywater for domestic users: an information guide*, May, http://publications.environment-agency.gov.uk/PDF/GEHO0511BTWC-E-E.pdf
18. Young, L. and Marys, G. (2000) *Water Regulations Guide*, Water Regulations Advisory Scheme, September [978–0–953–97080–3]
19. Institute of Plumbing (2002) *Plumbing Engineering Services: Design Guide*, Hornchurch [978–1–871–95640–5]
20. BS 6700: 2006 + Amendment 1: 2009 Design, installation, testing and maintenance of services supplying water for domestic use within buildings and their curtilages, BSI, London
21. CIBSE (2004) *Guide G: Public Health Engineering*, pp. 2-8 and 2-9, CIBSE, London [1–903–28742–1]
22. CIBSE (2004) *Guide G: Public Health Engineering*, pp. 2-11 to 2-31, CIBSE, London [1–903–28742–1]
23. RIBA (2006) *Approved Document H: Drainage and Waste Disposal*, RIBA Publishing, April [978–1–859–46208–9]
24. CIBSE (2004) *Guide G: Public Health Engineering*, pp. 3-1 to 3-16, CIBSE, London [1–903–28742–1]
25. BS 5572: 1994 Code of practice for sanitary pipework, BSI, London

14.1 | Mechanical engineering systems

Introduction

- After studying this section you should appreciate the importance of building services in the context of commercial buildings.
- You should understand the basic requirements for mechanical engineering systems.
- You should appreciate the regulations that govern the design of building services.
- In addition, you should understand the concept of thermal comfort and how buildings may be heated or cooled.

Overview

The term *Built Environment* describes the physical environment in which modern communities live and work. Large sections of the Built Environment were developed during a time when much less consideration was given to the finite nature of energy and materials than today. Our infrastructure must become more sustainable.

Building services are those parts of a building or infrastructure which actively use energy, delivering comfortable and safe internal environments, lighting, power, building transportation systems, hygienic systems for washing and waste disposal, and systems that monitor and protect us from fire and other risks.

Thermal comfort

The major function for building services systems is to control internal environments to either support some process or satisfy human thermal comfort. The notion of thermal comfort demonstrates how building services engineers must translate seemingly subjective concepts into measurable and achievable parameters that can be met by practical engineering systems [1]*.

CIBSE (Chartered Institution of Building Services Engineers) provides guidance for designers. The index of thermal comfort quoted by CIBSE is Dry Resultant Temperature (t_{res}) [2]. Provided that room air movement is below 0.1 m/s, t_{res} takes into account room air temperature and room surface temperature, and may be determined from:

$$t_{res} = 0.5t_{ai} + 0.5t_{r}$$

where t_{ai} is the inside air temperature and t_{r} is the room mean radiant temperature. For practical purposes t_{r} may be determined as an area weighted average room surface temperature.

$$\text{Mean radiant temperature } (t_{r}) = \frac{(A_1 t_1) + (A_2 t_2) + \dots}{(A_1) + (A_2) + \dots}$$

where t_1, t_2 etc. are the surface temperatures of walls, windows etc. and A_1, A_2 etc. are the areas of the corresponding room surfaces.

It is important to recognise that radiant (surface) temperatures affect comfort. This phenomenon means that the structure of buildings can contribute to internal environmental control. Other factors such as clothing, age, gender and activity will play a part in the comfort conditions within a space.

Building Regulations

Part L of the Building Regulations deals with 'Conservation of Fuel and Power' and so is an important area when considering building engineering services [3]. The Approved Document has four sections:

Part L1A – New dwellings
Part L1B – Existing dwellings
Part L2A – New buildings other than dwellings
Part L2B – Existing buildings other than dwellings.

---

* The numbers in square brackets throughout this chapter refer to the numbered references in the Info Point at the start of the chapter.

Compliance with Part L

Achieving limits for carbon emissions requires that building engineering services are efficient, correctly installed and commissioned, and that the building owner is provided with all necessary information to maintain engineering plant so that it can continue to be operated and controlled at optimum efficiency. *The Building Services Compliance Guide*, published by RIBA, provides further guidance in this area [4].

In order to limit building carbon emissions it may be necessary, in some instances, to deploy low or zero carbon technologies. The carbon emissions should be assessed by appropriate software such as SBEM (Simplified Building Energy Model) or by using an approved commercially available software.

For existing buildings L2B is the appropriate section. The type of work covered includes extensions, change of use and energy status, provision or extension of 'controlled fittings' (windows, roof-lights and doors), provision or extension of 'controlled services' (heating, hot water and other engineering services), and work on 'thermal elements' (walls, floors or roofs that can transmit heat).

For buildings with an existing floor area of 1000 m^2 or greater, the Regulations can require 'consequential improvements'. These are works that may be necessary to make an existing building more energy efficient, though the Regulations recognise that technical and financial implications can affect how practical it would be to undertake this work.

Heating

Commercial buildings tend to be heated by indirect systems which generate heat energy in a central plant from where it is distributed to the various rooms or spaces. The medium used to distribute the heat energy may be low-temperature hot water (around 85°C flow temperature), pressurised hot water, steam or air [5]. Pressurised hot water or steam may be used in large district heating schemes – for example, a large general hospital with several buildings located within the site. Heating by air may be part of a ventilation system or may be a requirement if space for radiators or other emitters is not available – for example, a library where all of the wall space is filled with bookshelves. Low-temperature hot water radiator systems are probably the most common heating arrangement in the UK.

Heating controls

The benefits of good design and installation can be wasted if the heating system operates when it is not required, overheats areas of a building, or does not operate as efficiently as possible. Control systems prevent energy being wasted in this way [6].

Time control can start systems at convenient times and prevent plant running unnecessarily. A time-clock is a simple device used to start and stop a heating system at fixed times, whereas an optimiser (often considered to be a 'clever' time-clock) containing microprocessors is capable of starting the heating earlier or later, depending on weather conditions. An optimiser can also stop the heating system early if the building is warm enough to remain comfortable until the end of occupancy.

Heating controls should be capable of recognising the heat requirement of different zones (Figure 14.1). Zonal heating loads can vary due to the movement of the Sun or perhaps because some areas will be occupied at certain times and while other parts of a building are empty.

It is common in commercial buildings for the load to be shared among several boilers. Although all boilers will fire in the coldest weather, outside temperatures can vary during the heating season and not all boilers may be needed. A sequence controller will arrange that, in multi-boiler systems, boilers will fire as needed to meet the demand.

For large non-domestic buildings, it can be difficult to determine which office or space provides a representative temperature that can be used to decide where a thermostat may be located. For commercial buildings, it is common practice for the heating system to respond to the feedback from an outside air temperature detector, so the temperature of the radiators can normally be varied in response to the outside temperature. This is known as *weather compensated control*.

Figure 14.1
Typical generalised multi-zone heating system.

The control system may also ensure that the building fabric is protected by maintaining a minimum temperature during periods of occupancy. This is known as *night set back*.

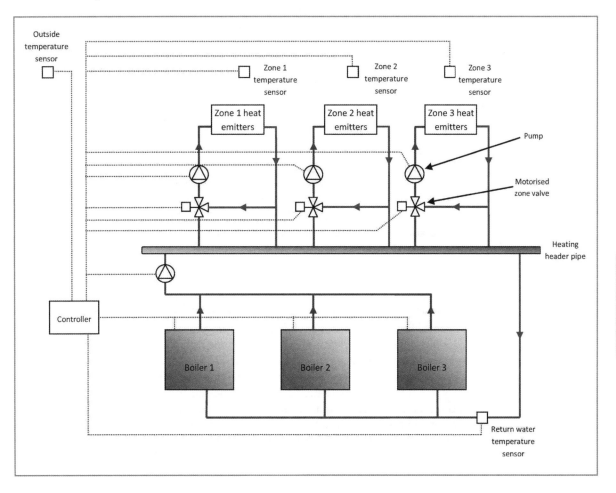

PART 4

Ventilation

Ventilation can have several functions in a building [7].

a. Provide sufficient fresh air for respiration
b. Dilute and remove odours or pollutants
c. Remove excess moisture
d. Provide some cooling or heating
e. Provide air/oxygen for fuel-burning appliances

Function (e) (air for combustion) must be arranged by experts because failure to ensure complete combustion in boilers and other fuel-burning appliances can have deadly consequences.

'Fresh air' is normally considered to be outside air, though the actual properties of outside air (and how 'fresh' it really is) depend upon a number of factors, including building location and surroundings. Ideally, the introduction of outside air should be controlled, though in older leaky buildings, quantities of heat energy tended to be lost because cold outside air was allowed simply to **infiltrate** into inside spaces.

Fresh air normally contains less moisture than indoor air, so as well as providing air for respiration, fresh air will absorb the moisture within a space and remove it as the air extracts from the building. Building Regulation Part F advises that an outside air supply rate of 10 litres per second per person will meet requirements for respiration and dilution of pollutants in office situations.

Natural ventilation

Natural ventilation is the most sustainable form of ventilation because it relies on naturally occurring pressure-differences to create air movement. These naturally occurring processes are stack and wind pressures. Commercial-type buildings can be designed to be satisfactorily ventilated by the stack and wind effects, however designers must be aware of the practical restrictions. The schematic diagrams shown in Figure 14.2 provide an indication of the limits of natural ventilation.

Mechanical ventilation

For certain buildings or processes natural ventilation may not be sufficient, and even buildings which are largely naturally ventilated must often have mechanically driven extraction for areas like toilets, kitchens or underground garages. For areas where natural ventilation does not generate sufficient pressure to overcome high resistances or cannot supply the volumes of air required, mechanical ventilation is necessary. By use of fans and ductwork, systems can be arranged to deliver and/or extract suitable volumes to all parts of a building.

Where mechanical ventilation supplies outside air to occupied spaces, it will require filtration and may also require heating. In a mechanical ventilation system the supply air temperature must be raised sufficiently to prevent discomfort. If the supply air temperature is raised sufficiently so that the space is actually heated this would then constitute a warm air heating system, which may be applicable in cases where other types of heat emitter cannot be used.

Figure 14.2
Commonly used natural
ventilation techniques:
(a) single-sided, single-
opening; (b) single-
sided, double-opening;
(c) cross-ventilation.

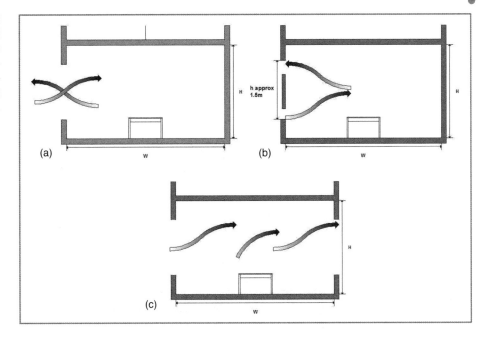

In some cases it is good practice to discharge 100 per cent of the extract air – for example, toilet or laboratory extract. However, energy may be saved if some of the extract air is recirculated. Other methods (outside the scope of this chapter) for reclaiming energy from extract air include the use of cross-plate heat exchangers, run around coils, thermal wheels and heat pipes. For further reading on these devices see *CIBSE Guide F*, Chapter 4 [8].

Night cooling

Ventilation can provide some cooling effect and *CIBSE Applications Manual AM10* discusses how moving air can have a cooling effect [9]. AM10 also describes a 'night ventilation strategy' in which lower external night-time temperatures pre-cool the structure and so help to prevent overheating of internal spaces during the daytime. The availability of this pre-cooling is affected by what is known as the 'thermal mass' of the building structure. The practice of exposing concrete soffits results in an increase in thermal mass, which in turn increases the heat-storage capability of a building structure. The inclusion of such techniques again demonstrates the benefit of cooperation between building services engineers and other construction disciplines.

Air conditioning

In some situations, mechanical ventilation may not be sufficient. This could be because:

- Heat gains to the internal space make the environment intolerable even with mechanical ventilation
- A building may require to be sealed for security reasons or because the outside environment is polluted or noisy
- A process requires close control of environmental conditions.

Air conditioning provides some of the functions of ventilation with the addition of control of humidity and an ability to deal with higher heat gains [10].

Whilst there are several different types of air conditioning (AC) systems, it is convenient to classify them under a few broad headings:

- All-air systems
- Air and water systems
- Unitary systems.

All-air AC systems

This method of air conditioning uses air as the medium to transfer cooling and heating energy around a building. All-air systems include constant volume systems, dual duct systems and variable air volume systems. Typically, these systems are served by central plant which may be located on a roof top or in a plant room. Because air is the heating/cooling medium they require large ventilation ducts, though having a central plant, usually away from building occupants, can be useful for maintenance purposes.

Constant volume systems tend to be specified for zones with similar loads and times of operation. They can be uneconomic when central plant, designed to serve large areas, must run fully if only some of the space is in use.

Variable air volume systems are better able to cope where cooling loads shift around a building – for example, solar gains which vary as the Sun travels from east to west. Because air volumes alter to match loads, it may be possible to apply an allowance for diversity to the sizing and selection of plant (see *CIBSE Guide B*, pp. 1-56 and 1-168).

Dual duct systems can control air conditions for spaces with different loads. Each zone is served by a hot duct and a cold duct from which various proportions are mixed to meet the zone loads. Because systems blend air from hot and cold ducts, they can be higher energy users. They also require lots of space to accommodate the two ducts.

Air and water systems

Fan coil systems (Figure 14.3) are probably the best known example of an air and water system. A fan coil system still has a central air handling unit, however this unit may be selected to deliver only sufficient air for ventilation and room moisture control. Therefore the air handling unit, fan and ductwork are much smaller than for an all-air system. Much of the cooling, or heating, in the space is carried out by the fan coil unit which resembles a fan convector. This unit is supplied with cooling or heating energy by cool or hot water delivered via pipework. Zone or room control is arranged merely by shutting down the room fan coil when it is not needed, and this very simple local control makes fan coil systems popular for air conditioning hotel rooms.

Other air and water systems include chilled ceilings, chilled beams and water-cooled concrete slab arrangements.

Chilled ceilings

These systems also exploit the radiant cooling effect of lower surface temperatures. The comfort equation means that with a lower mean radiant temperature, comfort can be achieved with higher air temperatures. Because much of the cooling heat transfer is

Figure 14.3
Typical fan coil
installation.

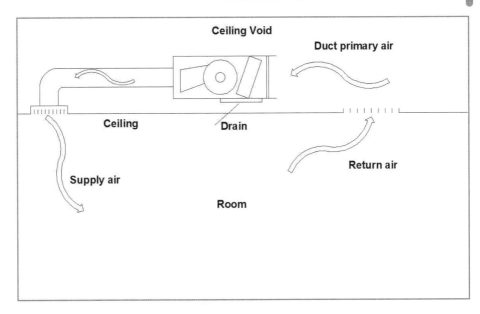

radiant, less convective cooling is necessary, so fan duties are smaller. The chilled water temperatures to cool chilled ceilings are higher than those normally used to cool air for all-air systems, so refrigeration duties may be reduced.

Chilled beams

Chilled beams operate in a similar fashion to chilled ceilings but have a greater cooling capacity. In both cases, a ventilation air supply is required. This is a separate supply for 'passive' chilled beams, though 'active' chilled beams will incorporate an air supply which is delivered through the beam. The chilling effect is achieved by circulating chilled water so space requirements are reduced.

Water-cooled concrete slab

By circulating chilled water through a concrete slab, cooling energy can be stored within the building fabric. This can be combined with night ventilation which also cools the slab by rejecting building heat into the cool night air. By careful design, the cooling energy stored in the slab will contribute to comfortable conditions in the space during occupied periods and can reduce the energy used for active cooling.

Unitary systems

There are several variations of air conditioning units, but probably the most familiar is the 'split system' which comprises an indoor unit and an outdoor unit linked by refrigerant pipework. In cooling mode the indoor unit acts as an evaporator and cools the space by absorbing heat into the refrigerant. This refrigerant is pumped to the outdoor unit where it is condensed and rejects its heat to atmosphere. A development of the split system is the variable refrigerant volume system (VRV). VRVs can have several indoor units and by controlling the refrigerant flow, they are capable of cooling a zone whilst simultaneously heating another.

PART 4

REFLECTIVE SUMMARY

- The comfort and well-being of building occupants depends upon a combination of conditions which building services engineers together with building designers attempt to control: the comfort temperature (which is a combination of ambient air and radiant temperatures), air quality, humidity and air movement.
- Most heating systems for commercial buildings are likely to be wet systems using low-temperature hot water, though to achieve sustainability and energy efficiency, careful thought must be given to the fuel used to heat the water and the strategy employed in the system controls.
- Buildings require ventilation for a variety of reasons, and for sustainability and energy efficiency should be naturally ventilated wherever possible. Since natural ventilation is often very difficult to control and may not even be sufficiently effective alone, mechanical systems must often be installed.
- Buildings tend to overheat and become humid during the summer months, and although natural or mechanical ventilation can be employed to remove some of the unwanted heat and moisture, this is not always sufficient to maintain comfort conditions so air conditioning systems are required.
- There are several strategies and methodologies available to control temperature and air quality in buildings: selection of the correct design solutions and integration of these into the overall building design, alongside an appropriate controls strategy, are vital for energy efficient operation of heating, ventilation and air conditioning systems.

REVIEW TASKS

- Observe the role of ventilation in your own home. Firstly, consider the importance of keeping your home fresh and comfortable, then identify where ventilation has another function such as removing moisture from bathrooms or kitchens. What causes air movement in and between rooms? How much of the ventilation in your home is natural and how much is mechanical?
- Consider a public or commercial building that you use and know quite well (choose a building that is a few years old if possible). Is the building ever too hot, too cold or draughty? Think about how it is heated and ventilated, and look to see if you can find out what systems are installed. Are these the best options with regard to energy efficiency and human comfort? Are there good control systems in evidence? Jot down some ideas about how the heating, ventilation and air conditioning systems might be improved.
- Visit the companion website at www.palgrave.com/engineering/riley2 to view sample outline answers to the review tasks.

14.2 | Electrical engineering systems

Introduction

- After studying this section you should understand the importance of electrical services in commercial buildings and how these services can contribute to the overall comfort of the occupiers.

- You should have developed an understanding of electrical services installations and controls for these systems.
- You should also have developed an appreciation of the issues related to lighting design and the importance of daylighting factors in these designs.

Overview

It would be difficult to overestimate the contribution that electricity makes to our daily lives. Electricity is a powerful, convenient and flexible power source that is a vital component of the built environment. It is so convenient that its benefits are, in many cases, taken for granted.

Traditionally, electrical power supplies are generated in central power stations and distributed to urban sites (Figure 14.4) [11]. Power stations and associated supply grids are operated by skilled engineers who can coax maximum efficiencies from generators and deliver power safely to users, but, despite the efforts of engineers and designers, power generators cannot overcome the thermodynamic limits set by nature.

Typical generating stations operate at around 40 per cent efficiency. This means that 60 per cent of the energy in the fuel is converted to heat and rejected to the atmosphere. Combined cycle power stations generate electricity using gas turbines and use the exhaust from the gas turbines to create steam to drive steam turbines, which also generate electricity. Combined cycle stations can achieve efficiencies of 50–55 per cent. The challenge for the built environment is to ensure the secure supplies of this vital power source whilst overcoming the problems of dwindling resources, poor efficiencies

Figure 14.4
Typical distribution
schematic.

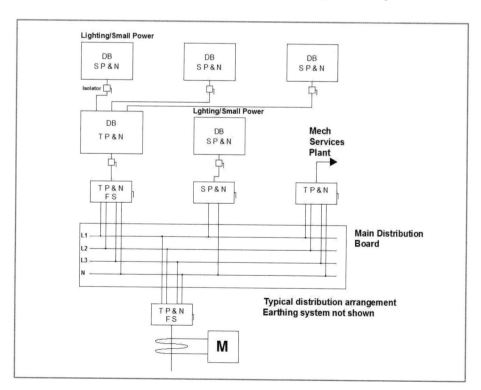

and carbon emissions. Carbon is emitted from the fossil fuels that are burnt to drive the generators. Of course, nuclear power generation does not create carbon but nuclear technology has considerable other environmental implications that must be factored into its consideration.

Grid-supplied electricity is generated and delivered from central power stations at high voltages, which are necessary to drive the electrical current over long distances. Substations on the system transform voltages down to levels to suit particular users. Industry may require electricity at 33 kV, and urban substations tend to receive electrical power at 11 kV. For commercial and domestic customers 11 kV is transformed to provide a 400/230 V low-voltage (LV) supply.

Typically, commercial-type buildings utilise a LV three-phase, 400 V supply. This allows commercial buildings the flexibility of single-phase systems for lighting, small power and similar loads, but allows three phases to drive machinery such as air conditioning plant.

An electrical supply is brought into a building where it is metered and fed into a main distribution board, from which subdistribution boards are supplied. Distribution boards should be strategically located for convenience and to limit the losses that can occur from long lengths of low-voltage cable.

This electrical distribution is arranged in circuits, each circuit being protected by a fuse or other protective device. This enables each circuit to be sized to meet a particular load and the fuse/circuit-breaker is selected to prevent damage to the wiring and switchgear, as well as protecting against electric shock. Selection of cabling systems must also ensure that the wiring is protected against heat, fire, explosion and any damp or corrosive elements.

Combined heat and power

Combined heat and power (CHP) (Figure 14.5) is a low, rather than a zero carbon technology. By generating electricity on-site the associated heat developed can be used to offset building heat demand. Generating electricity close to where it is used also reduces transmission losses. In comparison to photovoltaics, significant amounts of electrical power can be generated from CHP equipment that can be sited in a relatively small plant space.

Lighting

The Society of Light and Lighting (SLL) Code for Lighting describes good-quality lighting as 'lighting that allows you to see what you need to see quickly and easily and does not cause visual discomfort but does raise the human spirit' [12]. To achieve good-quality lighting that meets this definition, lighting designers must have knowledge of optics and electrical engineering as well as being able to cooperate with architects and others in matters of aesthetics and mood.

Economy and energy are also important considerations. The term *efficacy* describes how efficiently a lamp converts electrical power into light. The units for efficacy are lumens/watt.

However, an early and important step in achieving good-quality lighting is to interpret the client's requirements. Lighting levels are measured in lumens/m^2 or lux and *CIBSE*

Figure 14.5
CHP plant schematic.

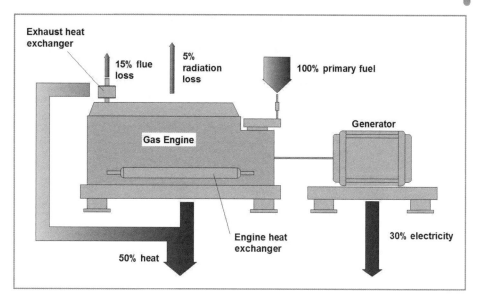

Guide A includes guidance for determining required lighting levels for different applications (Table 14.1) [13].

Lamps are normally included within a fitting or luminaire. The *SLL Code* defines luminaires as 'apparatus which distributes, filters or transforms the light transmitted from one or more lamps and which includes, except for the lamps themselves, all of the parts for fixing and protecting the lamps and, where necessary, circuit auxiliaries together with the means of connecting them to the electric supply'. Obviously, a luminaire will affect the light output from the lamps contained within it. The amount of useful light that is delivered from a luminaire, as a proportion of the total light flux that the lamps could give out is termed the *light output ratio* (LOR). In order to comply with Building Regulations which set limits for the amount of electrical power that can be used for lighting, designers must factor efficacies and LORs into their calculations.

Details of lumen outputs, efficacies etc. for different lamp types are available from manufacturers, who will also supply optical performance data for luminaires, including LORs.

Manufacturers' data will also include other necessary optical data:

- *Colour rendering index*: a numerical scale which indicates the ability of a light source to reveal the colours of a surface
- *Luminous intensity*: the power of a light source or illuminated surface to emit light in a given direction.

Because luminous intensity is related to direction, manufacturers' data will include information on intensity on a polar diagram. For lamps which do not have symmetrical intensities (for example, a fluorescent tube) there will be a different intensity when viewed 'side-on' from when viewed 'end on').

PART 4

Table 14.1 Comparison of different light sources.

Light source	Lumens	Power (watts)	Efficacy (lumens/watt)
Fluorescent			
T12	10 000–10 500	25–140	50–80
T8	650–6200	13–70	50–96
T5	120–8850	6–120	20–93
Compact (CLF)			
Non-integral control gear	250–9000	8–120	30–70
Integral control gear	100–1500	5–30	20–50
High-pressure mercury			
MBF/HPL	2000–58 500	60–1040	33–57
Metal halide			
Quartz tube	5200–200 000	85–2050	60–98
Ceramic tube	1600–26 000	20–250	65–97
Low-pressure sodium			
SOX, SOX-E	1800–32 000	26–200	70–180
High-pressure sodium			
Standard SON	4300–130 000	85–1040	53–142
Delux SON	12 500–37 000	165–430	75–86
White SON	1800–5000	45–115	40–44
Induction	2600–12 000	55–165	47–80
LEDs	20–220	1–5	30–100

Daylighting and controls

It is wasteful for electric lighting to be turned on when it is not required. Lighting may not be needed because a space is unoccupied, or because the space can be adequately illuminated by daylight. Whilst manual switching is incorporated into most lighting installations, this method of lighting control does not always ensure that lights are switched off when they are not needed. Lighting control systems can improve lighting energy management.

Lighting controls work automatically by dimming lighting or by switching lights on or off. Lighting controls may operate by timers, presence/absence detection, or by detecting the availability of daylight.

Lighting control systems should be specified after an analysis of the potential for daylighting and the likely behaviour and needs of the room occupants, and of course they must be safe. The BRE publishes guidance on space classifications and appropriate lighting control systems.

The reflectance values for room surfaces have an influence on lighting energy use. Effective daylighting requires careful design of building shape and form as well as glazing and shading arrangements.

REFLECTIVE SUMMARY

- Electricity is a convenient and flexible form of energy that fulfils a critical function in the modern built environment. The traditional methods for the generation of electricity are limited by thermodynamic principles which mean only around 40–55 per cent of the primary energy can be converted to electricity while most the remainder is converted to heat. Combined heat and power systems can improve the overall efficiency of power production by making use of this heat.
- Most electrical power is generated in central power stations, where it is generated at high voltage so that it can be delivered over long distances to where the demand is located. Transformer substations reduce the voltage to suit particular users. Industry may require electrical power at perhaps 33 kV. Local substations in urban locations receive electrical power at 11 kV and make it available to commercial and domestic customers at 400 V three-phase and neutral (tp&n) or 230 V single phase and neutral (sp&n)
- Commercial-type buildings are normally supplied with a tp&n supply. Electrical distribution arrangements within buildings are typically designed so that building services engineering plant and other equipment can operate on a three-phase supply whereas small power in rooms/offices/corridors (lighting, socket outlet circuits etc.) are supplied with a 230 V single-phase supply.
- Electrical distribution is conveniently arranged in circuits and subcircuits so that each circuit is protected by its designated protective device (fuse or circuit breaker). This means that should a fault occur in any particular circuit, only the designated protective device should operate. Wiring systems are selected so that they conduct electricity safely and are also suitably protected from any risks associated with the cable location or ambient conditions.
- Good-quality lighting design requires knowledge of optics, electrical engineering and aesthetics. To integrate these factors successfully means that lighting designers must work closely with architects and interior designers so that lighting combined with building structures and finishes can create mood and tone, as well as exploiting available daylight. Lighting controls can contribute to energy saving.

REVIEW TASKS

- Estimate how many times an average person, during a typical day, pushes a switch to use an electrical appliance or illuminate a room. Then further estimate how many times that appliance, or light fitting, fails to operate and determine an approximate percentage which would describe an average person's expectation for the reliability of the UK electricity supply. Given that the National Grid aims to match the electrical power that is generated in the UK to the electrical power required, consider what strategies the National Grid can use to make sure that everybody has a secure electrical supply without running expensive power stations unnecessarily.

- Consider how many lights are switched on in your home or workplace during the day. How does the weather affect when lights are switched on during daylight hours? Is it more likely that lights are switched on when the weather is gloomy or cloudy? Does the weather affect lights being switched off or are they left on, no matter how bright the day is? Make a study of how many lights are left on unnecessarily during the day.
- Visit the companion website at www.palgrave.com/engineering/riley2 to view sample outline answers to the review tasks.

14.3 | Public health engineering systems

Introduction

- After studying this section you should have developed an understanding of the need for public health engineering which involves predominantly the systems that allow for the provision of water in buildings.
- You should understand the different levels of cleanliness of water that are required for different activities and how this water can be provided in buildings.
- You should also understand how hot and cold water are supplied in buildings and how drainage systems are designed to remove waste water effectively.
- In addition you should appreciate how public health engineering systems can, if designed properly, contribute to the overall sustainability of a building.

Overview

Public health engineering services are usually considered to be the services which deliver clean water for human consumption and ablutions, and remove the products of human waste. Thus the general areas of hot and cold water supplies, and above- and below-ground drainage, are usually recognised as being within the province of public health engineering [14].

Clean water is required by the population primarily for drinking, washing and other domestic purposes. Water with a level of cleanliness making it suitable for human consumption is termed 'potable'. Water is, however, also necessary for many industrial processes and the requirement for purity may be more or less stringent than that for potable water.

Responsibilities of the local water companies

Each region of the UK is served by a local water company operating independently but under the terms imposed by a set of national water regulations [15]. The regulations stipulate installation standards for water supply systems and plumbing fittings that are designed to safeguard public health and prevent undue wastage of water.

Measures to avoid wastage of water in buildings

The water companies are charged with working to reduce consumption, and several measures are commonly employed, including the provision of water meters and planned preventive maintenance regimes to repair faults and leaks. The Water Supply (Water Fittings) Regulations 1999 [16] contribute to the imperative to conserve water by continually updating the list of permissible water fittings, particularly the specifications for WC cisterns and flushing mechanisms that use less water. Designers may also contribute by specifying items such as spray taps and automatically controlled taps, and by including techniques such as rainwater harvesting and grey water recycling into water supply systems. Detailed consideration of these various measures falls outside the scope of this chapter, though further information may be found in several publications, including:

- *Rainwater Harvesting: Regulatory Aspects* [16]
- *Greywater for domestic users: an information guide* [17]
- *Water Regulations Guide* [18].

Systems of cold water supply [19]

The **direct system** (Figure 14.6a) may be installed where the local water company permits it, however since the statutory requirement for water supply stipulates a pressure of 100 kPa (approximately 10 m head), a simple direct system is usually unworkable for taller buildings where water may be required at a height greater than 10 m from the ground.

The main disadvantage of the direct system is that if the local street main must be isolated for maintenance or repair then the building is completely without water until the street main is returned to service.

The system of cold water supply which maintains a reserve of cold water to cope with temporary disconnection of supply is termed the **indirect system** (Figure 14.6b). In this system all points of use are supplied by water from a storage cistern located at high level, except for the kitchen sink in dwellings, and specially designated drinking water taps in public buildings, these being connected directly to the water main. Drinking water outlets must be provided in all buildings and the quality of water from storage cisterns cannot be assured, whereas mains water always meets supply standards.

It is estimated conservatively that people in Britain use around 150 litres of cold and hot water (for drinking, washing and toilet flushing) per 24-hour period and therefore storage cisterns in public buildings are sized according to this requirement. There is some conjecture about this figure and different storage quantities are recommended for different types of building, usually depending upon whether people stay overnight in buildings or just work there. Guidance on sizing storage cisterns may be found in BS 6700 [20] and *CIBSE Guide G* [21].

In multi-storey buildings there is a requirement for boosted systems employing pumps to boost water up to high levels [21].

Water is very heavy: 1 litre has a mass of 1 kg, while 1000 litres has a mass of 1 tonne. Therefore a great deal of energy is required to raise water to high levels and thus the correct sizing of plant is vital to achieve energy-efficient system operation.

Figure 14.6
Schematics of
(a) typical direct cold
water supply system
and (b) typical indirect
cold water supply
system.

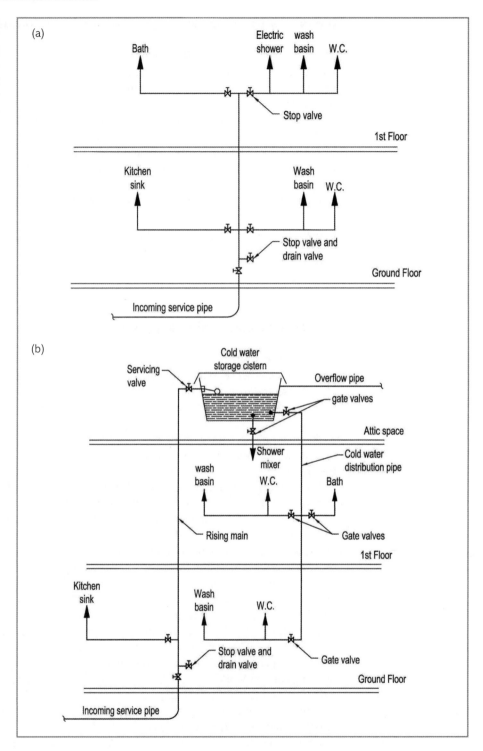

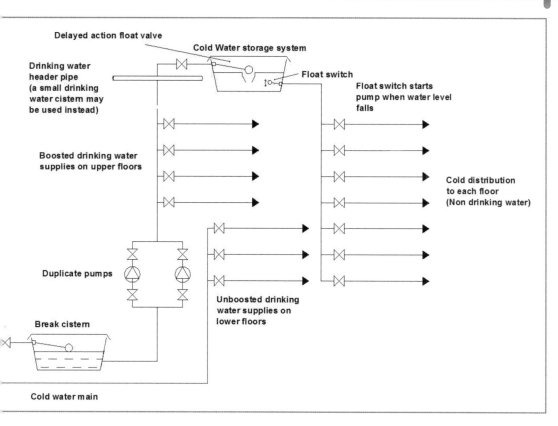

Figure 14.7
al boosted cold
supply system.

Boosted systems usually feature a low-level break cistern from which water is pumped, since to pump directly from the service pipe could overburden the public water main and mean shortages elsewhere in the locality. Pumps are normally installed in sets of two, one being the service pump, the other being a stand-by pump, which is brought into use when the service pump is undergoing maintenance or repair.

An inherent difficulty with the design of pumped systems is that in most buildings water tends to be drawn off intermittently in small amounts throughout the day, and this could potentially trigger a high number of pump starts. This is undesirable as it would result in excessive wear on pump components. Techniques such as combining delayed-action float valves and electronic float switches with extra storage cisterns can reduce pumping cycles. A typical generalised system is shown in Figure 14.7.

Provision of hot water

For public health considerations sufficient hot water for washing, bathing and domestic purposes must be provided in dwellings [22]. The two main approaches of providing hot water are to use localised heater units or centralised systems.

Localised units

Localised units normally heat the water instantaneously at the point of use, for instance

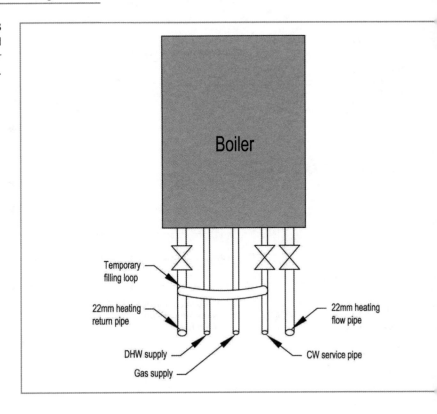

Figure 14.8
Typical gas-fired
combination boiler
arrangement.

above or adjacent to sinks or basins, and these often represent a cost-effective s
for small isolated cloakrooms, toilet areas or kitchens in public buildings. An
shower is a common example of a localised water heater.

Centralised systems

There are two main types of centralised systems, termed *storage* and *non-storage* s

- **Non-storage systems** include an instantaneous water heater with no storage c
  the most common type being the gas- or oil-fired combination boiler (Figur
  These are increasingly the preferred solution in modern dwellings and
  commercial buildings since they allow space heating and hot water to be heat
  the same appliance without the need for a storage cistern.
- For commercial buildings, **storage systems** represent the most usual app
  economies of scale apply. A suitable building maintenance regime will protec
  health.

At the centre of such a system is a hot water storage vessel. In small buildings or d
an upright copper cylinder with a capacity of around 100–150 litres serves as the
vessel, however in larger commercial and public buildings storage vessels of varic
are used. Such storage vessels, usually termed 'calorifiers', are often cylindrical i
though tend to be designed to be positioned on end rather than to stand upright.
of the higher system pressures, greater mechanical strength is required and ca

are therefore most often constructed of steel. For energy efficiency of hot water supply systems, storage vessels and associated plant must be carefully sized: design guidance for commercial and public buildings may be found in BS 6700 [20] and *CIBSE Guide G* [22].

The calorifier serves the purpose of providing a store of hot water from which quantities may be drawn off intermittently, thus there is a danger of heat loss and consequently energy wastage. To combat this, all hot water storage vessels are well insulated, and many are supplied incorporating layers of factory-fitted insulation foam which cannot be removed.

Public health considerations

Since human body temperature is around 38°C, the optimum temperature for washing and bathing is one or two degrees higher than this. It thus may seem sensible to store and supply domestic hot water at around 40–45°C, and this would undoubtedly be the best energy-saving option. Plumbing systems, however, harbour various species of micro-bacterial life, many of which are harmless, but some of which, such as the Legionella bacterium, can cause serious illness.

Harmful microbes are inactive at temperatures below 20°C and are killed at temperatures above 55°C, and they tend to breed and multiply most successfully in conditions close to human body temperature. It is thus considered good practice for cold water to be kept at as low a temperature as possible (without allowing it to freeze). Cold water pipes should be insulated and not be located above hot water or heating pipes. Best practice for hot water storage is to maintain a minimum storage temperature of 60°C. There are various techniques to limit cooling of pipes in distribution networks, such as pipework insulation, electrical trace heating of pipes and secondary return pipework. Storing and circulating water at temperatures too high for human use may seem counter to the notion of energy efficiency, but public health considerations must be addressed.

Expansion of water

Water expands by around 4 per cent when heated from 10°C to 60°C, so in a standard domestic cylinder this equates to more than 4 litres of extra volume when hot. Hot storage vessels could be damaged or rupture due to excessive pressures caused by expanded water. Hot supply systems must therefore incorporate some sort of device to accommodate expansion. The traditional method is to use a cold water cistern to feed the cylinder and act as an expansion vessel, and to provide an open vent pipe to allow air (and steam, if necessary) to escape, as shown in Figure 14.9.

Unvented hot water systems

Increasingly unvented systems are preferred, since these offer substantial savings in plant room space, materials and labour. These systems (Figure 14.10) allow the hot storage vessel to be filled from the cold water main, and an expansion vessel and relief valves remove the need for an open vent and feed cistern.

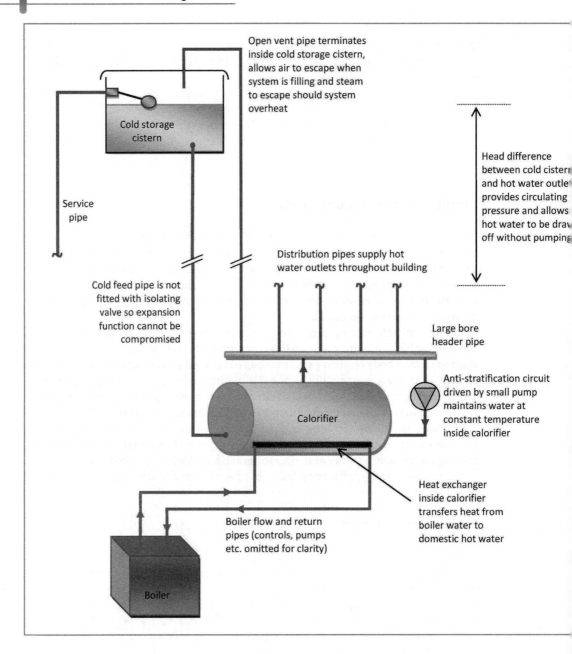

Open vent pipe terminates inside cold storage cistern, allows air to escape when system is filling and steam to escape should system overheat

Cold storage cistern

Service pipe

Head difference between cold cistern and hot water outlet provides circulating pressure and allows hot water to be drawn off without pumping

Cold feed pipe is not fitted with isolating valve so expansion function cannot be compromised

Distribution pipes supply hot water outlets throughout building

Large bore header pipe

Anti-stratification circuit driven by small pump maintains water at constant temperature inside calorifier

Calorifier

Heat exchanger inside calorifier transfers heat from boiler water to domestic hot water

Boiler flow and return pipes (controls, pumps etc. omitted for clarity)

Boiler

Figure 14.9
Typical generalised hot water supply schematic for multi-storey building.

Unvented systems for larger buildings would tend to be composed of separate nents, which must be carefully sized by the design engineer, though for smaller bu they are often supplied as packaged units, particularly in the case of domestic sy

Clearly for multi-storey buildings, as with cold water supply systems, the hc must be lifted up to higher levels of the building, and this would require the pumps. For multi-storey buildings there is merit in supplying the calorifier from level cistern, where the resulting head of water from the cistern would supply the sary circulating pressure.

Figure 14.10
Unvented hot water
system schematic.

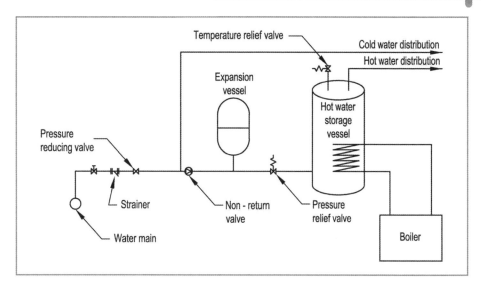

Minimising system heat losses

A particular problem for large centralised systems is that these may contain lengths of hot water pipework where the water stands still for long periods of time. Such sections are known as 'dead legs'. Excessively long dead legs are undesirable for three main reasons:

- It can take a long time for water to run hot at the tap, and users tend to allow water simply to run to drain while they wait. Apart from causing annoyance, this wastes potable water.
- Water standing still for a long period of time in pipework will, even though hot water pipework is always well insulated, give up its heat to the surroundings. There is thus an unwanted heat loss and this renders such systems energy inefficient.
- Long sections of hot water pipework containing static water allowed to cool down close to body temperature represent a serious public health risk, since ideal conditions can exist for dangerous bacteria to colonise. For this reason in hospitals a maximum dead leg of 1.5 m is permitted, though this figure is relaxed in other public buildings where usually around 3 m is permitted.

To prevent excessively long dead legs, secondary circulation is often employed as shown in Figure 14.11. Continuous circulation is achieved by pumping. This kind of solution protects public health but does have an energy cost. The desire for sustainability must be balanced against the requirement to protect public health.

Alternatively, electrical trace heating tape can be applied to sections of hot water pipework so that the required water temperature can be maintained while the water is static.

PART 4

Figure 14.11
Hot water distribution
incorporating
secondary return.

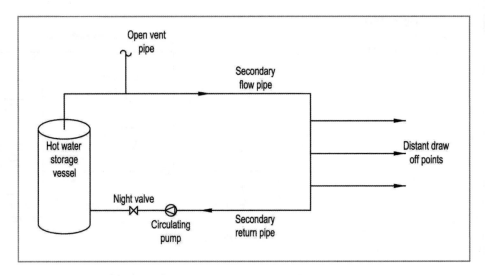

Building drainage [23, 24]

Drainage systems are required to remove waste water, usually including human bodily waste, and rainwater from buildings. Drainage systems are integral to the building and must perform their function efficiently for public health reasons, though for aesthetic reasons they are generally expected to be unobtrusive.

Systems are categorised as above ground or below ground: in simple terms, the above-ground system collects rainwater and all waste water from sanitary appliances and transports this to the below-ground system, which in turn then transports it away from the building.

Above-ground sanitary drainage

Sanitary appliances are generally connected via waste pipes which are laid to a slight fall so that gravity assists the discharge, to vertical pipes, termed discharge stacks. In modern installations the discharge stack connects directly to the below-ground drainage system.

Appliances are classified according to the type of waste matter that they collect. WCs, urinals and any appliances that remove human toilet waste are classified as 'soil appliances'. All others are termed 'waste appliances'.

This division may seem moot since in most systems nowadays all appliances discharge into the same pipework systems, however this distinction is a legacy from a time when the discharge from waste appliances was not considered harmful and might have been allowed simply to discharge onto the ground, or run to a soakaway or a water course, while only the discharge from soil appliances would have entered the below-ground drainage system. The distinction continued into modern times because in larger public buildings it is sometimes expedient to install a 'two pipe system' (i.e. to provide a waste discharge system separate from the soil system) since pipe diameters can be smaller when they convey discharge from waste appliances only. It is also helpful to continue with this classification since grey water recycling is becoming ever more popular as a method of reducing fresh water consumption.

Figure 14.12
Typical waste traps.

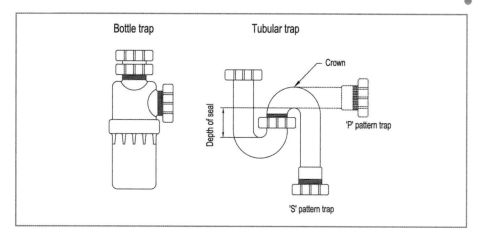

Waste traps

Every sanitary appliance must be provided with a trap ('U bend') either as an integral part of the appliance (such as in WC pans), attached directly to the appliance outlet, or fitted in close proximity to the appliance (Figure 14.12).

The purpose of the trap is to provide a barrier to prevent foul air from the drainage system entering the building, but in addition the water reservoir can act to cool very hot discharge or dilute chemically aggressive discharge. In can also intercept small accidentally dropped objects and serve to prevent the ingress of vermin from the drain.

Waste traps are required by the Building Regulations Part H [23] to retain a minimum water seal of 50 mm under normal working conditions. The flow of waste water as appliances discharge tends to cause variations in the air pressure inside the pipework, and this in turn leads to trap seals being drawn out due to syphonic action (a gurgling noise is often heard when discharging appliances). Older systems attempted to prevent trap seal loss by connecting additional ventilation pipework to the system, thus stabilising pressures within the pipework. This was of course an expensive solution, hence the later introduction of the single stack system.

Single stack above-ground drainage system

It may seem obvious to use a single vertical stack to collect the discharge from all appliances, but this was a relatively recent innovation based on quite extensive research, and was first advised by BS 5572 in 1974 (updated periodically, latest version published in 1994) [25]. Prior to this, two-pipe systems and one-pipe ventilated systems were common. Clearly one-pipe systems where waste and soil discharge are mixed would normally require less pipework and would thus be cheaper and more sustainable in today's terms. The single stack system is similar to the one-pipe system except that if a set of special installation rules stipulating pipe diameters, gradients and lengths of connection pipes are followed, then no ventilation pipework is required (Figure 14.13). Thus the system became very quickly the preferred option.

Since there is no ventilation of the sanitary pipework in the single stack system, there are still pressure fluctuations and it is still common for traps to gurgle. However, if the

PART 4

Figure 14.13
Single stack system
showing installation
rules.

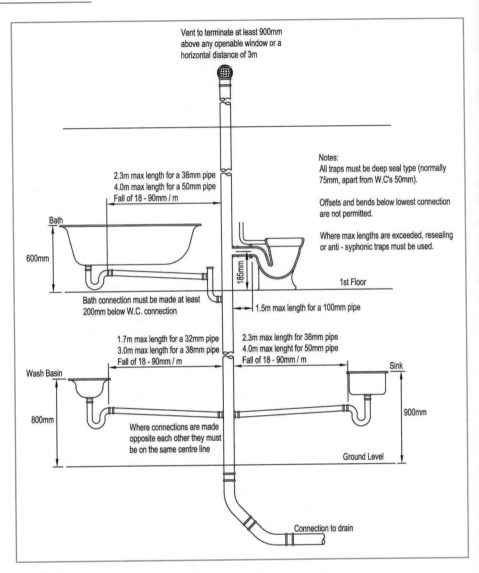

Vent to terminate at least 900mm
above any openable window or a
horizontal distance of 3m

Notes:
All traps must be deep seal type (normally
75mm, apart from W.C's 50mm).

Offsets and bends below lowest connection
are not permitted.

Where max lengths are exceeded, resealing
or anti - syphonic traps must be used.

2.3m max length for a 38mm pipe
4.0m max length for a 50mm pipe
Fall of 18 - 90mm / m

Bath

600mm

185mm

1st Floor

Bath connection must be made at least
200mm below W.C. connection

1.5m max length for a 100mm pipe

1.7m max length for a 32mm pipe
3.0m max length for a 38mm pipe
Fall of 18 - 90mm / m

2.3m max length for 38mm pipe
4.0m max lenght for 50mm pipe
Fall of 18 - 90mm / m

Sink

Wash Basin

800mm

Where connections are made
opposite each other they must
be on the same centre line

900mm

Ground Level

Connection to drain

installation rules are correctly followed, a water seal of 50 mm would still be retained in the trap under normal working conditions.

The top of the stack is left open to atmosphere to ventilate the stack as far as possible and to provide ventilation for the underground drainage system.

The single stack system represents the simplest, cheapest and, it could be argued, most sustainable above-ground drainage option, since it requires the absolute minimum amount of pipework possible. However, as grey water recycling becomes more common, a return to two-pipe sanitary plumbing seems likely. Systems incorporating grey water recycling and rainwater harvesting are of course more complex and consequently more expensive, but their use results in lower use of potable water and smaller volumes of waste entering the below-ground drainage system.

Below-ground drainage

Systems below ground are of course vital for the maintenance of public health, and are required to be unobtrusive and virtually maintenance free. Pipes are therefore laid in straight lines as far as possible, at a gradient to enable effluent to flow at a 'self-cleansing' velocity. Where changes of direction are necessary or connections are made, access must be provided so that blockages can be easily cleared.

The term 'drain' is used to describe a pipework system serving and belonging to one premise, whereas 'sewer' is used to describe a system serving two or more premises. In urban areas most building drainage systems convey effluent from the building to a network of public sewers, which are normally located under publicly owned land and are maintained by the local water company. Occasionally, however, local conditions dictate that private sewers running beneath private property must be provided. The upkeep and maintenance of public sewers are funded by the water rates and water metering charges, though there are various different arrangements in place to pay for public sewers.

Urban sewerage systems convey foul effluent to large treatment plants where a series of processes are utilised to remove foul and toxic constituents before the water can be released into nearby watercourses.

In rural locations where there is no public sewerage system available for connection, small community sized or individual treatment plants may be provided. These may be managed by the local council, by the local water company or by a private contractor, though the users would pay for their upkeep and maintenance. Small treatment plants treat the effluent in much the same way as do major plants, though on a much smaller scale, and treated water may be released to a local stream, river or soakaway.

Combined system

The first below-ground drainage systems were combined, that is they consisted of one set of pipes to collect all rainwater and waste water from buildings and convey them to a combined public sewer. The obvious point here is that clean rainwater entering the drainage system would immediately become polluted by foul effluent. When combined drainage and sewerage systems were first introduced however, this was not considered to matter much, since effluent was often released to watercourses with minimal treatment. As the UK's population grew and medical science became more advanced, the importance of not releasing anything into watercourses that could potentially contaminate drinking water sources was realised. Thus nowadays toxins and dangerous microbiological life are routinely removed from sewage such that water returned to watercourses cannot cause public health risks.

Separate system

Separate below-ground drainage systems consist of two sets of pipes, one for foul water and one for rainwater (termed 'surface water' drains). This means that, unlike the combined system, during periods of rainfall there is no change to the quantity of effluent to be conveyed to the foul water sewerage system and onto the treatment plant. The provision of two sets of drainage pipework would seem to contradict the drive for economy and sustainability, but this is more than balanced by the need for much smaller sewerage treatment plant. Surface water drains convey the water either directly to a watercourse or soakaway, or connect to a surface water sewer.

In all new building developments separate systems of drainage must be installed, though in urban areas where sewers were installed in the late 19th or early 20th century combined sewerage systems may well be in evidence. Thus a separate drainage system for a new building might well connect to a combined sewerage system! The logic of this approach is that it is expected at some future time that the combined sewerage system would be replaced with a separate system.

Figure 14.14 shows examples of combined and separate systems.

Partially separate system

During the 1960s and 1970s there was a fashion for a while to install partially separate systems. These are essentially separate systems where occasional rainwater connections could be connected to the foul water drain where convenient. The logic of this approach is that occasionally rainfall will clean out the inside of the drain, which is without doubt an argument with merit. The setback with partially separate systems was that building owners could mistakenly make cross-connections which could result in surface water drains being contaminated: a common error for instance is for kitchen sinks to be allowed to discharge over a rainwater collection grid connected to a surface water drain.

Partially separate systems are no longer permitted, though many are still in existence.

Access to drains

Access to drains is required for maintenance and cleaning at changes of direction, gradient and pipe size, and at junctions. The traditional method for providing access to drains and sewers is the access chamber – in general terms an access chamber is classified as an 'inspection chamber' if it less than 1 m deep and it could not be accessed by a person, whereas it is considered to be a 'manhole' if it is deeper than 1 m and a person could climb inside if necessary.

Inspection chambers and manholes traditionally were rectangular or square and were built *in situ*. Since they were constructed of dense engineering brick with a concrete base and cast iron covers, they tended to be expensive and labour-intensive items. Later the trend was to use precast concrete rings and construct the chamber from these on top of the concrete base. Nowadays a great many inspection chambers and manholes are made off-site from plastics or glass fibre reinforced plastics, thus making procurement and installation far cheaper.

At certain points of drainage systems where blockages are unlikely, for reasons of economy it is common to use rodding points instead of inspection chambers to provide access.

Treatment of waste water

Safe return of properly treated waste water to the natural environment is key to ensuring that water sources remain suitable for public supply.

Figure 14.14
Examples of combined
and separate below-
ground drainage
systems.

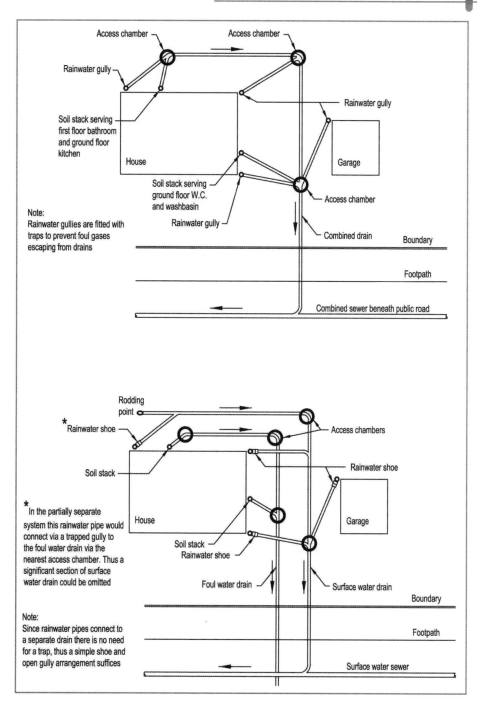

Early sewerage systems simply allowed waste water to discharge into the nearest water-course, be it a stream, a river or the ocean. The ecosystems present in large watercourses are naturally able to break down organic pollution like human excreta and food waste in reasonable quantities. If, however, the concentration of organic pollution becomes excessive this has a detrimental effect on the local ecosystem and kills animal and plant life.

As well as human and food waste, waste water may contain industrial or chemical waste. Of particular environmental concern are heavy metals, which are not biodegradable and are toxic, and can accumulate in plants, insects and fish, and thus can compromise the food chain. Industrial discharges to sewers are consequently well regulated in the UK and most industrialised countries (though not necessarily in third world nations).

REFLECTIVE SUMMARY

- Access to clean, wholesome water is a basic human need that large numbers of people worldwide do not enjoy. Strategies are required in the UK to assure continuity of water supply to all populated areas, and people in developed countries need to be educated to recognise the worth of potable water and develop more sustainable habits in their water use. Designers can assist by specifying appropriate water-saving technologies in buildings.
- There are two main types of cold water supply system, the *direct system* is relatively simple and saves on materials and installation costs, while the *indirect system* maintains a 24-hour supply of cold water in the event of mains failure, but is more expensive and complex to install and maintain.
- Larger multi-storey buildings require boosted systems where water is lifted to higher levels of the building using pumps and/or pneumatic lifting devices.
- Localised hot water systems include items like electric shower heaters, while centralised hot water supply systems may utilise instantaneous heaters such as combination boilers or hot storage cylinders. There are advantages and disadvantages inherent in all approaches in respect of sustainability, energy efficiency and public health.
- Above-ground drainage systems are largely based upon the single stack system, which is the least expensive to install, though with technologies like grey water recycling gaining popularity a return to two-pipe systems is likely in some buildings.
- Foul water and surface (rain) water must normally be kept separate in below-ground drainage systems. Though this means two sets of drains and sewers must be provided, there is a saving in the volume of sewage to be treated.

REVIEW TASK

- Knowing how to act in an environmentally friendly way and actually behaving that way are two very different things. Take the simple concept of conserving water. Consider a public or commercial building that you use and know quite well (choose a building that is a few years old, if possible). Do you think there is any cold or hot water wastage? How could you save water in this building? Write down your ideas. How could you communicate these ideas to the other building occupiers and educate people to be more circumspect with their water usage? Is there any scope for capturing and using rainwater or recycling waste water in your building?
- Visit the companion website at www.palgrave.com/engineering/riley2 to view sample outline answers to the review task.

Index